STP 1310

Tribology of Hydraulic Pump Testing

*George E. Totten, Gary H. Kling, and Donald J. Smolenski,
Editors*

ASTM Publication Code Number (PCN):
04-013100-12

ASTM
100 Barr Harbor Drive
West Conshohocken, PA 19428-2959
Printed in the U.S.A.

Library of Congress Cataloging-in-Publication Data

Tribology of hydraulic pump testing / George E. Totten, Gary H. Kling,
and Donald M. Smolenski, editors.
      p.   cm. — (STP ; 1310)
    "Papers presented at the symposium of the same name, held on 4-5
December 1995 . . . sponsored by ASTM Committee D-2 on Petroleum
Products and Lubricants and its Subcommittee D02.N on Hydraulic
Fluids.
    Includes bibliographical references and index.
    ISBN 0-8031-2422-8
    1. Tribology—Congresses.   2. Oil hydraulic machinery—Testing—
Congresses.   3. Pumping machinery—Testing—Congresses.
I. Totten, George E.   II. Kling, Gary H., 1931-   .   III. Smolenski,
Donald M., 1955-   .   IV. ASTM Committee D-2 on Petroleum Products
and Lubricants.   V. Series: ASTM special technical publication ;
1310.
TJ1075.A2T56   1997
621.2'52—dc21
                                                    96-40253
                                                      CIP

Photocopy Rights

Peer Review Policy

Each paper published in this volume was evaluated by two peer reviewers and at least one of the
editors. The authors addressed all of the reviewers' comments to the satisfaction of both the technical
editor(s) and the ASTM Committee on Publications.

To make technical information available as quickly as possible, the peer-reviewed papers in this
publication were printed "camera-ready" as submitted by the authors.

The quality of the papers in this publication reflects not only the obvious efforts of the authors and
the technical editor(s), but also the work of these peer reviewers. The ASTM Committee on
Publications acknowledges with appreciation their dedication and contribution to time and effort on
behalf of ASTM.

Printed in Philadelphia, PA
January 1997

Foreword

This publication, *Tribology of Hydraulic Pump Testing,* contains papers presented at the symposium of the same name, held on 4–5 December 1995. The symposium was sponsored by ASTM Committee D-2 on Petroleum Products and Lubricants and its Subcommittee D02.N on Hydraulic Fluids. George E. Totten of Union Carbide Corporation in Tarrytown, New York; Gary H. Kling of Caterpillar Inc. in Peoria, Illinois; and Donald J. Smolenski of General Motors Corporation in Warren, Michigan presided as symposium chairmen and are editors of the resulting publication.

Contents

Overview

Traditionally, numerous tests have been used to determine the lubrication properties of hydraulic fluids. These tests have included both pump tests and bench tests. However, none of these tests has achieved consensus acceptance within the fluid power industry. This lack of consensus has affected everyone in the industry.

Fluid users are confronted with a myriad of data obtained from different tests, if any at all, and almost all of the tests are conducted differently with no assurance that there is any correlation with specific types of wear that may be encountered in their hydraulic pumps.

Hydraulic pump OEMs (original equipment manufacturers) face a similar dilemma in that they are continually being asked to approve the use of new fluids on the basis of test data, if it exists, that may be conducted under conditions that may have no applicability to normal hydraulic pump usage or to their pumps.

Fluid suppliers are also confronted with obtaining lubrication data that illustrates that their fluids will exhibit the expected lubrication properties in every manufacturer's pumps of all designs and bearing configurations and used in widely varying conditions, which are often severe. This problem is compounded by the fact that OEMs will not accept any test data except a use test in their own particular pump, often under unique evaluation conditions that may not correlate to the acutal use conditions of the pump. Furthermore, it is impossible to evaluate every fluid in every pump under numerous evaluation conditions.

Therefore, there is a need to develop a hydraulic fluid testing protocol that will provide the desired insights into the lubrication properties of hydraulic fluids in a widely varying array of pumps and use conditions. This testing protocol should provide the user a method of specifying fluids for particular uses and use conditions. The OEM should be able to apply the data obtained from standard tests to predict the lubrication properties that would be attained with different pumps, pressures, rotational speeds, wear surfaces, and bearings. Ideally, the fluid supplier should have available standard tests accepted by everyone in the industry that can be applied cost-effectively to determine fluid lubricity in hydraulic pumps and motors. Furthermore, these lubrication data could be correlatable to the expected performance in any manufacturer's pumps and use conditions.

Thus far, the above stated objectives are only a dream. The Tribology of Hydraulic Pump Testing conference was held in Houston, Texas on December 4–5, 1995 as a first step in addressing this very complex issue. The objective of this conference was to obtain an overview both of testing procedures that have been applied and new tests that are currently being developed to successfully evaluate hydraulic fluid lubricity.

The topics addressed at this conference include the potential application of fundamental tribological principles in solving and generalizing the lubrication problems in hydraulic fluid lubrication of a broad range of pumps and motors. An overview of the most commonly encountered pump tests and new pump testing proposals was provided. The predictability of fluid contamination on pump wear was addressed. Finally, a number of bench testing procedures that are currently under evaluation to supplement or replace current pump testing procedures were discussed. This book provides a collection of the papers presented at this conference.

The tests and recommendations made by the speakers at this conference will be carefully analyzed by the newly formed "Hydraulic Pump Testing Task Force" within the ASTM D.02N committee. The task of this committee is to recommend and develop both bench test and pump tests as appropriate to address the pressing need within the fluid power industry for more effective fluid lubrication evaluation.

Finally, *Tribology of Hydraulic Pump Testing* represents the first focused global conference addressing only the subject of hydraulic pump lubrication testing that has been held. This conference was attended by leaders in this technology from Europe, Asia, and North America. A further objective of the conference will be to facilitate continued global technical information.

George E. Totten

Union Carbide Corporation
Tarrytown, NY
Symposium chairman and editor

Gary H. Kling

Caterpillar Inc.
Peoria, IL
Symposium chairman and editor

D. J. Smolenski

General Motors Corporation
Warren, MI
Symposium chairman and editor

Lubrication Fundamentals

Lavern D. Wedeven[1], George E. Totten[2] and Roland J. Bishop, Jr.[2]

TESTING WITHIN THE CONTINUUM OF MULTIPLE LUBRICATION AND FAILURE
MECHANISMS

REFERENCE: Wedeven, L. D., Totten, G. E. and Bishop Jr., R.J., **"Testing Within the Continuum of Multiple Lubrication and Failure Mechanisms,"** *Tribology of Hydraulic Pump Testing, ASTM STP 1310*, George E. Totten, Gary H. Kling, and Donald J. Smolenski, Eds., American Society for Testing and Materials, 1996.

ABSTRACT: The inherent difficulty of bench testing for the tribological performance of hydraulic fluids is the interaction of multiple lubrication and failure mechanisms. The engineer judges the performance limits in descriptive terms relating to what the load bearing surfaces have experienced. The lubrication and failure pathway that leads to the final surface condition is at the mercy of what lubrication and failure mechanisms have been invoked. Lubrication mechanisms, such as hydrodynamic, elastohydrodynamic, and boundary can be isolated with specialized testing, along with failure mechanisms, such as those described in general terms of wear, scuffing, and pitting. The interaction and competitive nature of these mechanisms, which exist in hardware, makes bench testing a nightmare.

A rational approach using a highly flexible and computerized test machine, WAM3, is described. The approach demonstrates how performance attributes of fluids and materials can be evaluated as they are made to travel through multiple lubrication and failure pathways. The testing protocol is terminated when the test specimen's surface reaches the same failure condition the engineer uses to judge performance limits of component hardware. Testing pathways are demonstrated that lead to wear, scuffing and micro-pitting. Along the testing pathway, viscous film-forming attributes and chemical boundary lubrication attributes of the fluid are characterized. Tests conducted with a range of fluid types, including two hydraulic fluids, demonstrate a wide range of traction, viscous film-forming and boundary film attributes. The continuum approach, which maps out performance in terms of hardware relevant criteria, provides a means to determine the impact of development strategies based on fluid and material technologies.

KEYWORDS: hydraulic fluid, performance mapping, hydrodynamic, elastohydrodynamic, boundary, bench testing, lubrication.

There has been an ongoing interest in the development of "bench tests" as alternatives to hydraulic pump testing to model hydraulic fluid lubrication performance. Examples of bench tests include: Shell 4-ball, Falex pin-on-V-block and the FZG gear test. The standard ASTM versions of these tests do not adequately model lubrication performance; although, with considerable experimentation, test conditions can be identified that appears to correlate with hydraulic fluid lubrication performance in a hydraulic pump [1]. In some cases, where standard bench tests are inadequate, custom bench tests are constructed.

---

[1]Wedeven Associates Inc., 5072 West Chester Pike, Edgement, PA 19028-0646.

[2]Union Carbide Corporation, 777 Old Saw Mill River Road, Tarrytown, NY 10591.

For example, Jacobs, et. al. developed a bench test to model sliding wear regimes that occur in hydraulic pumps as a function of material pairing and also fluid contamination [2].

Instead of using one or more of these bench tests to characterize the lubrication performance of hydraulic fluids, it is also possible to evaluate a fluid over a broad range of fundamental performance variables such as contact velocity, load and fluid viscosity. Tessmann and Silva [3,4] characterized hydraulic fluid performance using bench tests and the classic Stribeck-Hersey curve,[5] as shown in Fig. 1.

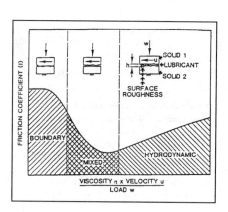

FIG.1--Stribeck-Hersey Curve.

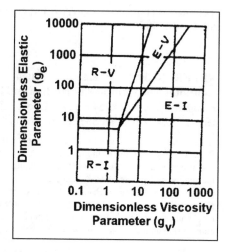

FIG.2--EHL Lubrication Map From Ref.10.

Mizuhara and Tsuya have thoroughly examined the ability of an ASTM block-on-ring test (ASTM D 2714-68) to model wear at various contact points in three different hydraulic pumps (vane, piston and gear) with antiwear oil, water-glycol, water-in-oil emulsion, oil-in-water emulsion, phosphate ester and polyolester hydraulic fluids [1]. The results of their work showed: it was necessary to evaluate fluids under a broad range of loads, speeds and material pairs in order to obtain any correlation with each different wear contact, load-carrying capacity has nothing to do with antiwear properties of the hydraulic fluids evaluated, accelerated tests usually provide the wrong results, materials for the wear contact must be similar to those of the wear contact in the pump, and to successfully evaluate the wear characteristics of hydraulic fluids, a wear test in a hydraulic pump must be conducted.

Ueno, et. al. have performed a detailed theoretical analysis of wear occurring at the vane on ring contact in both fixed and variable displacement vane pumps [6,7]. Some of the parameters examined included: delivery pressure, type of hydraulic fluid, pump rotational speed, vane contact eccentricity, hardness of the vane and ring materials of construction, radius of curvature of the vane tip and the length of contact force action range, ratio of frictional work and wear, and occurrence of abnormal wear. The conclusions of these extensive studies showed [6,7]:

1. Cam ring wear predominated.
2. Wear could not be simply represented as wear work even when accounting for pump delivery pressure and rotational speed.
3. The occurrence of abnormal wear was related to the temperature of the frictional surface.
4. No generalized quantitative conclusions could be drawn due to the hydraulic pump complexity

and the specific nature of the wear phenomena.

Ueno and Tanaka calculated the film thickness of different classes of hydraulic fluids under various load conditions encountered during the operation of a hydraulic pump [6]. Their results showed that film thickness was an important parameter controlling wear in hydraulic systems. Others have also reported that film thickness is an important hydraulic fluid characterization parameter [8-13].

In a more recent study, Kunz, et. al. modeled the wear process in a vane pump with subsequent experimental validation studies [14]. Their various conclusions were:

1. A good correlation between the load and wear location on the cam ring of the Vickers V-104 vane pump was found.
2. A mathematical wear model produced good correlations with respect to wear mass and wear rate with experimentally derived data.
3. A mathematical model based on shear energy hypothesis retraced the wear behavior in friction regimes with boundary lubrication, with the exclusion of additives.
4. Good input data for the model can be obtained only with great effort.
5. Wear prediction is not possible since some of the input data is actually derived from the modelling experiment.

Runhua and Caiyun have described in detail the use of elastohydrodynamic (EHD) theory to aid in the design of the optimal vane to cam ring geometry [15,16]. These vane profiles can be calculated from the luorication map shown in Fig. 2, the corresponding EHD film thickness equation from Table 1 and the following parameters:.

The film parameter Λ for two contacting surfaces is calculated from the minimum film thickness (h) from [17]:

$$\Lambda = \frac{h}{(H_{j1}^2 + H_{j2}^2)^{1/2}} \tag{1}$$

where: H_{j1} and H_{j2} are the surface finishes of the two contacting surfaces.

If the film parameter Λ is equal to 1, boundary lubrication and asperity contact occurs, mixed lubrication occurs when $\Lambda = 1$-3, and EHD lubrication occurs when $\Lambda = 3$-10 [18].

The minimum film thickness (h) is dependent on, α (pressure-viscosity coefficient), η_0 (kinematic viscosity), v' (entraining velocity), R' (effective radius), E' (effective elastic modulus), and w = load of a unit line contact. These are interrelated by the following dimensionless parameters:

Dimensionless speed parameter:

$$\overline{U} = \frac{\eta_0 \, v'}{E' \, R'} \tag{2}$$

Dimensionless load parameter:

$$\overline{W} = \frac{w}{E' \, R'} \tag{3}$$

Dimensionless materials parameter:
Dimensionless elastic parameter:

$$\overline{G} = \alpha \, E'$$
(4)

$$g_\epsilon = (\overline{WU})^{-\frac{1}{2}}$$
(5)

Dimensionless viscosity parameter:

$$g_\nu = (\overline{GW})^{3/2} \, \overline{U}^{-1/2}$$
(6)

where:

$$R' = \frac{R \, R_1}{R - R_1}$$
(7)

where: R = Radius of pump stator.
       R_1 = Radius of vane profile.

$$v' = \frac{1}{2} \, (v_1 + v_2)$$
(8)

$$E' = \frac{1}{2} \, [\frac{1 - \nu_1^2}{E_1}] + \frac{1}{2} \, [\frac{1 - \nu_2^2}{E_2}]$$
(9)

TABLE 1--Film thickness equations for wear regimes represented in Fig. 2.

Regime	Name of Equation	Equation
R-I	Martin	$h/R' = 4.9 \, (\overline{UW})^{-1}$
R-V	H.Block	$h/R' = 1.66 \, (\overline{GU})^{2/3}$
E-V	Dowson-Higginson	$h/R' = 2.65 \overline{G}^{-0.54} \overline{U}^{0.7} \overline{W}^{-0.13}$
E-I	K. Herrebrugh	$h/R' = 2.32 \overline{U}^{-0.6} \, \overline{W}^{-0.2}$

An alternative strategy to the use of the various bench or pump testing protocols described above, either those commercially available or custom tests, is to characterize hydraulic fluid performance with respect to fundamental lubrication performance variables which can be related to lubrication and failure modes in pump hardware [19]. This can be accomplished by examination of the variation of lubrication performance relative to the entraining velocity which generates an oil film and a sliding velocity which is a major component for oil film breakdown [19,20,21]. Fluid evaluation over a sufficient range of entraining and sliding velocities permits the characterization of the performance of the hydraulic fluid with respect to safe and unsafe lubrication and failure regions.

This paper describes the construction of performance maps to characterize the lubrication performance of hydraulic fluids. Experimentally measured EHD film thickness, along with an analysis to determination the viscous film-forming properties of a fluid is described. The analysis is used to derive an "effective" pressure-viscosity coefficient over a temperature range.

DISCUSSION

A. Continuum Concept of Lubrication and Failure

When the lubrication of a hydraulic fluid is stretched to the limit its viscous and chemical attributes are involved in several mechanisms of lubrication. These would include hydrodynamic elastohydrodynamic and boundary lubricating mechanisms. The failure of a component, which is judged by a criteria defined by the user, involves several failure mechanisms. These would include plastic flow, fatigue, adhesion and chemical reactions. The deterioration of the surfaces is generally expressed in descriptive terms, such as wear, scuffing and contact fatigue. In real mechanical systems, lubrication and failure mechanisms do not operate alone, or even in an orderly sequential process. They operate instead in a continuum with multiple pathways that can potentially lead to failure. The performance of the hydraulic fluid should always be considered with respect to its participation within a tribo-contact system. It helps to consider a tribo-contact system in terms of its structural elements. The major structural elements are:

1. A hydrodynamic or "full-film"
2. Surface films (boundary lubricating films)
3. Near-surface region (including surface topography)
4. Subsurface region

The life and durability of two contacting bodies is a function of the dynamic response of the lubrication and failure mechanisms associated with these structural elements. While the attributes of the hydraulic fluid are particularly associated with the interfacial lubricating films, the ultimate performance of the fluid cannot be disassociated from the other structural elements which are material dependent. If one considers the multiple lubrication and failure mechanisms and their dependence on the operating conditions that are imposed, it is easy to see why simple bench tests are difficult to correlate with hardware performance.

An opening strategy, which can reduce the complexity of testing, is to address the failure modes that are known or anticipated. A testing protocol that replicates the same failure features found on component hardware is a prerequisite for bench test correlation with field hardware. To do this requires an ability to properly simulate the structural elements of a lubricated contact and to invoke the same lubrication and failure mechanisms that control the performance of field hardware. For most practical applications, there is limited information to provide a detailed simulation. The best that can be done is to map performance over a range of conditions which is thought to be operative in field hardware or to characterize fluid properties that are believed to be important for performance. To do this requires a "smart" tribology test machine that has the flexibility and precision to control the key parameters that invoke a multitude of lubrication and failure mechanisms. In addition, a test methodology must be developed that follows an operating pathway that ultimately leads to the replication of the failure features found on hardware surfaces.

B. Test Methodology

1. WAM3 Test Machine - Traction coefficients and scuffing behavior of two model hydraulic fluids and two reference oils were determined using the "WAM3" test machine, shown in Fig. 3 [20,21]. The test machine is designed for precision control and monitoring of a single contact in three-dimensional space [19,20].

The test machine provides a tribo-system consisting of a ball-on-disk specimen configuration. The contacting specimens can be operated over a wide range of loads, rolling and sliding contact velocities, temperatures and environmental conditions. The flexibility of the machine allows simulation of specific contact conditions and performance mapping over a wide range of contact conditions.

The ball drive is positioned at an angle to control the amount of spin within the contact. The ability to control the tribological input variables provides an opportunity for wear measurement and the development of tribological performance maps in terms of R and S. The range of R and S available with the WAM3 are shown in Fig. 4. For comparison, the operating conditions for a 4-ball test, FZG test and Ryder Gear test are also shown.

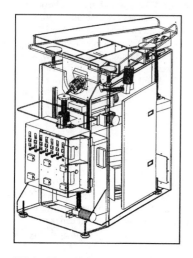

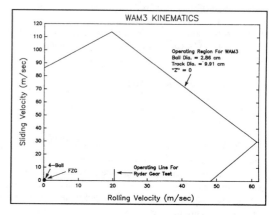

FIG.3---The WAM3 Machine uses a ball-on-disk contact configuration.

FIG.4--Map of entraining velocity (R) and sliding velocity (S) available in the WAM3 machine.

Precise control of ball and disk surface speeds and their direction (velocity vectors) is provided. This allows the introduction of a large range of rolling and sliding velocities which can be independently controlled. The independent control of R and S provides a "decoupling" of entraining velocity in the inlet region from the sliding velocity in the Hertzian region. This decoupling provides a method for separation, or at least controlling, the interaction between the physical and chemical properties of a hydraulic fluid. It also provides a direct connection to specific rolling and sliding velocities encountered across the face of a contacting gear mesh or across the contact ellipse of rolling element bearings. Heating or cooling of individual specimens provides thermal control of test conditions. Continuous recording of traction, load and sliding speed allows real time calculation of flash temperature.

2. EHD Film Thickness Measurements by Interferometry

The WAM3 machine provides the capability for the measurement of elastohydrodynamic (EHD) film thickness between the two contacting surfaces. This is accomplished with a dual chromatic optical interference fringe system. The interference fringes are calibrated to supply the optical film thickness between a transparent (pyrex) disc and a smooth M50 steel ball. A typical interference fringe pattern around the contact is shown in Fig. 5.

FIG.5--Dual chromatic interference fringe pattern used to measure EHD oil film thickness between ball and disc contact.

The EHD contact is monitored with a microscope, closed circuit TV and a video recording system. The tests are conducted with a load of 44.48 N (10 lbs) which gives a Hertzian contact stress of 0.59 GPa (86,000 psi). The EHD film thickness measurements are performed by recording the optical fringe color in the center of the contact as a function of rolling velocity. Optical film thicknesses are measured at the "center" of each fringe and at the transition between each fringe. The optical film thickness data is converted to actual film thickness by making corrections for the refractive index of the test fluid, including the effect of pressure on density under the Hertzian contact. The Lorenz equation [22] is used to correct the refractive index for density and the Hartung's empirical formula [23] for hydrocarbons is used to correct the density for the Hertzian pressure used in the test.

Film thickness data are generated by determining the entraining velocity (rolling velocity) corresponding to each fringe color. Film thickness tests were conducted at nominal temperatures of 23, 40, 70 and 100°C. Tests were limited to 23 and 40°C, for the fluids containing water, due to the loss of water by vaporization at higher temperatures. The tests were conducted with a recirculating fluid supply in a heated chamber. A trailing thermocouple was used to measure the ball temperature. A computer controlled peristaltic pump was used to recirculate the fluid.

The test results are graphically displayed using three types of plots: (1) EHD film thickness (h_o) vs.

entraining velocity (U_e, m/s), (2) dimensionless film thickness parameter (h_o/R) vs. dimensionless speed parameter ($\eta_o U_e/E'R$) and (3) pressure-viscosity coefficient (Alpha, GPa^{-1}) vs. temperature (°C).

B. Performance Map Construction

Lubrication performance is reflected in both physical and chemical attributes of the hydraulic fluid. Designers, in general, use physical properties to predict in-use performance. Yet, most conventional bench tests, focus on the chemical and metallurgical properties of the contact pair under specific operating conditions. The tribological conditions being studied are frequently unrelated to those actually encountered in a specific pump wear regime. A single wear or scuffing test number does not adequately model the multi-dimensional character of a lubricant. Multi-dimensional characterization of a hydraulic fluid should also include such attributes as "antiwear", EHD film-forming ability, and traction coefficient. Non-pitting performance is also an attribute of interest. The synergism between the viscous film forming ability and the boundary film forming ability of a fluid determines its quality. The attributes of both are necessary for a proper performance determination.

A more comprehensive approach to lubricant evaluation is to model or "map" the performance of a contact system over a range of entraining and sliding velocities. Performance maps for a phosphate ester hydraulic fluid (Fluid 1), a polyolester (Fluid 2) and a perfluoropolyalkyl ether (Fluid 3) are shown in Figures 6-8, respectively [20]. These figures show that Fluid 1 has substantially greater mixed-film or boundary lubricating performance than Fluid 2, and especially Fluid 3. The viscous film-forming attributes of each fluid are judged by the entraining velocity necessary to bring the contact into the EHD region. While Fluid 3 has limited boundary lubricating properties, its EHD film-forming ability is superior to the other fluids. Fluids with good viscous film forming properties may sometimes hide their chemical or boundary lubricating deficiencies.

The relative size of the mixed-film lubrication region reflects the synergism between the viscous and boundary film lubricating ability of the fluid. Fluid 1 shows good performance in this regard. The scuffing or severe wear boundary reflects the fluid and material attributes with respect to their ability to prevent a sudden loss of surface integrity, or a transition into a severe wear mode. The location of this boundary does not necessarily reflect good wear resistance, since a sacrificial wear process can prevent the onset of a sudden scuffing event.

To further understand the physical meaning of these transitions, it is first necessary to understand the tribological concepts depicted by these maps. Performance maps are generated in terms of rolling (R) and sliding (S) velocity vectors. The generation of an EHD film is primarily a function of the entraining or rolling velocity (R) in the inlet region of the Hertzian contact region. In this region, the lubricating film generation, is primarily a function of the physical properties of the fluid (viscosity and pressure-viscosity coefficient). The sliding vector (S) determines the shear strain within the high-pressure Hertzian contact region. This region is important with respect to heat generation, surface film formation, wear and scuffing within the tribo-contact. The magnitude of the sliding velocity in the Hertzian region, along with the degree of surface interaction, invoke the chemical properties of the fluid, e.g., adsorbed films, chemical reaction films, tribochemical reactions and thermal/oxidative stability.

To create a performance map, a series of 10-minute tests are conducted over a wide range of rolling and sliding velocities. The M50 steel specimens are run with a contact stress of 2.07 Gpa. [15] The ball specimen has a "hard grind" surface finish of Ra = 0.25 μm. The disk specimen is ground to a finish of Ra = 0.076 μm. Although M50 steel specimens were used here, other materials can be used for specific hardware simulation.

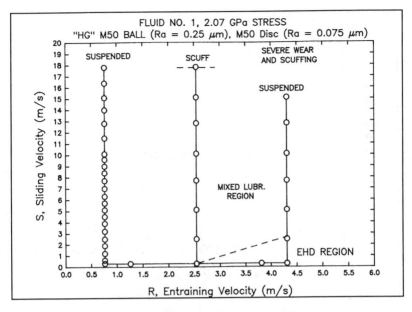

FIG.6--Performance map for hydraulic Fluid No.1.

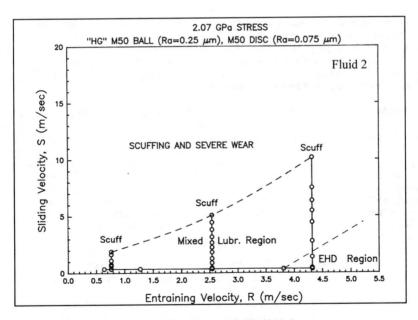

FIG.7--Performance map for Fluid No.2.

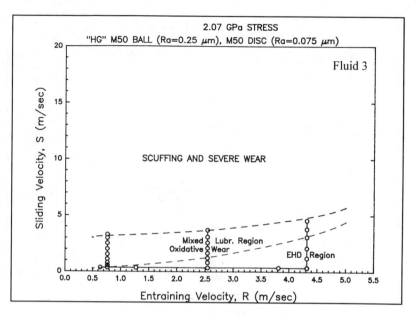

FIG.8--Performance map for Fluid No.3.

Lubrication and failure regimes are identified by the presence or absence of wear on the specimens, as well as transitions into a severe wear or scuffing failure. Specimen temperatures and the traction (friction) between the specimens are measured for each test. To identify the transition between "full-film" EHD lubrication and mixed-film lubrication, the running tracks on the specimens are microscopically examined and photographed as the sliding velocity is increased. The flexibility of the test machine allows the sliding velocity to vary while the rolling velocity remains constant.

The transition point from full-film EHD lubrication is defined as the operating condition where local surface polishing on an asperity scale is observed. As the sliding velocity increases, the "hard grind" finish (Ra = 0.25 μm) on the ball becomes smoother. The polishing of surface features can be considered as the initiation of a run-in process, at least from a topographical viewpoint. The formation (and removal) of surface films, are expected to become more predominant as "S" increases. Also, there is greater heat generation which results in a temperature rise as the sliding velocity increases. The transition from mixed-film lubrication to scuffing, or in some cases to severe wear, is determined by running each fluid at a constant velocity (R = 80, 250 and 430 cm/s) with increasing sliding velocities until a scuffing event is observed. A scuffing transition is characterized by a sudden increase in traction coefficient in response to a complete breakdown of the lubricating film and gross smearing of the contact surfaces.

The traction derived from an EHD fluid film is a function of the limiting shear strength of the pseudo-solid fluid in the Hertzian contact region for highly loaded contacts. The traction coefficient decreases with increasing temperature. The decrease in traction is usually a linear function with respect to temperature. Monitoring the traction coefficient as a function of increasing sliding velocity reflects the nature of the sheared material within the contact region. In addition to bulk hydraulic fluid, the sheared material is likely to be composed of wear particles and breakdown products. The traction coefficient is sensitive to surface roughness, especially with high sliding velocities.

For most normal operating conditions, the bulk fluid traction plays a predominant role in the overall traction at the contact. The contribution of "asperity" friction is only significant under severe operating conditions. This is shown in Fig. 9, which is a plot of the traction coefficient vs. rolling velocity (R) for each fluid at a constant sliding velocity (S = 0.36 m/s). According to Figures 6-8, a rolling velocity over 4 m/s puts each fluid in the full-film EHD region. As the rolling velocity

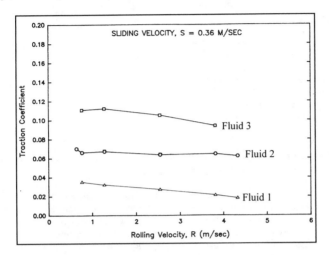

FIG.9--Effect of varying rolling speed at a constant sliding velocity.

decreases, the contact enters the mixed-film lubrication region. The four fluids have considerably different traction behavior. Fluid 1 has only 30% the traction coefficient of Fluid 3, even with its operation deep into the mixed-film lubrication region. Traction behavior is an important design consideration in view of its direct link to heat generation and its influence on the tangential shear stress at the contact surface. Traction is generally missing from fluid property characterization.

These data show that the determination of the location of the boundaries for lubrication and failure regions provide only a starting point for a more comprehensive performance characterization of the fluid. The fluid can be further characterized for pressure-viscosity properties that impart EHD film-forming ability. Testing along specific pathways through the mixed-film lubrication region can provide wear and micro-pitting characterization.

D. EHD Film Thickness Results

In the above discussion, it was shown that performance maps identify the regions of full-film EHD lubrication, mixed film lubrication and scuffing. The location or size of the EHD region reflect the viscous film-forming properties of the fluid. These are usually divided into viscosity-temperature and viscosity-pressure properties. The viscosity (η_o) and pressure-viscosity coefficient (α) are required for EHD film thickness calculations. While viscosity-temperature properties are generally available, the pressure-viscosity coefficient is frequently missing. As part of a comprehensive characterization of fluid performance with the WAM3 machine, the pressure-viscosity coefficient can be determined from the measurement of EHD film thickness. The viscosity and refractive index data for the test fluids are provided in Table 2. Independently measured pressure-viscosity coefficients are given in Table 3 [24].

The film thickness data in Fig. 10a-c show the effect of rolling speed and temperature on film thickness for each fluid. The log-log plots in these figures show that the relationship between film thickness and speed is exponential. From EHD theory, film thickness varies with entraining velocity to the power of 0.67 as seen in EQ(10).

$$(10) \quad h_o = 1.9 \frac{(\eta_o U_e)^{0.67} \alpha^{0.53} R^{0.397}}{E^{0.073} w^{0.067}}$$

where: h_o = Film thickness, center of contact.
R = Combined radius of curvature.
η_o = Viscosity @ atm. press. & test temp.
α = Pressure-viscosity coefficient.
U_e = Entraining vel., $U_e = 1/2(U_1+U_2)$.
E = Combined elastic modulus.
w = Load of the unit line contact.

Film thickness decreases with increasing temperature because the viscosity (η_o) and the pressure-viscosity coefficient (α) decrease with temperature.

The three test fluids are plotted in dimensionless form for two temperatures in Fig. 11a,b. The only EHD lubricant characteristic missing from the two dimensionless parameters is the pressure-viscosity coefficient (α) and load (w). The film thickness measurements were conducted under the same load. If the α-value for each fluid measured were the same, the film thickness data would all fall on a single line. Higher values of (h_o/R) for the same ($\eta_o U_e/E'R$) reflect higher pressure-viscosity coefficients.

These data are very interesting since they show that the optically measured pressure-viscosity coefficient is always lower than the value obtained by capillary viscometry, and by approximately the same amount. Of the various potential reasons for this, perhaps one of the most significant is the temperature-viscosity effect. These data suggest that the pressure-viscosity and temperature-viscosity coefficients do not vary independently of each other. This would be reasonable since the traction coefficients for these lubricants are different, the contact temperatures are also probably different. If the contact temperature varies between the various fluids, the actual temperature of measurement of the film temperature may vary somewhat. Therefore, these results reflect the "effective" pressure-viscosity coefficient actually experienced.

Using Fluid No. 3 as the reference fluid (since the pressure-viscosity of this fluid is well characterized), it is possible to calculate the "relative" pressure-viscosity coefficient from the equation below which is derived from Eq(10).

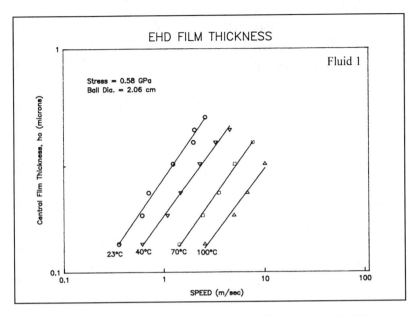

FIG.10a--Effect of rolling speed and temperature on film thickness for Fluid No.1.

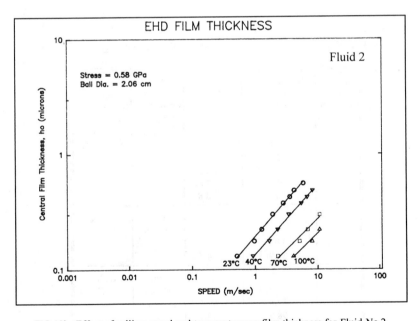

FIG.10b--Effect of rolling speed and temperature on film thickness for Fluid No.2.

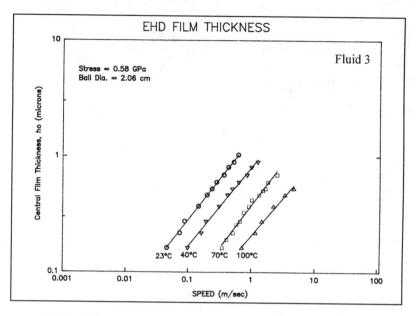

FIG.10c--Effect of rolling and temperature on film thickness for Fluid No.3.

TABLE 2--<u>Viscosities and Refractive Indices of Experimental Hydraulic Fluids.</u>

Fluid No.	Temperature (°C)	Viscosity (cSt)	Pressure (psi)	Refractive Index
1			Atm.	1.4690
			92,000	1.5799
	-17.8	1202		
	0	325		
	25	82.0		
	50	29.6		
	98.9	8.8		
2			Atm.	1.4500
			92,000	1.5590
	37.8	21.4		
	98.9	4.5		
3			Atm.	1.2985
			92,000	1.3620
	37.8	69.1		
	98.9	9.1		

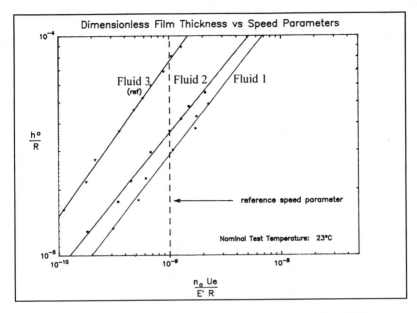

FIG.11a--Dimensionless film thickness vs. speed for each fluid at 23°C.

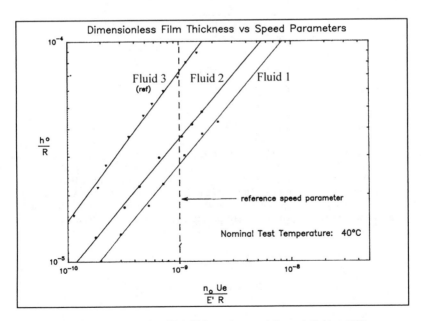

FIG.11b--Dimensionless film thickness vs. speed for each fluid at 40°C.

1. Calculation of Pressure-Viscosity Coefficient - For each test fluid, the film thickness data was plotted in dimensionless form. The dimensionless "film thickness parameter" (h_o/R), where:

h_o = Film thickness in the center of the contact.
R = combined radius of curvature (0.3968 in.).

$$\alpha = \alpha_{ref\ oil} \left[\frac{h_o/R}{(h_o/R)_{ref\ oil}} \right]^{1.887}$$ (11)

The above equation assumes that h_o is proportional to $\alpha^{0.53}$ according EHD theory EQ(10). The relative pressure-viscosity coefficients for the test fluids were determined at a selected speed parameter of 1×10^{-9}. This value was selected because it corresponds with film thickness data for all the test fluids. The calculated α-values are shown in Table 3.

TABLE 3--**Pressure-Viscosity Coefficients at 65°C**
Obtained by Capillary Viscometry and
EHD Film Measurements (GPa^{-1}).

Fluid No.	Capillary Viscometry	Optically Measured From EHD Contact
1	11.5	7.8 (68°C)
2	10.0	11
3	37.0	37 (ref. fluid)

These data suggest that the pressure-viscosity and the temperature-viscosity coefficients do not vary independently of each other. This would be reasonable since the traction coefficients of these lubricants are different. If the contact temperature varies between these various fluids, the actual temperature of the film may vary somewhat. Therefore, these results reflect an "effective" pressure-viscosity coefficient actually experienced by the fluid in the process of film generation. However, more work is required to definitively explain this behavior. Therefore, the variation of pressure-viscosity coefficient with temperature is shown Fig. 12. Interestingly, this data shows that the pressure-viscosity coefficient of Fluid No. 3 is more sensitive to temperature than Fluid No. 2. It is generally known that fluids with sensitive viscosity-temperature properties also have sensitive viscosity pressure characteristics.

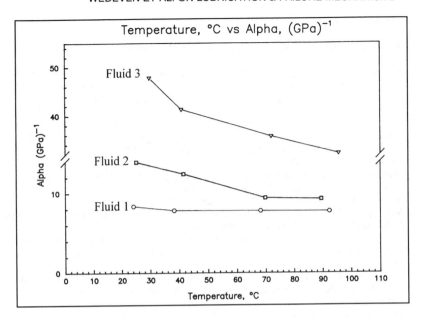

FIG.12--Pressure-viscosity coefficient vs. temperature.

CONCLUSION

An important, and often little understood, function of hydraulic fluids is to lubricate the hydraulic pump where either sliding, e.g. vane on ring in a vane pump, or rolling, e.g. rolling element bearings, where contact velocities and related wear may result. There are various methods of studying these effects which include wear tests in hydraulic pumps, the use of custom designed bench tests or the use of a test methodology which permits the study of potential wear in its most fundamental form by modeling traction and scuffing as a function of rolling and sliding speed and load.

The use of performance maps to identify the boundary regions of various wear regimes along with fluid traction behavior within these regions has been demonstrated using a new and comprehensive test approach. The test approach includes a capability to measure the oil film thickness within the contact region. Film thickness measurements conducted over a range of temperatures can be used to calculate an "effective pressure-viscosity coefficient." Although the "effective pressure-viscosity coefficient" is consistently less than when measured by capillary viscometry, they are expected to reflect the dynamic conditions invoked by the film generating process within the EHD contact.

This work demonstrates the value of examining hydraulic lubrication and wear from a fundamental, first-principles approach. This approach does not require numerous specialized tests, nor does it require the frequent use of relatively expensive, large volume hydraulic pump tests. With the availability of the appropriate information, it is possible to use such fundamental lubrication calculations in the design phase of hydraulic pump development and subsequent fluid qualification. This will be the subject of a future report.

REFERENCES

[1]   Mizuhara, K. and Tsuya, Y., <u>Proc. of the JSLE Int. Tribology Conf.</u>, (1985), Tokyo, Japan, p. 853-858.

[2]   Jacobs, G., Backe, W., Busch, C., and Kett, R., "A Survey on Actual Research Work in the Field of Fluid Power", Orally presented at the 1995 STLE National Meeting, Chicago, IL.

[3]   Tessmann, R.K. and Hong, I.T., <u>SAE Technical Paper Series</u>, Paper Number 932438 (1993).

[4]   Silva, G., <u>SAE Trans</u>, Vol. 99, 1990, pp. 635-652.

[5]   Stribeck, R., <u>Zeit V.D.</u>, Vol. 46, 1902, pp. 1.

[6]   Ueno, H. and Tanaka, K., <u>Junkatsu (J. Jap. Soc. Lubr. Eng.)</u>, Vol. 33, 1988, pp. 425-430.

[7]   Ueno, H., Tanaka, K., and Okajima, A., <u>Nippon Kikai Gakkai Ronbunshu, Bhen</u>, Vol. 52, No. 480, 1986, pp. 2990-2997.

[8]   Wan, G.T.Y., Kenny, P., and Spikes, H.A., <u>Tribology Intl.</u>, Vol. 17, 1984, pp. 309-315.

[9]   Spikes, H.A., <u>J. Synth. Lubr.</u>, Vol. 4, pp. 115-135.

[10]   Wan, G.T.Y., <u>ASLE Trans.</u>, Vol. 27, 1984, pp. 366-372.

[11]   Veresnyak, V.P., Zaretskaya, L.V., Imerlishvili, T.V., Kel'bas, V., Lukashvili, N.V., Sedova, L.O., and Ryaboshapka, V.M., <u>Trenie Iznos</u>, Vol. 10, 1989, pp. 919-27.

[12]   Veresnyak, V.P., Zaretskaya, L.V., Imerlishvili, T.V., Kel'bas, V., Lukashvili, N.V., Sedova, L.O., Ryaboshapka, V.M., Shvartsman, V. Sh., and Shoikhet, V. Kh., <u>J. Frict. Wear</u>, Vol. 10, pp.120-126.

[13]   Veresnyak, V.P., Zaretskaya, L.V., Imerlishvili, T.V., Kel'bas, V., Lukashvili, N.V., Sedova, L.O., Ryaboshapka, V.M., Shvartsman, V. Sh., and Shoikhet, V. Kh., <u>Trenie Iznos</u>, Vol. 12, pp. 144-153.

[14]   Kunz, A., Gellrich, R., Beckmann, G., and Broszeit, E., <u>Wear</u>, Vol. 181-183, 1995, pp. 868-875.

[15]   Runhua, T. and Caiyun, Y., <u>J. Hebei Institute of Technology</u>, Vol. 20, No. 3, 1991, pp.43-50.

[16]   Runhua, T. And Caiyun, Y., <u>Proceed. Of the Int. Fluid Power Appl. Conf.</u>, March 24-26, 1992, pp. 475-480.

[17]   Dowson, D. and Higginson, G.R., <u>Elasto-Hydrodynamic Lubrication</u>, Pergamon Press, 1977.

[18]   Zuengze, W., <u>Advanced Machine Design</u>, Qinghua University Press, 1991.

[19]   Wedeven, L.D., "Thin Film Lubrication and Tribological Simulation", <u>Adv. in Eng. Tribology</u>, STLE SP-31, April, 1991, pp. 164-184.

[20]   U.S. Patent Application Number: 07/963,456.

[21]   Wedeven, L.D., Totten, G.E., Bishop, R.J. Jr., <u>SAE Technical Paper Series</u>, Paper Number 941752, 1994.

[22]   Glasstone, S., <u>Textbook of Physical Chemistry</u>, Macmillan and Co. Ltd., London, 1948.

[23]   Chu, P.S.Y. and Cameron, A., "Compressibility and Thermal Expansion of Oils", <u>J. Inst. Petrol</u>, May 1963, Vol. 49, No. 473, pp. 140-5.

[24]   Bair, S. and Winer, W.O., <u>Tribol. Trans.</u>, Vol. 31, 1987, pp. 317.

Monica Ratoi-Salagean[1] and Hugh A. Spikes[1]

THE LUBRICANT FILM-FORMING PROPERTIES OF MODERN FIRE RESISTANT HYDRAULIC FLUIDS

REFERENCE: Ratoi-Salagean, M. and Spikes, H. A. **"The Lubricant Film-Forming Properties of Modern Fire Resistant Hydraulic Fluids,"** *Tribology of Hydraulic Pump Testing, ASTM STP 1310,* George E. Totten, Gary H. Kling, and Donald J. Smolenski, Eds., American Society for Testing and Materials, 1996.

ABSTRACT: Fire resistant hydraulic fluids tend to show significantly poorer tribological performance in hydraulic systems than conventional mineral oil-based fluids. There have recently been performance problems associated with increases of operating temperatures of mining hydraulics. This paper describes measurements of the elastohydrodynamic and boundary film-forming properties of a range of different hydraulic fluid types at temperatures up to 80°C. These are compared with friction and wear results obtained using the same fluids.

KEYWORDS: elastohydrodynamic lubrication, film thickness, hydraulic fluid, fire resistant, wear

INTRODUCTION

Fire resistant hydraulic fluids are widely used in mining and other safety-critical applications where there is a significant risk from fire. Many of these fluids contain a high proportion of water which confers their resistance to ignition, although others are based upon synthetic materials which are either less volatile or possess intrinsically more resistance to combustion than mineral oils.

One limitation of water-based hydraulic fluids and also of some synthetics is that their lubricating properties tend to be significantly poorer than conventional, mineral oil-based fluids. In some cases this is because they form thinner hydrodynamic films than mineral oils of equivalent viscosity. This is particularly the case with some water-based fluids. Some

---

[1]Research student and Reader, respectively, Tribology Section, Department of Mechanical Engineering, Imperial College, London SW7 2BX, England

fire resistant hydraulic fluids may also possess less effective boundary lubricating properties than mineral oil blends. Because hydraulic pumps generally rely on the fluid being pumped for their lubrication, the inferior lubrication properties of many fire resistant fluids mean that pumps have to be de-rated, and also often give performance problems such as high wear or low fatigue life in service.

A recent trend which has exacerbated this problem in mining applications in the United Kingdom is a significant increase in the temperature at which hydraulic fluids are operating. Whereas maximum hydraulic fluids temperatures used to be about 60 to 70°C, temperatures as high as 80°C are now common. This results from higher ambient temperatures, either because of deeper seams or from the large quantities of energy-dissipating, electrical equipment required for increased mine mechanisation. This rise in temperature is believed to have contributed to an increase in pump failure rates in the last two years. One particular problem has been the failure of gear pumps with needle roller bearings. Failures have taken place within "days or weeks" and have been concentrated in mine areas where ambient temperatures are high, at about 38°C.

A useful, fundamental measure of the lubricating ability of fluids is the film thickness that they form in concentrated contacts. This indicates the extent to which they are likely to form separated films and thus limit wear in high pressure contacts such as are present in gear pumps. This paper describes measurements of the film-forming properties of a range of different commercial fire resistant hydraulic fluids in concentrated contacts at temperatures up to 80°C, with a special emphasis on water-based fluids. The friction and wear performance of some of the fluids have also been determined over the same temperature range. The main aim was to compare the behavior of different types of fluids at the high temperatures now being reached in some mining applications and thus to indicate appropriate fluids for such applications.

BACKGROUND

According to ISO standard 6743/4 there are four main categories of fire resistant fluid, HFA to HFD, as indicated in table 1 (1). This specification is not normally considered to include some aviation hydraulic fluids based on synthetic hydrocarbons which possess moderate fire resistance (2), although the family HDFU may encompass these as well as silicone-based fluids and recently developed water-free glycol-based materials.

Most of these fluids, and in particular the water-based types, are poorer lubricants than conventional mineral oil-based hydraulic fluids, whose evolution has been charted by Grover and Perez (3). Of the fire resistant fluids, HFD phosphate ester fluids are generally considered to have the best lubricating characteristics, followed by the HFB class. HFAE oil in water emulsions are usually regarded as the poorest lubricants whilst HFAS and HFC fluids are intermediate (4).

This order of effectiveness tends to be reflected in their applications. Thus HFB fluids are employed, together with some HFCs or even mineral oils, in demanding applications such as mobile and static hydrostatic power transmissions whilst HFAE fluids are confined to low contact pressure, static or slow moving applications such as longwall powered roof supports (5).

TABLE 1--Classes of fire resistant fluid

Class	Sub-class	Compositional requirements	Typical composition
HFA	HFAE	Oil in water emulsions, less than 20% wt. combustible materials	Typically 5% wt. oil
" "	HFAS	Solutions of chemical in water, less than 20% combustible materials	Typically 1 to 5% wt. hydrocarbon phase as micro-emulsion or solution
HFB		Fire resistant water in oil emulsions	Typically 40% wt. water
HFC		Fire resistant polymer solutions, at least 35% wt. water	Typically 15% high MWt polyglycol, 40% monoglycol, 45% water
HFD	HFDR	Fire resistant synthetic fluids based on phosphoric acid esters	Typically trialkyl and/or triaryl phosphate
" "	HFDS	Fire resistant synthetic fluids based on chlorinated hydrocarbons	
" "	HFDU	Fire resistant synthetic fluids of other types	e.g. silicate esters

The performance problems associated with water-based fluids have been discussed by a number of authors and include low fatigue life, high wear rate and cavitation erosion (5)-(8). There appear to be three main reasons for the inferior lubricating performance of fire resistant fluids as compared to mineral oils;

(i)   poor hydrodynamic or elastohydrodynamic film formation, related either to the low viscosity and pressure viscosity coefficient of water or to the low pressure viscosity coefficient or shear thinning behavior of some synthetics

(ii)   poor boundary lubrication associated either with the difficulty of finding boundary additives effective in water systems or the lack of suitable additives soluble in some synthetics

(iii)   corrosion, erosion or embrittlement caused by the presence of water

One particular problem associated with the use of water-based fluids is their relatively poor film-forming properties in high pressure, elastohydrodynamic-type contacts. Low film thickness results in high levels of solid/solid contact which has been shown to lead to high wear (9), reduced fatigue (10) and increased likelihood of seizure (11). Water itself has very little elastohydrodynamic film-forming ability due to its very low pressure-viscosity coefficient and this is partially reflected in the film-forming properties of most water-based fluids.

The film-forming properties of water in oil emulsions have been studied using optical interferometry by Hamaguchi (12) and later by Wan (13). These authors showed that for most emulsions, the EHD film thickness formed is that predicted from the emulsion base oil. In practice this is a good deal less than that predicted from the emulsion itself which is generally more viscous than the component base oil. Wan noted, however, that W/O emulsion having very small water droplet size tend to form slightly thicker films than predicted from the viscosity of the component oil (13).

Wan also examined the EHD film-forming properties of HFC-type polyglycol solutions (14). He showed that the presence of even modest proportions of water in monoglycol and polyglycol greatly reduces the pressure viscosity coefficient of the blend, leading to poor film formation.

The film-forming properties of water in oil emulsions have been extensively investigated (15-17). The key to the performance of this type of fluid appears to be the ability of the oil phase to wet the rubbing surfaces and thus form a localised pool of oil in the rubbing contact inlet. At low sliding speeds, many oil in water emulsions are able to do this and thus form lubricant films close to those produced by the water-free oil component. However at some critical speed, which depends strongly upon the emulsifier type and concentration, the supply of oil due to wetting cannot keep up with the contact requirements and severe starvation and film collapse occurs (18). It has recently been shown how the composition of O/W emulsions can be optimised to maximise surface wetting properties and thus film formation at high speeds (19).

Very little work has been carried out on the film-forming properties of HFAS-type fluids although their stability and rheological properties have been outlined by Rasp (20).

All the above work was carried out at relatively low temperatures, between room temperature and about 50°C, and it would be of considerable interest to determine and compare the film-forming properties of fire resistant hydraulic fluids at higher temperature, for reasons outlined in the introduction to this paper. One cause of the lack of previous high temperature measurements is that it was not possible until recently to measure film thicknesses in contact below about 50 nm. In practice, at temperatures above 50°C, because of the reduced viscosity of the fluids, film thickness can fall considerably below this limit. In the last few years, however, a new technique has been developed which is able to measure film formation in rolling contact down to less than 2 nm (21). This means that it is not only possible to measure the film-forming properties of fire resistant fluids at high temperatures but also to examine some of their boundary film forming properties.

The aim of the work described in this paper was to make such film thickness measurements on a range of fire resistant hydraulic fluids in the high temperature range in order to compare their performance under these conditions.

EXPERIMENTAL TECHNIQUES

Elastohydrodynamic/Boundary Film Thickness Measurement

Elastohydrodynamic (EHD) film thickness was measured in a rolling concentrated contact using ultrathin film interferometry (21). The principle of the technique is shown in figure 1. A high pressure contact is formed between the flat surface of a glass disc and a reflective steel ball. The glass disc is coated with a thin, semi-reflective layer of chromium, on top of which is a layer of silica about 500 nm thick. White light is shone into the contact and some is reflected from the chromium layer whilst some passes through the spacer layer and any lubricant film present to be reflected from the steel ball. The two beams recombine and interfere and the spacer layer ensures that interference will occur even if no oil film is present.

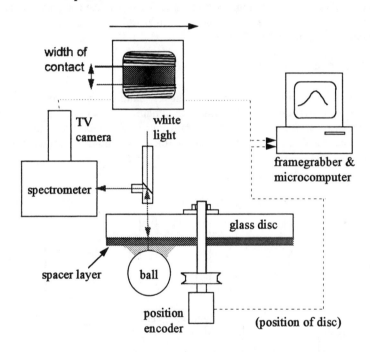

FIGURE 1--Schematic diagram of ultrathin film method

The interfered light from a strip across the contact is then passed into a spectrometer where it is dispersed and detected by a solid state black and white TV camera. A frame grabber is used to capture this image at a pre-selected moment and a microcomputer program determines the wavelength of maximum constructive interference in the central region of the contact. The lubricant film thickness is then calculated from the difference between the measured film thickness and the thickness of the silica spacer layer at that position.

In the test rig, the ball is loaded upwards against the underside of the glass disc and both ball and disc are held within a temperature-controlled, stainless steel chamber. A slotted perspex cover with a side rim is sealed against the top of the glass disc so that the fluid level can be higher than the glass disc to produce a fully-immersed contact. This is important when using some water-based fluids so as to avoid water evaporation from the otherwise very thin out-of-contact track on the glass disc.

In each test, the silica spacer layer film thickness is initially recorded using a position encoder, at a series of set positions around the disc using a dry contact. Fluid is then circulated and a series of film thickness measurements are made at different rolling speeds. In doing this, the position encoder is used to ensure that interference images are detected and grabbed from these same positions where the spacer layer thickness has already been determined.

In the work described in this study the disc was rotated and itself drove the ball in nominally pure rolling. A fresh 19 mm diameter, steel ball of surface roughness 11 nm RMS was used for each test. The load applied was 20 N which corresponds to a maximum contact pressure of 0.52 GPa.

Wear and Friction Test Method

To compare friction and wear properties of the test fluids, a high frequency reciprocating rig (HFRR) was used. This is shown in figure 2. A steel ball is loaded and reciprocated against a steel flat. The contact is fully immersed in lubricant whose temperature is controlled to 0.5°C. The friction coefficient is monitored continuously and the average diameter of the wear scar is measured at the end of a test.

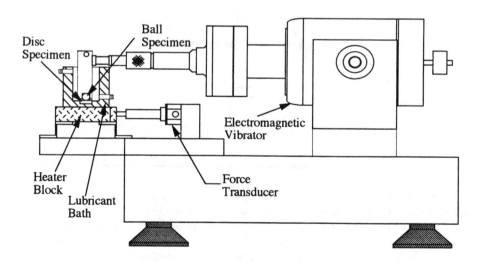

FIGURE 2--HFRR friction and wear tester

This test was designed to compare the wear properties of diesel fuels and is thus suited to studying low lubricity fluids. In the context of the current work it had two other advantages. Firstly, because the motion is reciprocating, the test will operate for a significant proportion of time in the mixed lubrication regime. Thus it is likely to be more responsive to lubricant film-forming ability than a test operating primarily in the boundary regime. Secondly, because of the relatively mild load/speed conditions, there is very little flash temperature heating in the contact, so the test should not be dominated by antiwear or extreme pressure additive effects as other, more severe tests might be (22). The HFRR test conditions used in this study are summarised in table 2.

TABLE 2--HFRR test conditions used in this study

Stroke length	1000 μm
Stroke frequency	20 Hz
Load	400 g
Temperature	40°C or 80°C
Test duration	60 minutes
Ball properties	AISI 52100
Disc properties	AISI 52100

FLUIDS TESTED

The fluids tested are listed in table 3. All were fully-formulated, commercial hydraulic products, spanning a range of different classes of fluids obtained from three different suppliers.

TABLE 3--Hydraulic Fluids Tested

Designation	Viscosity Grade	Composition
HM-46	46	Mineral oil based, fully-formulated
HM-68	68	Mineral oil based, fully-formulated
HM-100	100	Mineral oil based, fully-formulated
HFA-1		5% O/W emulsion
HFA-2		5% O/W micro-emulsion
HFA-3		5% O/W micro-emulsion
HFA-4		5% O/W emulsion (high biodegradability)
HFB-68	68	W/O emulsion
HFB-100	100	W/O emulsion
HFC-1	46	Polyglycol/glycol/water type
HFC-2	46	Polyglycol/glycol/water type
HFC-3	46	Polyglycol/glycol/water type
HFD-46	46	Phosphate ester
HFD-68	68	Phosphate ester

Three were mineral oil-based hydraulic fluids having different viscosities and two were phosphate ester-based synthetics. Four were O/W HFA fluids, two were invert HFB fluids and three were glycol-based HFCs

FILM THICKNESS RESULTS AND INTERPRETATION

Mineral Oil-Based Fluids

Figure 3 shows film thickness results for the three mineral oil-based fluids. These, and all other film thickness results in this paper, are shown as log(film thickness) versus log(speed). According to EHD theory, the central film thickness, h_c is given by (23);

$$h_c = k(U\eta_o)^{0.67}(\alpha)^{0.53} \tag{1}$$

where U is the mean rolling speed, η_o is the viscosity, and α is the pressure viscosity coefficient of the lubricant. Thus so long as it obeys EHD theory, a given fluid should give a straight line of gradient approximately 0.67 on a log(h_c) versus log(U) plot. Any deviation from such behavior is indicative of non-classical EHD behavior such as starvation, non-Newtonian fluid behavior or the presence of boundary films.

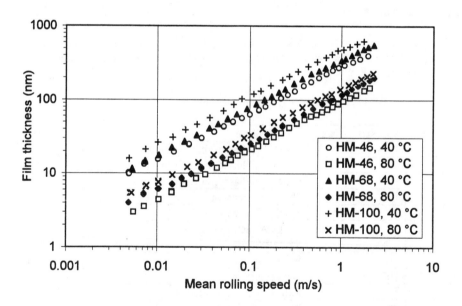

FIGURE 3--Film formation by mineral oil-based hydraulic fluids

All of these mineral oil-based fluids give straight line plots at both temperatures, indicative of conventional EHD behavior down to 3 nm. The film thicknesses at 80°C are lower than at 40°C primarily because the viscosities, η_o of the fluids are reduced at the higher temperature and also, to a small extent because the pressure viscosity coefficients, α are reduced.

These mineral oil results will be used as reference values with which to compare the other types of hydraulic fluid.

Type HFD Fluids

Figure 4 shows the film-forming behavior of the two HFD fluids. These also obey EHD theory over most of the speed/film thickness range although there is a slight deviation at very thin films at 80°C. The plots appear to flatten at film thicknesses below 20 nm, which may indicate a surface film of enhanced viscosity (24).

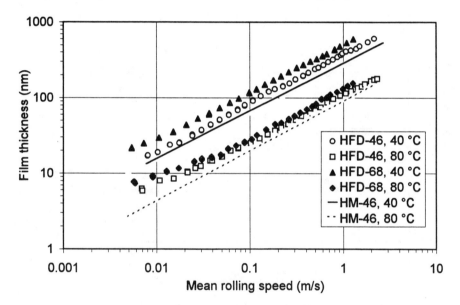

FIGURE 4-- Film formation by HFD hydraulic fluids
(mineral oil shown as solid/dashed lines)

In figure 4, the performance of the mineral oil-based fluid HM-46 is also shown as solid/dashed lines. The corresponding phosphate ester gives a significantly thicker film than the mineral oil at 40°C. This is probably because the phosphate ester has a larger pressure viscosity coefficient than the mineral oil. The difference at 80°C is less striking than at 40°C. The same trend is present with the other HFD fluid. Phosphate esters generally have quite low viscosity indices (VI) and thus thin more rapidly with temperature than mineral oils. At 80°C this will tend to nullify the effect of the higher pressure viscosity coefficient of the phosphate esters compared to the mineral oils.

Type HFB Fluids

Figure 5 shows results for the two HFB invert emulsions. Both fluids give reasonably linear plots except at very high or at slow speed. Interestingly, both fluids form similar film thicknesses at 40°C and the HFB-68 is the thicker of the two at 80°C, even though the HFB-100 has higher bulk viscosity. Previous work on model W/O emulsions has suggested that EHD film thickness depends upon the viscosity of the base fluid rather than on the generally higher viscosity of the bulk emulsion (12). The current results thus imply that the two fluids contain similar viscosity base oils. It should be noted that in low pressure, hydrodynamic contacts the benefits of the more viscous fluid might be more evident.

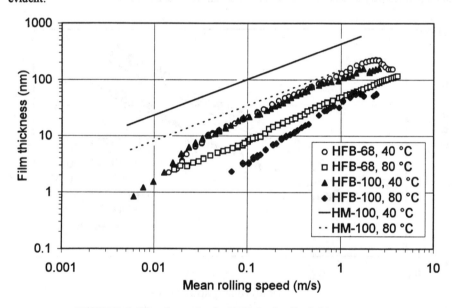

FIGURE 5--Film formation by HFB hydraulic fluids
(mineral oil shown as solid/dashed lines)

Figure 5 also includes the mineral oil-based HM-100 film thicknesses as solid/dashed lines. It can be seen that HFB-100 fluid gives a much lower film thickness than the corresponding mineral oil

The origin of the fall in film thickness at high speed with HFB-68 at 40°C is not known. It may be produced by some starvation in the inlet. One interesting feature is that both fluids show a steepening of the film thickness/speed plot at 40°C; to yield thinner than expected films in this region. This type of behavior has been noted previously when a mixed base fluid consisting of a low viscosity polar ester and a higher viscosity non polar

hydrocarbon was tested (25). The more polar component concentrated near the polar
solid surfaces to produce a low viscosity surface film.

<u>Type HFC Fluids</u>

The three HFC fluids tested are plotted separately in figure 6, together with the film
thickness behavior of HM-46 shown as solid (40°C) and dashed (60°C) lines. All three
fluids show an increase of film thickness with speed, with gradient very approximately
0.67. However within this general trend, two of the fluids show quite complex behavior.
Fluid HFC-1 shows a marked plateauing at a film thickness of about 25 nm. Fluid HFC-2
forms an irregular boundary film of about 5 nm thickness and also exhibits a pronounced
plateauing and even a small, temporary fall in film thickness at about 25 to 20 nm. Fluid
HFC-3 showed far more regular behavior and gives thicker films.

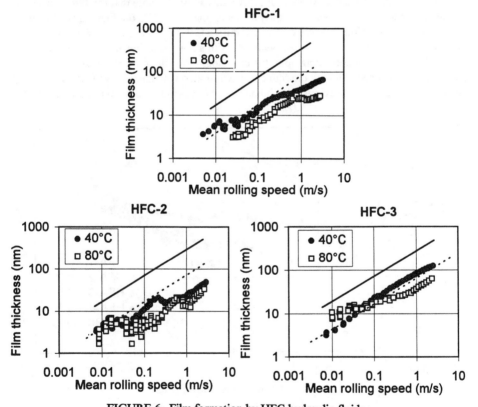

FIGURE 6--Film formation by HFC hydraulic fluids

This type of intermediate levelling-out of film thickness with speed at about 10 to 30
nm has also been observed with solutions of polar VI improver polymers in mineral oils,
where it was ascribed to the presence on the solid surfaces of a layer of polymer solution

about one random coil diameter in thickness (i.e. about 20 nm) which had enhanced polymer concentration and thus viscosity (26). Since HFC fluids contain some high molecular weight polyglycol polymer this may also be the origin of the observed behavior with these two fluids. Alternatively it may reflect some other form of fluid structuring close to the rolling solid surfaces. Polyglycol/monoglycol/water mixtures are known to form quite complex internal fluid structures (14). These structures and consequently their viscosities and pressure viscosity coefficients are likely to be modified in the vicinity of solid surfaces.

All of the HFC fluids give much thinner EHD films than their mineral oil counterparts.

Type HFA Fluids

Figure 7 shows results for the four HFA fluids at 40°C. Note that the film thickness scale on this figure is different from the preceding ones. At 80°C the films formed were so thin that the rubbing surfaces were damaged during tests so it was not possible to acquire reliable results. All four give similar behavior, with a boundary film at low speeds which was 1.5 to 5 nm thick depending upon the emulsion. This film thickness was broadly independent of speed although the film collapsed to zero when motion was halted. Similar behavior has been seen previously for boundary additives in hydrocarbon-based oils (27).

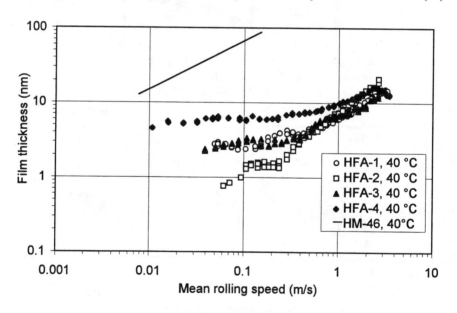

FIGURE 7--Film formation by HFA hydraulic fluids at 40°C

At higher speed, elastohydrodynamic films start to form for all fluids but these do not reach more than 20 nm thickness within the speed range studied.

This film thickness behavior is quite poor. Previous studies using O/W emulsions for application as rolling oils have tended to show much thicker elastohydrodynamic film formation in the slow speed range. Figure 8 shows the results for a commercial rolling oil emulsion where it can be seen that the film thickness first rises with a gradient of about 0.67, then collapses and then starts to rise once more. As outlined in the background section of this paper, the slow speed EHD film formation is due to the build up of a reservoir of oil in the contact inlet caused by wetting. The film collapse at higher speed then occurs when this reservoir cannot be replenished fast enough. For the emulsions tested in the current study there appears to be negligible slow speed EHD film formation. This may be because the emulsions have been formulated in such a way that the oil droplets are unable to wet the solid surfaces.

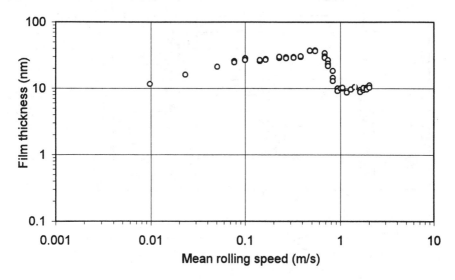

FIGURE 8--Film formation by a commercial O/W rolling emulsion

WEAR RESULTS

Friction coefficient and wear test results are listed in table 4. For most of the fluids tested, the wear was less at 80°C than at 40°C, suggesting that antiwear additives or other surface chemical behaviour was being activated at the higher temperature. The exception was HFC-3 which showed very poor wear performance at 80°C.

Table 4 also lists the measured film thickness at 0.1 m/s rolling speed. There was quite good correlation between the wear scar diameters and the film thicknesses at 40°C, as shown in figure 9. The solid line is a regression-fitted exponential curve and matches the data with $R^2=0.95$. The correlation at 80°C was poorer, presumably because of additive effects. Friction coefficient also decreased with increasing film thickness over the range of fluids except that HM-68 gave considerably higher friction and HFC-3 lower friction than expected from their film-forming properties. The high friction for the mineral oil-based fluid suggests that, whilst the fluid almost certainly contains antiwear additives, these are not at all effective at controlling friction in the boundary and mixed lubrication regimes.

TABLE 4--Wear scar and friction coefficient results

Fluid	Temperature, °C	Wear scar diameter, μm	Friction coefficient	Film thickness at 0.1 m/s, nm
HM-68	40	109	0.141	80
	80	106	0.148	23
HFA-1	40	218	0.146	2.5
HFA-2	80	228	0.164	1.5
HFB-68	40	161	0.128	20
	80	111	0.129	7
HFC-1	40	192	0.132	11
	80	193	0.117	7
HFC-3	40	173	0.099	20
	80	260	-0.133	18
HFD-68	40	97	0.121	110
	80	93	0.126	30

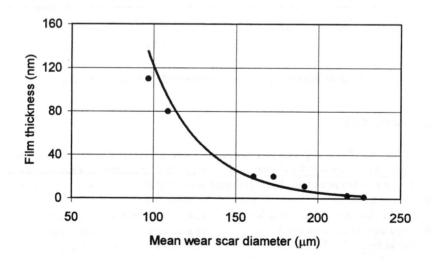

FIGURE 9--Correlation between wear and film thickness measurements at 40°C

DISCUSSION

The results show a wide diversity of behavior from conventional EHD performance for mineral oils and phosphate esters, to a combination of boundary film formation with only a very thin EHD film for type HFA emulsions.

The film thickness results are broadly in agreement with known pump performance; i.e. mineral oil/phosphate esters are the best lubricants, W/O emulsions somewhat less effective, polyglycols/glycols much less effective and O/W emulsions the worst. The wear results correlate quite well with the film thickness measurements at 40°C but not at 80°C, possibly because additive effects control wear at the higher temperature. This raises the question of whether wear or film thickness measurements are more relevant in determining real pump performance at high temperatures. The authors suspect that film thickness may be more relevant under the complex combination of conditions found in pumps. The fact that failures tend to occur at higher temperatures whilst the measured wear performance is much the same at both temperatures may support this.

There were quite striking differences in behaviour between the various HFC fluids and some fluids showed very interesting non-linear film-forming behavior. Clearly further work needs to be carried out on these fluids. It also appears that there may be scope for improving the lubricating properties of some HFA hydraulic fluids

CONCLUSIONS

(i)   The EHD and boundary film-forming properties of a range of commercial hydraulic fluids has been measured. Some wear properties have also been determined.

(ii) There are large differences in the thickness of films formed by different classes of fluid, spanning two orders of magnitude from mineral oil-based fluids to HFA O/W emulsions.

(iii) The relative film formation correlates with known relative performance in lubricating hydraulic pumps.

ACKNOWLEDGEMENTS

The authors wish to thank Dr David Yardley of E.D. Yardley Associates for much valuable advice and discussion and also those companies who kindly provided hydraulic fluids for testing.

REFERENCES

1.   ISO standard 6743/4, Intern. Standards Organisation, Geneva, Switzerland
2.   Snyder, C.E. and Geschwender, L.J. "Fire Resistant Hydraulic Fluids", STLE, Presented at STLE Annual Meeting, Chicago, May 1995, preprint 95-AM-4D-1, to be published in Tribology Transactions.

3. Grover, K.B. and Perez, R.J. "The Evolution of Petroleum Based Hydraulic Fluids", Lubrication Engineering 46, pp. 15-20, (1990).
4. Spikes, H.A. "Wear and Fatigue Problems in Connection with Water-Based Hydraulic Fluids." J. Synthetic Lubrication 4, pp. 115-135, (1987).
5. Myers, M.B. and Townsend, F. "Fire-resistant Fluid and Lubricant Trends Within the Mining Industry", Proc. Intern. Conf., Tribology in Metal Extraction, Sept. 1994, Instn. .Mech.Engnrs. C336/84, pp. 95-104,
6. Wymer, D.G. "The Use of Water Hydraulics in Machines in Mechanised Gold Mining in South Africa", Proc. Intern. Conf. Mining and Machinery, Brisbane, July 1979, pp. 320-328.
7. Rudston, S.G. and Whitby, R.D. "Recent Developments in the Performance Properties of Water Based Cutting and Hydraulic Fluids", J. Synthetic Lubrication 2, pp. 183-212, (1985).
8. Rubery, A.M. "Compatibility of Performance and Fluid Properties", Inst. Petr. Intern. Symp. Performance Testing of Hydraulic Fluids, Oct. 1978, London, ed. R Tourret and E.P. Wright, publ. Heyden and Son Ltd., London.
9. Sastry, V.R., Sethuramiah, A. and Singh, B.V. "A Study of Wear Under Partial EHD Conditions", Proc. 11th Leeds-Lyon Symposium on Tribology, Leeds, 1984, Mixed Lubrication and Lubricated Wear, Ed. D Dowson et al. Publ. Butterworths, 1985.
10. Life Adjustment Factors for Ball and Roller Bearings, An Engineering Guide Sponsored by the Rolling-Elements Committee, The Lubrication Division of ASME, 1971.
11. Dyson, A. "The Failure of Elastohydrodynamic Lubrication of Circumferentially Ground Discs", Proc. Inst. Mech. Engnrs. 190, (1976), pp. 52-76.
12. Hamaguchi, H., Spikes, H.A. and Cameron, A. "Elastohydrodynamic Properties of Water in Oil Emulsions." Wear 43, pp. 17-24, (1977).
13. Wan, G.T.Y., Kenny, P. and Spikes, H.A. "Elastohydrodynamic Properties of Water Based Fire-Resistant Hydraulic Fluids." Tribology International 17, pp. 309-315, (1984).
14. Wan, G.T.Y and Spikes, H.A. "The Elastohydrodynamic Lubricating Properties of Water-Polyglycol Fire Resistant Fluids." ASLE Transactions 27, pp. 366-372, (1984).
15. Kimura, Y. and Okada, K. "Lubricating Properties of Oil in Water Emulsions." Tribology Transactions 32, pp. 524-532 (1989).
16. Nakahara, T. M., Makino, T. and Kyogoku, K. "Observations of Liquid Droplet Behavior and Oil Film Formation in O/W Type Emulsion Lubrication" Trans ASME J. of Tribology 110, pp. 348-353 (1988).
17. Zhu, D., Bireshaw, G., Clark, S.J. and Kasun, T.J. "Elastohydrodynamic Lubrication with O/W Emulsions", ASME Preprint 93-Trib-4, (1993).
18. Barker, D.C., Johnston, G.J., Spikes, H.A. and Bünemann T. "EHD Film Formation and Starvation of Oil in Water Emulsions." Tribology Transactions 36, pp. 565-572, (1993).
19. Ratoi Salagean, M., Spikes, H.A. and Rieffe, H.L. " Optimising Film Formation by Oil in Water Emulsions" to be presented at STLE Annual Meeting,. Cincinnati, May 1996.
20. Rasp, C.R. "Water Based Hydraulic Fluids Containing Synthetic Components", J. Synthetic Lubrication, 6, pp. 233-251, (1989),.

21. Johnston, G.J., Wayte, R. and Spikes, H.A. "The Measurement and Study of Very Thin Lubricant Films in Concentrated Contacts." Tribology Transactions 34, pp. 187-194, (1991).
22. Spikes, H.A., Bovington, C., Caprotti, R., Meyer, K. and Kreiger, K. "Development of a Laboratory Test to Predict Lubricity Properties of Diesel Fuels and its Application to the Development of Highly Refined Diesel Fuels." Tribotest 2, pp. 93-112, (1995).
23. *Ball Bearing Lubrication: the Elastohydrodynamics of Elliptical Contacts*, B T Hamrock and D Dowson, publ. J Wiley, New York, 1981.
24. Guangteng, G. and Spikes, H.A. "Boundary Film Formation by Lubricant Base Fluids". Accepted for publication, Tribology Transactions, May 1995.
25. Spikes, H.A. "Boundary Lubricating Films", Proceeding of Intern. Tribology Conf., Yokohama 1995, Satellite Forum on Tribochemistry, publ. Jap. Soc Trib., Tokyo, October 1995.
26. Smeeth, M., Gunsel, S. and Spikes, H.A. "Boundary Film Formation by Viscosity Index Improvers". Accepted for publication in Tribology Transactions, preprint no. 95-3B-TC-1.
27. Spikes, H.A. and Cann, P.M. "The Thickness and Rheology of Boundary Lubricating Layers", presented at International Tribology Conference, Yokohama, October 1995. Accepted for publication in the Proceedings.

Kazuyuki Mizuhara,[1] and Makoto Tomimoto[2]

THE EFFECT OF REFRIGERANTS IN THE MIXED LUBRICATION REGIME

REFERENCE: Mizuhara, K. and Tomimoto, M. **"The Effect of Refrigerants in the Mixed Lubrication Regime,"** *ASTM STP 1310,* George E. Totten, Gary H. Kling, and Donald J. Smolenski, Eds., American Society for Testing and Materials, 1996.

ABSTRACT: Because of environmental concerns, CFC (chlorofluorocarbon) refrigerants must be replaced with HFCs (hydrofluorocarbons.) As a result, many tribological problems are caused especially in rotary piston compressors. To solve the problem, the effects of refrigerants on friction and wear characteristics of the oil and refrigerant mixtures at the mixed lubrication regime are investigated. The difference in refrigerants are clearly observed not only in boundary but also in the mixed lubrication regime. The effects of operating conditions on sliding conditions and experimental results are also discussed. It is concluded that for practical application where long life is essential, experiments must be conducted under the mixed lubrication regime. Also, the importance of defining the lubrication regime in terms of film parameter is emphasized.

KEYWORDS: refrigerants, mixed lubrication, film parameter, friction, wear, rotory piston compressor

To solve the problem caused by the replacement of CFC refrigerants by HFCs, many rig- and bench-tests have been conducted to screen the lubricants and it has been found that the test methods used for CFCs are not appropriate for HFCs. This discrepancy occurs because the test conditions are generally much more severe than the practical conditions, thus the effect of extreme pressure (EP) additives dominates [1]. It is well known that CFCs have lubricity [2] but HFCs don't, and that the lubricity of CFC is that of EP additives [3]. Then increasing the severity of the test would be somewhat advantageous to the CFCs but not to the HFCs. Also an "accelerated" test in a CFC system is not as severe as that compared with an HFC system. This might have enhanced the disagreement between the bench- and rig-tests results in the HFC system.

Analytical studies have predicted that the lubrication conditions at the vane tips are in the boundary condition [4], although it is somewhat suspicious that compressors operating under

[1]Senior Research Official, Tribology Division, Mechanical Engineering Laboratory, Tsukuba 305 Japan

[2]Research Engineer, Hoya Corporation, Kitaoka, 403 Japan

TABLE 1--Test specimen

	Material	Surface roughness* (μm)	Hardness (MPa)
Vane	Ni$_3$P	0.15	2.8
Disk	aluminum-silicon alloy	0.15	1.5

*Ra(Center line average)

TABLE 2--Lubricants

Type of Oils	Viscosity (mm^2/s)		Total Acid Values (mgKOH/g)	Additives	Water (ppm)
	40°C	100°C			
Naphthenic	33	4.45	0.01>	None	25>
PAG	32.3	6.66			100>

Test Procedure

Preceeding the test, specimens and the holders were cleaned ultrasonically by using acetone : hexane (2:1) for 5 minutes and dried in the air. A small quantity (1.25 cm^3) of the refrigerant oil was dropped on the disk. Then, the chamber was evacuated down to 1 Pa, and refrigerant was fed into the chamber up to a pressure of 0.1 MPa. This procedure was repeated twice, and later on, the refrigerant was introduced into the chamber up to the test pressure and kept at that pressure for at least 30 minutes. Since the oil layer thickness is only 1 mm, this process generally ensures the quasi equilibria of refrigerant in oil and no pressure change was observed during the experiments. Table 3 shows the test conditions employed.

TABLE 3--Test conditions

Duration	Speed	Load	Oil temperature	Oil amount	Refrigerant pressure
hr	m/s	N	°C	cm^2	MPa
2, 4	0.02-1.37	9.8	75	1.25	0.1- 0.5

Wear of the test specimens was measured after the test and coefficient of friction was measured at the end. Wear rates were calculated as the wear volume in m^3 divided by a sliding distance in m. Wear volumes were calculated from wear scar widths on the vanes after the tests.

boundary condition have been achieving a life of 10 years. To estimate the lubricating condition in actual compressors, the authors have conducted short term bench tests. As a result, it is estimated that vane pumps start in the boundary lubrication regime but enter in the mixed lubrication regime within ten minutes [5]. This transition is caused by the formation of a flat area at the vane tip during the initial wear process.

In this work we attempt to clarify the reason for this discrepancy, to find appropriate screening test conditions, and investigate the effect of refrigerants in relatively mild conditions (mixed lubricaton regime).

EXPERIMENTS

Test Apparatus

The schematic diagram of the pin on disk type friction and wear tester used is shown in Fig.1. The machine was equipped with a hermetic chamber, which contained the disk and the contact pins (vanes). The range of refrigerant pressure was from 0.1 to 0.5 MPa. The drive shaft was sealed by a mechanical seal to prevent any leakage of the refrigerant. The static normal load was applied within the chamber. The temperature of the oil and the disk was controlled by circulating water under the disk through the driving shaft.

Fig. 2 shows the test configuration of two vanes and a disk. Each vane was made of an actual vane (or blade) which is utilized in a rotary compressor. The rotating disk was made with the material used for a cylinder in a vane type compressor (see Table 1.) Table 2 summarizes the chemical properties of the oils used. Generally, naphthenic oil is used in a rotary compressor for CFC-12. PAG (polyalkyleneglycol) oil was developed for a rotary compressor for HFC-134a. No additives were used in these oils.

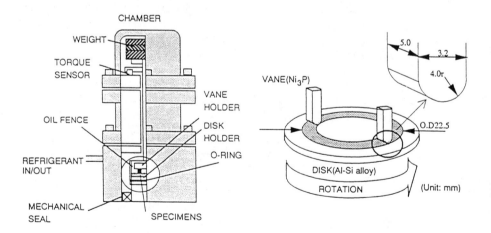

Fig. 1--Schematic of vane on disk tester   Fig. 2--Configuration of test specimens

Evaluation of Oil Film Thickness Parameter

As mentioned above, the test conditions were selected to produce the mixed lubrication regime. In this regime, since it is well known that the friction and wear phenomena strongly depend on the oil film parameter (h_{min}/Ra or λ), friction and wear data were interpreted using this parameter [8]. During the tests, the cylindrical surface of the vane is partially flattened by wear. Therefore, to calculate the oil film thickness (h_{min} or h_0), the model as shown in Fig.3 was employed and Reinolds' equation (1) was applied to the equation (2). The oil film thickness (h_{min} or h_0) was calculated by equation (2) using Newton's method.

$$\frac{dp}{dx} = 6\eta V\left(\frac{h - h_0}{h^3}\right) \qquad (1)$$

$$W = l\left\{\int_{-b/2}^{-a} p(x)dx + 2ap(-a)\right\} \qquad (2)$$

where

V=velocity
W=load
h_0=film thickness
h=distance from the surface at x
$p(x)$=pressure at x
l=vane width

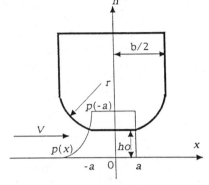

Fig. 3--Model for film thickness calculation

Viscosities of oil-lubricant mixtures at each test condition were calculated from the data provided by the oil suppliers and are shown in Figs.4 and 5. As shown in Fig. 4, at the pressure of 0.5 MPa, CFC-12 dissolves in the naphthenic oil more than HFC-134a does in the PAG oil, although at the pressure of 0.1 MPa the refrigerant concentration are almost the same. As shown in Fig. 5, the viscosity of the mixtures decreases with increasing refrigerant concentration.

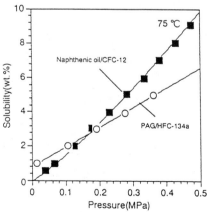

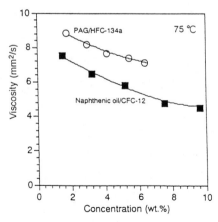

Fig. 4--Relation between solubility and refrigerant pressure

Fig. 5--Relation between refrigerant concentration and viscosity

RESULTS AND DISCUSSIONS

Effect of Refrigerants in Boundary Lubrication Regime

Fig. 6 shows the effect of refrigerant pressure in the boundary condition. In this condition, the effectiveness of EP additives increases with increasing μPV value, which is equivalent to the energy consumed at the contact [1]. Although the μPV values are relatively small (0.01-0.1 Nm/s,) the friction coefficient (a) and wear rate (b) are both reduced with increase of the refrigerant pressure. From these figures, it could be concluded that CFC-12 has better lubricity than HFC-134a. In this regime, since the viscosity of oils has little effect on friction and wear phenomena, the effect of refrigerant pressure is mainly responsible for increasing the refrigerant concentration. As CFC-12 is known to have EP capability so, the increase of the refrigerant concentration is equivalent to increasing the additive concentrations. Then, the data is sorted by refrigerant concentration.

As can be seen in Fig. 7 (a), reduction in friction coefficient is almost the same for the two refrigerants. It should be noted that the supremacy of CFC-12 in reduction of wear rate is partly responsible for a higher solubility of refrigerants (additives) in naphthenic oil (see Fig. 4.) It is interesting to see even HFC-134a shows almost the same performance as CFC-12 which suggests that HFC-134a also has EP capability under this condition on the aluminum-Ni_3P interface.

The EP effect of HFC-134a is observed only in dry friction using copper alloy and steel but not with lubricant [2] so this result is a bit surprising, although this will be attributed to relatively high testing temperature and higher reactivity of aluminum. It is reported that halogenides which have a higher heat of formation value are more easily formed [3] and that of aluminum fluoride (AlF_3) is -376.6 Kcal/g mole which is much higher than those of iron (-177.8) and copper (-129.6) fluorides [9]. On the wear scars of the aluminum disks tested under 5 MPa of HFC-134a, trace amounts of fluorine were detected by means of the XPS (X-ray photoelectron spectroscopy.)

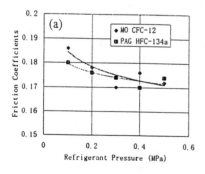
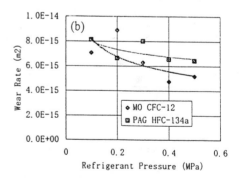

Fig. 6--Effect of the refrigerant pressures on friction (a) and wear (b) in boundary lubrication regime

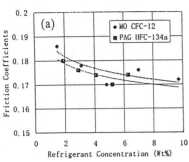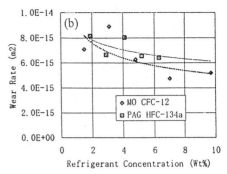

Fig. 7--Effect of the refrigerant concentration on friction (a) and wear (b) in boundary lubrication regime

Effect of Refrigerants in Mixed Lubrication Regime

Fig. 8 shows the relationship between the frictional coefficient and film parameter at 0.1 and 0.5 MPa refrigerant pressure. At 0.1 MPa (a), the difference in refrigerant had no effect and the friction coefficient was reduced with increasing film parameter. This indicates the lubricating condition is in mixed lubrication (ML) regime. At 0.5 MPa (b), frictional coefficient is also reduced with increasing film parameter. At film parameters below 0.05, friction coefficients are not dependent on the film parameter so that film parameters below 0.05 can be defined as boundary lubrication (BL) regime. Thus, the film parameters between 0.05 to 1 are considered to be in the mixed lubrication regime in these experiments. As can be seen in Fig 8 (b) the combination of CFC-12 and naphthenic oil show lower friction coefficient than the HFC-134a and PAG combination in the ML regime. It must be noted that the differences in friction coefficient in ML regime are larger than in BL regime (see Fig . 6 and 7.)

Fig. 9 shows the relationship between the film parameter and wear rate. Wear rates also reduced with increasing film parameter. Again, the difference in refrigerant is not clear at 0.1 MPa (Fig. 9 (a)), but the advantage of CFC-12 over HFC-134a is prominent at 0.5 MPa (Fig. 9 (b)). In contrast to the frictional results, below the film parameters of 0.05, wear rates continued to increase with decreasing film parameter.

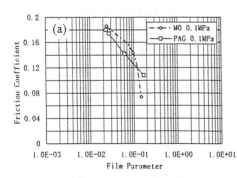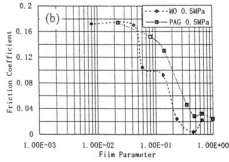

Fig. 8--Relationship between the frictional coefficient and film parameter at (a) 0.1 and (b) 0.5 MPa of refrigerant pressures

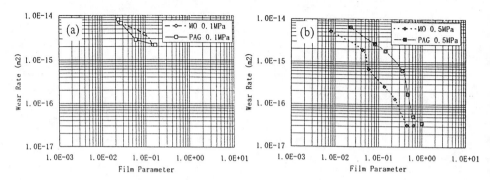

Fig.9--Relationship between wear rate and film parameter at (a) 0.1 and (b) 0.5 MPa of refrigerant pressure

Fig.10 shows the effect of pressure on wear rate. As shown in Fig.10 (a), in CFC-12, increasing the pressure reduced the wear rate dramatically even in the mixed lubrication regime. In contrast, the pressure increase of HFC-134a only affected the reduction in film thickness parameter, and showed no effect on wear rate reduction (see Fig. 10 (b).)

It can be concluded that CFC-12 has lubrication capability in the mixed lubrication regime where even the μPV (equivalent to the energy consumed at the contact) or PV values are quite low. Also the lubricity of CFC-12 over cames the disadvantageous effect of reducing the viscosity and thus the film thickness. On the other hand HFC-134a which shows some advantageous effects in boundary lubrication has no effect in the mixed lubrication regime if the data is presented in terms of film parameter. This means that HFC-134a has a disadvantageous effect in mixed lubrication since the film parameter is reduced by dissolution of refrigerant and the friction and wear increases with decreasing film parameter.

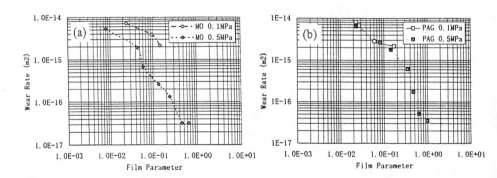

Fig. 10--Effect of refrigerant pressure on wear rate in mixed lubrication regime.
(a) CFC-12, (b) HFC-134a

Calculating Film Thicknesses Based on EHL Theory

When looking at these results bends are observed in Fig. 10 (b) both of which are at a sliding speed around 0.5 m/s. These bends are caused by the model used for film thickness calculation, since the lubrication below 0.5 m/s is estimated to be in the 'elastic-isoviscous' (E-I) regime in elasto-hydrodynamic lubrication (EHL) conditions by means of elasticity and viscosity (g_3) parameters[10-13]. Therefore film parameters below 0.5 m/s were also calculated by Herreburg's formula (3) proposed for the E-I regime [14], using the Young's modulus for vane and disk as 98 and 79 GPa, Poisson's ratio of 0.35 and assuming no wear.

$$\bar{h} = 3.10 \ g_3^{0.8} \tag{3}$$

$$g_3 = \frac{w}{(\mu_0 uE\ 'R\ )^{1/2}} \tag{4}$$

$$h_{min} = \frac{\bar{h} \ \mu_0 uR}{w} \tag{5}$$

where

$\bar{h}$=dimensionless film thickness
μ_0=viscosity
u=mean velocity
E'=equivalent elastic modulus
R=equivalent curvature radius
w=load per unit width

The results using the film parameter based on EHL theory are shown in Fig. 11 as circles. From Fig. 11 (a) the flatting out of friction coefficient occurred at a film parameter around 0.08 for both refrigerants. It should be noted that the transition of friction coefficient values from boundary to hydrodynamic lubrication (HL) takes place over a relatively narrow range of film parameter, say between 0.08 and 0.7.

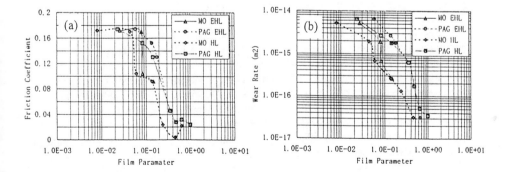

Fig. 11--Data sorted using film parameters based on EHL and HL lubrication

In contrast to the friction coefficient, wear rate did not flatten out but kept on increasing as film parameter decreased. Concerning the bends, as can be seen in Fig. 11 (b), the relation between wear rate and film parameter becomes more straight. Further investigation is needed to confirm this linear relationship on a logarithmic scale.

Effect of Experimental Period

Fig. 12 shows the friction and wear data at 2 and 4 hour periods of testing. Friction coefficients for HFC-134a at 4hrs are lower than that at 2hrs but that of CFC-12 are scarcely changed after 2hrs. This suggests that CFC-12 achieves a steady state much faster than HFC-134a with less wear and at lower film parameter values.

As can be seen in Fig. 12 (a), friction coefficients in HFC-134a after 4hrs are similar to those in CFC-12 although wear rates in HFC-134a at 4hrs are still higher than in CFC-12 (b). Thus it can be concluded that the effect of CFC-12 is more relevant to wear than to friction.

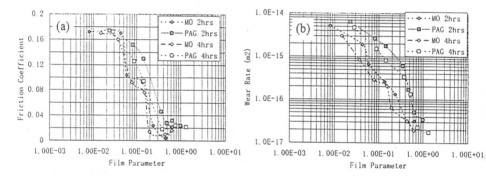

Fig. 12--Friction and wear results at 2 and 4 hours

Advantage of Film Parameter over Sribeck's Parameter

Fig. 13 shows the friction and wear data in terms of the Stribecks' parameter. This parameter has similar tendencies to the film parameter if the film thickness is calculated assuming the fluid dynamic lubrication. The advantage of this parameter is that it can be readily calculated from the experimental conditions, although the physical meaning of the parameter value is vague as compared with the film parameter. Thus it is difficult to judge the lubrication regime from these parameter values. On the other hand, film parameter values give clear images of the lubricating conditions; below 1 solid-solid contact is expected. But we must also not forget that the values could differ when different roughness and film thickness calculation methods are used. As can be seen in Fig. 9, calculated film thickness value (assuming EHL and HL) differed by 3 times at most. So, we can not rely too much on precise values of film parameter, but the tendency must be observed. Then, as discussed above, the film thickness based on combination of EHL and HL lubrications should be used for more detailed analysis of data.

Concerning the effect of worn area, changes to the film thickness of only 30% were observed in these tests. It is thus feasible to calculate the film thickness ignoring the configuration change because of wear. Then, no convergent calculation such as Newton's method is necessary and all the film thicknesses could be calculated on conventional spread sheet software.

This result appears somewhat different from the conclusion obtained for an actual compressor, where the flat area formed by the wear increased the film parameter as much as 8 times [5]. This is because the pressure at the inlet of the contact (which controls the film thickness) is mostly decided by the rough configuration of the contact.

The other advantageous feature of film parameter can be seen in Fig 11 (a) the difference in refrigerants is not clear in Stribeck's diagram even at 0.5 MPa. Considering that in mixed lubrication regime, some solid-solid contact must be involved, then data presented in terms of film parameter is more understandable.

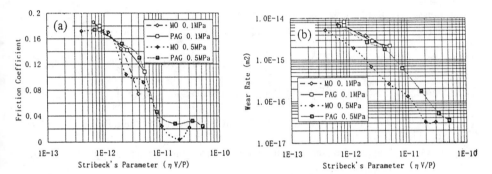

Fig. 13--Experimental results sorted by Stribeck's parameter

CONCLUSIONS

The following points are concluded:

HFC-134a has some lubricating capability at an aluminum/Ni$_3$P interface under relatively mild boundary lubrication regime.

CFC-12 shows lubrication capability not only in the boundary but also in the mixed lubrication regime.

The lubrication effect of CFC-12 in the mixed lubrication regime is more relevant to wear reduction than to friction reduction.

HFC-134a shows a disadvantageous effect in mixed lubrication regime by reducing the film thickness.

The friction and wear phenomena in mixed lubrication regime in an HFC-134a atmosphere are dependent on film parameter if the load is unchanged.

It is suggested that the presentation of data in terms of film parameter is valuable in analyzing the friction phenomena in the mixed lubrication regime.

It is also suggested that film thickness be calculated based on EHL and HL lubrication theory ignoring the wear.

REFERENCES

[1] Mizuhara, K., Tsuya, Y., "Investigation of a Method for Evaluating Fire-Resistant Hydraulic Fluids by means of an Oil Testing Machine," Tribology International, Vol. 25, No. 1, 1992, pp 37-43

[2] Murray, S.F., Johnson, R.L. and Swikert, M.A., "Difluorodichloromethane as a Boundary Lubricant for Steel and Other Metals," Mechanical Engineering, Vol. 78, March 1956, pp 233-236

[3] Mizuhara, K. and Akei, M., "The Friction and Wear Behavior in Controlled Alternative Refrigerant Atmosphere," Tribology Trans., Vol.37, 1, pp120-128, 1994.

[4] Yanagisawa, T, Shimizu, T., Chu, I. and Ishijima, K. " Motion Analysis of rolling Piston in Rotary Compressor," Proc. 1982 Purdue Compressor Tech. Conf., 1982, pp185-192

[5] Mizuhara, K., Tomimoto, M. and Nishizawa, T., "Initial Wear of a Rotary Compressor Vane in HFC-134a", Proc. of 6th Nordic Symposium on Tribology '94 Uppsala, Vol.2, Jun. 1994, pp 473-480

[8] Czichos, H., Habic, K.H., in Dowson.D, Taylor CM, Godet, M., Berthe, D., "Mixed Lubrication and Lubricated Wear", Proc. 11th Leeds-Lyon Symposium on Trib., Butterworths, 1985, pp135-147

[9] Handbook of Chemistry and Physics, Weast, R.C., ed., CRC Press, Cleveland, 53rd Ed., 1972

[10]Johnson, K.L.,"Regimes of Elastohydrodynamic Lubrication,"J. Mech. Eng. Sci., 12, 1970, pp9-16

[11] Greenwood, J.A.,"Presentation of Elastohydrodynamic Film-Thickness Results," J. Mech. Eng. Sci., 11, 2, 1969, pp128-132

[12] Moes, H.,"Elastohydrodynamic Lubrication," Proc. Instn mech. Engrs, 180, (Pt 3B) communication, 1965-66, pp244

[13]Hooke, C.J.,"The Elastohydrodynamic Lubrication of Heavily Loaded Contacts," J. Mech. Eng. Sci., 19, 4, 1977, pp149-156

[14]Herrebrugh, K.,"Solving the Incompressible and Isothermal Problem in Elasthydrodynamic Lubrication Through an Integral Equation," Trans. ASME, Ser. F, 90, 1968, pp262-270

Atsushi Yamaguchi[1]

TRIBOLOGY OF HYDRAULIC PUMPS

REFERENCE: Yamaguchi, A., **"Tribology of Hydraulic Pumps,"** *Tribology of Hydraulic Pump Testing, ASTM STP 1310,* George E. Totten, Gary H. Kling, and Donald J. Smolenski, Eds., American Society for Testing and Materials, 1996.

ABSTRACT: To obtain much higher performance than that of alternative power transmission systems, hydraulic systems have been continuously evolving to use "high-pressure". Adoption of positive displacement pumps and motors is based on this reason. Therefore, tribology is a key technology for hydraulic pumps and motors to obtain excellent performance and durability.
    In this paper the following topics are investigated:
    1. The special feature of tribology of hydraulic pumps and motors.
    2. Indication of the important bearing/sealing parts in piston pumps and effects of the frictional force and leakage flow to performance.
    3. The methods to break through the tribological limitation of hydraulic equipment.
    4. Optimum design of the bearing/sealing parts used in the fluid to mixed lubrication regions.

KEYWORDS: tribology, mixed lubrication, fluid lubrication, power loss, sliding parts, pump, motor, fluid power

1. INTRODUCTION

    Machine systems do useful work for human beings by using power of prime movers. The prime mover has inherent characteristics, and the work or the load, which human beings want to do, has its own characteristics also. Therefore, to harmonize with each other, various types of power transmissions are adopted. In brief, it is to make an adjustment in the force (torque) and velocity (angular speed).
    There are three types of typical power transmissions, which have infinitely variable drive characteristics under a wide range of operating conditions. That is, hydrostatic(hydraulic), pneumatic and electric power transmissions(or drives).
    Here, we define hydraulic systems as the systems which use a liquid as a medium of hydrostatic energy and are composed of hardware and software. Since these hydraulic systems use a highly-pressurized liquid, it is possible to make hydraulic equipment or systems small in size and light in mass. This means that hydraulic systems make an important feature of the ease of control, because the mass of the moving

[1]Professor, Department of Mechanical Engineering and Materials Science, Yokohama National University, 156, Tokiwadai, Hodogaya-Ku, Yokohama, 240 Japan

element (such as valve elements and fluids) and the distance of motion
are small. Based on the feature hydraulic systems are favorably used for
high performance and automation of various machine systems.

2. TRIBOLOGY OF HYDRAULIC EQUIPMENT

To keep the merits of hydraulic systems to electric drives in the
speed of response, the system pressure is continuously increased. Here,
we examine problems of development of high-pressure pumps and motors,
because they are in charge of energy conversion.

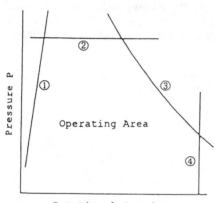

Fig. 1 Limits of Operation of hydraulic pumps and motors
(1: Fluid film formation, 2: Mechanical strength,
3: Heat balance or life of rolling bearing, 4: Cavitation)

Hydraulic pumps, which belong to positive displacement pumps, are
basically suitable to obtain a highly-pressurized liquid. This means
that tribology plays important roles in their performance. Figure 1
shows the limit of operation of typical hydraulic pumps. Besides the
limits based on the strength of materials (2) and cavitation (4), other
two limits are closely related to tribology. Curve (1) is the limit of
the formation of fluid films on the sliding parts, and (3) the limit of
the heat balance of fluid films and/or the life of rolling bearings.
The feature of tribology of hydraulic equipment (mainly pumps and
motors) is as follows:
(1) the working fluids (hydraulic fluids) are used as lubricants
also.
(2) the operational conditions vary under a wide range. Although
the loads on the sliding parts are basically determined by a fluid
pressure of the high-pressure side, the pressure varies widely according
to the power transmitted. The sliding speed changes widely (including
the change in rotational direction) due to the change in the shaft
speeds. For hydraulic motors, the ratio of the maximum to the minimum
speed reaches up to the 3rd or 4th power of 10.
(3) the sliding parts are also in charge of seals.
(4) the working elements as pumps and motors constitute the
sliding parts, and support forces and moments of force. Therefore,

freedom in the design is limited. Hydraulic motors should be much less in size and mass compared with electric motors and internal combustion engines of the same power.

(5) base on the item (2), the concept of hydrostatic lubrication is ease to apply to the sliding parts. Piston type machines are superior to the other types for this reason.

(6) to guarantee reliability as hydraulic control systems, filters to remove contaminants are generally adopted in the circuits.

To solve the tribological problems on hydraulic equipment, there are 3 basic answers:

(a) improvement of hydraulic fluids,
(b) improvement of materials and/or surface treatments,
(c) improvement of mechanisms (designs).

We can recognize that development of antiwear fluids has contributed to improvement of the characteristics and performance of hydraulic equipment and systems. Progress in today's hydraulic technology has been indebted to improvement of hydraulic fluids. The sliding parts in hydraulic equipment operate usually under a mixed lubrication region, then the item (b) is closely related to hydraulic fluids. New materials are occasionally tried, however, from the viewpoint of cost and reliability, they are not adopted widely.

The item (c) is not regarded as of major importance by nowadays. This appears directly in "hydraulic balance", that is, a design concept of the sliding parts in pumps and motors. This concept means to balance the forces and moments of force, but dose not include the stiffness and damping of the fluid films.

3. TRIBOLOGY OF PISTON PUMPS AND MOTORS

We use positive displacement type fluid machines as hydraulic pumps and motors, because this type of machines is basically suitable to the high-pressure operation. In JIS B 0142 (Japan Industrial Standard) the positive displacement pump is defined as follows: based on the movement and/or change in the sealed space formed by the casing and moving parts inscribed with it, the pump presses out a liquid from the suction to the delivery side.

Therefore, tribological conditions of the sliding parts have direct effects on the performance of hydraulic pumps.

From the historical point of view, subjects of the studies were the pump itself. There were a lot of studies on the performance formulas [1], experimental results [2] which delivered the conclusion shown in Fig. 1, and effects of solid contaminants on the performance [3], and so on. These studies related to tribological conditions of the sliding parts, directly or indirectly.

Figure 2 shows the test results [4] of frictional torque (ΔT) at the starting and low-speed conditions. The tested fluids were petroleum based hydraulic fluids of the same viscosity . Fluid A was ordinary , B with antiwear additives, and C with friction modifiers. The test was carried out with a balanced type vane motor, so it was seen that the line contact between the vane and cam-ring was the major reason for the results.

As the next stage, there were elaborated studies on each sliding part.

Here, let us examine tribology of piston pumps and motors. The reason is that these pumps and motors are suitable for the high-pressure operation and the variable displacement type, and have the high over-all efficiencies.

Figure 3 illustrates a piston pump and motor (swash plate type).

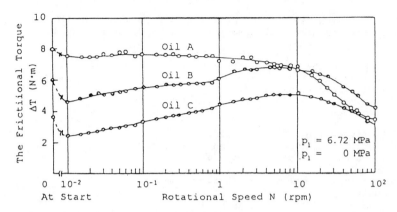

Fig. 2 Effects of fluid properties on frictional torque of
vane motor (p_1: Input pressure, p_2: Output pressure)

The major sliding parts are between valve plate and cylinder block,
between piston and cylinder wall, and between piston slipper and swash
plate. For bent axis type there is no last sliding part.

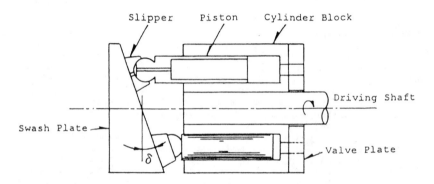

Fig. 3 Typical sliding parts of swash plate type piston pump

3.1 Between Valve Plate and Cylinder Block

Since this sliding part affects on the characteristics of piston
pumps and motors regardless of a swash plate type or a bent axis type,
there have been much work. References [5] and [6] studied the
performance with a specially designed valve plate, [7] gave a basic
investigation of a hybrid bearing, and [8] analyzed theoretically the
effect of elastic deformation of the shaft and casing. References [9]
and [10] studied the friction and leakage under starting and low-speed
conditions in the case of a swash plate type motor. Figure 4 shows the
leakage flow rate at a test rig modelling on a bent axis type [11]. In
Fig. 4(a), the case with hydrodynamic pads, the leakage flow rate
increases with decreasing fluid viscosity. On the contrary in Fig. 4(b),
without hydrodynamic pads, the effects of viscosity except the petroleum

base hydraulic fluid are not so clear. Considering the results of the measured film thicknesses, the reason of the difference in two cases is seen that an excess of the hydrostatically pushing force of the cylinder block to the valve plate is mainly supported by the wedge film effect for (a), and the squeeze film effect for (b). Moreover, the frictional torque and temperature of the sliding parts in the high water content fluids were less than those for the petroleum based hydraulic fluid.

Now, we discuss theoretical treatments. To obtain the fluid film

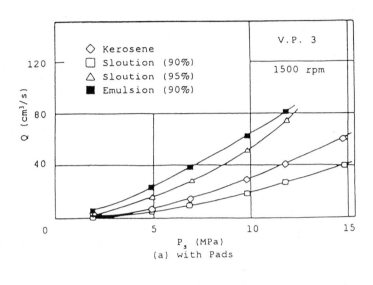

(a) with Pads

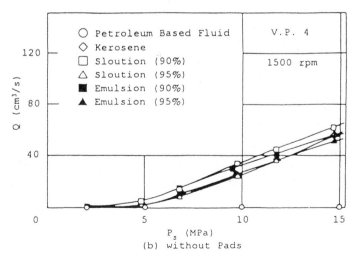

(b) without Pads

Fig. 4 Leakage at sliding part between valve plate and cylinder block —effect of hydrodynamic pad and fluid properties

shape between the valve plate and cylinder block, it is not enough to
analyze the sliding parts only. AS shown in Fig. 5, the cylinder block
is supported by the shaft and film pressure on the valve plate and in
some cases the outer journal bearing. The loads consist of the reaction
force of the piston mechanism (including the frictional force), besides
the force due to cylinder pressure. Under these forces and moments of
force, the cylinder block moves. This means that the motion of the
cylinder block is 5 degrees of freedom, even if the rotational speed of
the shaft is a constant.

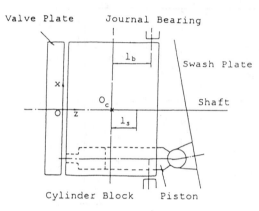

Fig. 5 Illustration of force   Fig. 6 Model of valve plate
       acting on cylinder block   with hydrodynamic pad

We consider a valve plate with a conventional hydrodynamic pad on
its outside as shown in Fig. 6. For a pump p_{r1} is the suction pressure
and p_{r2} the delivery. We assume the following: the radial length (width)
of the seal lands and pad is small enough compared to the
circumferential length; the angular span of kidney ports is π; the
hydrodynamic pad is a ring; and the sliding surfaces are geometrical
planes.

Since the inclination angle of the cylinder block is small (Fig.
7), then the fluid film thickness is given by [12]

$$\bar{h} = \bar{h}_0 + \bar{a}\bar{r}_{in}\cos(\theta - \varphi), \qquad (i = 1 \sim 3) \qquad (1)$$

where $\bar{h} = h/H$ (H: characteristic clearance),
         $\bar{a} = aR_C/H$ (R_C: pitch circle radius of valve ports),
         $\bar{r}_{in} = r_{in}/R_C$ (r_{in}: characteristic radius of valve plate land
                and pad).
         $i = 1$: inside seal land,
         $i = 2$: outside seal land,
         $i = 3$: pad.
Reynolds equation is

$$\frac{\partial}{\partial \bar{r}}(\bar{r}\bar{h}^3\frac{\partial \bar{p}}{\partial \bar{r}}) = \bar{r}(\frac{\partial \bar{h}}{\partial \theta} + 2\frac{\partial \bar{h}}{\partial \tau}) \qquad (2)$$

where $\bar{p} = p/[6\mu\omega(R_C/H)^2]$,

$\tau = \omega t$,

μ: viscosity,

t: time,

p: pressure.

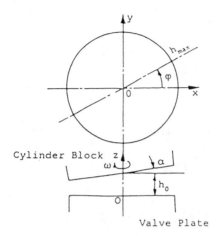

Fig. 7 Fluid film shape between valve plate
and cylinder block

The pressure on the sliding surface can be divided into three parts: $\bar{p}_1$ due to hydrostatic effect, $\bar{p}_2$ wedge film effect and $\bar{p}_3$ squeeze film effect. That is,

$$\bar{p} = \bar{p}_1/S_0 + \bar{p}_2 + \bar{p}_3 \tag{3}$$

The boundary conditions are

$$
\left.
\begin{array}{ll}
\text{at} \quad \bar{r} = \bar{r}_{i1}, \bar{r}_{i2}, & \bar{p}_2, \bar{p}_3 = 0 \\[4pt]
\qquad = \bar{r}_{11}, \bar{r}_{22}, & \bar{p}_1 = 0 \\[4pt]
\qquad = \bar{r}_{12}, \bar{r}_{21}, & \bar{p}_1 = 0 \quad (0 \le \theta < \pi) \\[4pt]
& \bar{p}_1 = 1 \quad (\pi \le \theta < 2\pi)
\end{array}
\right\} \tag{4}
$$

Subscripts are shown in Fig. 6. By introducing $S_0 = [6\mu\omega(R_C/H)^2]/p_{r2}$, the above conditions are made applicable to an arbitrary p_{r2}.

The load carrying capacities are given by

$$
\left.
\begin{array}{l}
\left(\overline{F}_z\right)_v = \iint \bar{p}\,\bar{r}\,d\bar{r}\,d\theta \\[6pt]
\left(\overline{M}_x\right)_v = \iint \bar{p}\,\bar{r}^2 \sin\theta\,d\bar{r}\,d\theta \\[6pt]
\left(\overline{M}_y\right)_v = -\iint \bar{p}\,\bar{r}^2 \cos\theta\,d\bar{r}\,d\theta
\end{array}
\right\} \tag{5}
$$

where force $\overline{F} = F / [6\mu\omega R_c{}^2(R_c / H)^2]$, and moment of force
$\overline{M} = M / [6\mu\omega R_c{}^3(R_c / H)^2]$. $(\ )_v$ denotes the force and moment due to the pressure on the valve plate. As we assume a infinitely short bearing, we can express $(F_z)_v$ and so on as analytical form. Similarly, we can obtain the load carrying capacities of the outside journal bearing $(F_x)_b$, $(F_x)_b$. And $(\ )_s$ denotes the force and moment due to the spring effect of the shaft.

Let the mass and moment of inertia of the cylinder block be
$\overline{m} = m\omega / [6\mu R_c(R_c / H)^2]$ and $\overline{I} = I\omega / [6\mu R_c{}^3(R_c / H)^2]$, respectively, then the equations of motion of the cylinder block

$$
\left.
\begin{array}{l}
\overline{m}\dfrac{d^2\overline{h}_0}{dt^2} = \left(\overline{F}_z\right)_v + \left(\overline{F}_z\right)_l \\[2mm]
\overline{m}\dfrac{d^2\overline{x}_c}{dt^2} = \left[\left(\overline{F}_x\right)_s + \left(\overline{F}_x\right)_b + \left(\overline{F}_x\right)_l\right]\Big/\overline{C} \\[2mm]
\overline{m}\dfrac{d^2\overline{y}_c}{dt^2} = \left[\left(\overline{F}_y\right)_s + \left(\overline{F}_y\right)_b + \left(\overline{F}_y\right)_l\right]\Big/\overline{C}
\end{array}
\right\}
\tag{6}
$$

$$
\left.
\begin{array}{l}
\overline{I}_x\dfrac{d^2\overline{\varPhi}_x}{d\tau^2} - \overline{I}_z\dfrac{d\overline{\varPhi}_x}{d\tau} = \left(\overline{M}_x\right)_v + \left(\overline{M}_x\right)_s + \left(\overline{M}_x\right)_b + \left(\overline{M}_x\right)_l \\[2mm]
\overline{I}_y\dfrac{d^2\overline{\varPhi}_y}{d\tau^2} + \overline{I}_z\dfrac{d\overline{\varPhi}_y}{d\tau} = -\left(\overline{M}_y\right)_v - \left(\overline{M}_y\right)_s - \left(\overline{M}_y\right)_b - \left(\overline{M}_y\right)_l
\end{array}
\right\}
\tag{7}
$$

where $\overline{x}_c$: x co-ordinates of center of mass of cylinder block, $\overline{x}_c = x_c/C$, $\overline{y}_c = y_c/C$, $\overline{C}$: tolerable displacement of cylinder block (radial clearance for case with an outside journal bearing), $\overline{C} = C/H$. $\overline{\varPhi}_x$: angle between centerline of cylinder block and xz plane, $\overline{\varPhi}_x = \varPhi_x R_c/H$, $\overline{\varPhi}_y$: angle between centerline of cylinder block and yz plane, $\overline{\varPhi}_y = \varPhi_y R_c/H$. $(\ )_l$ denotes the loads due to the cylinder pressure and piston mechanism. Finally we solve equations (6) and (7) to obtain the film shape $(\overline{h}_0, \overline{\alpha}, \varphi)$ between the valve plate and cylinder block and the center of mass $(\overline{x}_c, \overline{y}_c)$ of the cylinder block.

The stable region of the film shape on the valve plate with a hydrodynamic ring pad is much narrower than that with hydrostatic pads. The difference mainly depends on the magnitudes of spring and damping

Table 1 Change in fluid film shape due to rotational speed

$\overline{h}_0$	$\overline{\alpha}$	φ (rad)	S_0
1.047	0.175	1.79	500
1.063	0.268	1.46	50
1.058	0.274	1.41	5
1.058	0.274	1.41	0.5
1.058	0.274	1.41	0.005

coefficients of the films.

As already mentioned in chapter 2, the loads of sliding parts are basically determined by the pressure of the high-pressure side. Then, the adoption of the concept of hydrostatic bearings is very effective [13]. For example, Table 1 shows the change in the film shape with the rotational speed of the pump whose valve plate has three hydrostatic pads. In this case the swash plate angle is 20° and the friction coefficient between piston and cylinder wall is 0.05.

It is seen that, regardless of the change in rotational speed up to the 5th power of 10, film changes hardly.

3.2 Between Piston and Cylinder Wall

In a sense the sliding parts belong to the same category as spool valves [14], which are widely used as one of the typical hydraulic control valves. The difference is present as a lateral force acting on the piston. We can point out that grooves on the piston work on reduction of the load-carrying capacity of the fluid film. Fluid lubrication can be realized with an exponential-function-type piston whose diameter is small at the head [15]. The piston is equivalent to tapered or stepped ones. For swash plate type machines, the lateral force acting on pistons is so large that the metallic contact between the piston and cylinder wall does not ease to be avoided, especially under a large swash plate angle. Therefore, in this case it is effective to decrease the friction acting on pistons to add the hydrostatic effect to the sliding parts [16].

As to the experiments, there have been continuously a lot of works as follows: study on frictional forces acting on the pistons [17], study on the ratio of the period with metallic contact to the whole period under a constant pressure [18], and study on frictional forces of rotating-swash plate type pumps and motors [19].

3.3 Between Slipper and Swash Plate

This sliding part is inherent in swash plate types, and affects largely on their performance. Especially, the tribological condition on this part is taken notice in connection with their starting and low-speed characteristics.

As shown in Fig. 4, the cylinder pressure is connected to the slipper recess by way of a restrictor. This means that the film between the slipper and swash plate may have a stiffness as a bearing. The slipper is connected with the piston by a spherical bearing. Due to the limit in size, the spherical bearing has to operate under a mixed lubrication. So, it is necessary to take this point account in the case of analysis for the slipper motion.

For theoretical studies, there have been some simplified works. During the piston stroke, the cylinder pressure changes from the suction to the delivery values trapezoidally. Then, we should examine dynamic behaviors of the fluid film as the case of the fluid film around pistons.

Under the assumption that both surfaces of the sliding parts are geometrical planes, to clarify the behavior of the slipper, we should determine the film thickness at the center of sliding surface, inclination and azimuth angle of the film shape. Basically, the shape of film is given by the equations similar to Eq. (1) [20].

Deriving the loads (force due to cylinder pressure, inertia and frictional force of pistons, etc.), we obtain the equation of motion of the slipper (3 degrees of freedom).

Figure 8 shows the limit of fluid lubrication region, that is, the area where the minimum film thickness is equal to zero. The area 1

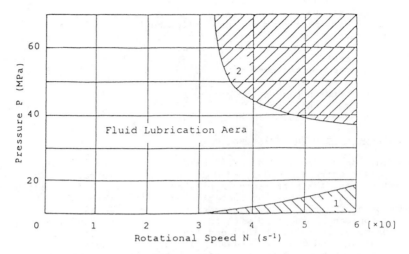

Fig. 8 Limits of fluid lubrication in slipper bearing

corresponds to the metallic contact during the pump delivery stroke. The area 2 corresponds to the case during the trapping period from the suction to the delivery stroke. The power-losses of the sliding part were also obtained in connection with the slipper shape [21]. In the case of a swash plate type motor, friction and leakage characteristics under starting and low-speed conditions were studied [22], [23]. There was work on overdamped slipper in a piston pump [24].

4. A MIXED LUBRICATION MODEL AND OPTIMUM DESIGN

The sliding parts of hydraulic pumps and motors are to operate under a mixed lubrication region, especially for the case of low-speed conditions or low viscous working fluids. To forecast the characteristics of such equipment, a lubrication model, which can unify from a fluid film to a mixed lubrication region, is essential. It should be emphasized that the model can be applied to both the friction and flow characteristics in a mixed lubrication.

Recently, such a model has been proposed [25], [26], and Figure 9 shows the power loss of the disk type hydrostatic thrust bearing which may be an effective simulation especially for the piston slipper bearing. The loss $\overline{L}$ is defined by the summation of the power losses due to the leakage flow rate and those due to the frictional torque. Here,

$$\overline{L} = L \Big/ \Big[6\mu\omega^2 R_2^3 (R_2/H)^2 \Big], \quad R_2: \text{ outer radius}, \quad \zeta = 2W\ln a \Big/ \Big[\pi(1-a^2)p_s R_2^2 \Big], \quad a = R_1/R_2:$$

ratio of recess radius, p_s: supply pressure. The parameter ζ implies the ratio of the load to the maximum hydrostatic load-carrying capacity. The loss in the range $\zeta < 1$ (equivalent to fluid lubrication) is mainly due to the leakage flow rate and that in the range $\zeta > 1$ (mixed lubrication) the frictional torque.

Figure 10 shows the test results for the effects of the speed of rotation N on L [27]. The theoretical results based on the mixed lubrication model agree well with the experimental results, especially

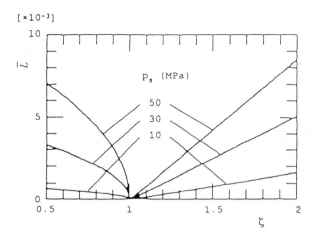

Fig. 9 Effects of supply pressure on power loss of hydrostatic
bearing in fluid and mixed lubrication (Theory)

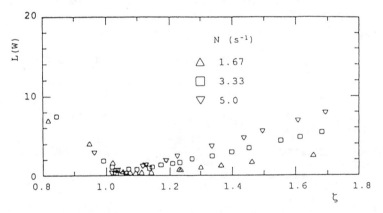

Fig. 10 Effects of rotational speed on power loss of hydrostatic
bearing in fluid and mixed lubrication (Experiment)

for large surface roughness. We anticipate a practical application of
the model to the CAE/CAD of hydraulic equipment.

From the viewpoint of the optimum design, it is worth while to
examine such conditions as the minimum power loss and maximum load-
carrying capacity besides the maximum stiffness [28]. For hydrostatic
bearings the maximum stiffness is usually adopted as the optimum
condition, and then the ratio of the recess pressure to the supply
pressure is derived. However, this pressure ratio is the parameter not
directly determined at the design. Selecting such direct parameters as
recess radius, feed-restrictor and bearing radius, we can determine the
parameters based on the condition, for example, the minimum power loss.
Including the mixed lubrication region, we can derive the optimum
surface roughness realizing the minimum power loss [27].

Here, let us discuss of the effects of elastic deformation of the
sliding parts on the power loss. A test on the disk type hydrostatic

thrust bearings, with four pieces of disk of different thickness, shows that there is the minimum leakage flow rate at certain thickness [29]. However, because of the probability of the local metallic contact and the reduction of the over-all bearing stiffness, the condition of the minimum flow rate due to elastic deformation may not be the optimum design point of such sliding parts.

5. CONCLUSIONS

We should recognize that the tribology of hydraulic equipment, especially hydraulic pumps and motors, is much severer and more difficult than that of other machinery. It is based on the reasons as the following: the sliding parts have to work as bearings and seals simultaneously; the sliding parts have to function over a wide range of operating conditions; the elements composed the sliding parts have to support not only forces but also moments of force; the materials of the sliding parts have to tolerate a heavy stress.

Therefore, to obtain experimentally the solution and development in connection with tribology of hydraulic pumps and motors, it is effective to use the test rigs simulated the specified sliding parts. To this step the theoretical analysis is helpful.

As a new system, it is widely trying to adopt a water as a working fluid. For this case tribology plays the dominating role for accomplishment of the system.

REFERENCES

[1]   Wilson, W. E., Positive Displacement Pumps and Fluid Motors, (1950), Pitmam.

[2]   Schösser, M. W. J., Mathematical Model for Displacement Pumps and Motors (Part two), Hydraulic power Transmission, 7 (1961), 324/328.

[3]   Fitch, E. C., An Encyclopedia Of Fluid Contamination Control for Hydraulic Systems, (1979), Hemisphere.

[4]   Hibi, A. and Ichikawa, T., Influences Of hydraulic Fluid on Toque Characteristics of Hydraulic Motor at Start and in Low Speed Range, J. JHPS, 9-6 (1978), 418/422.

[5]   Kosodo, H., Studies on the Intermittent Lubrication System Of Axial Piston Pumps and Motors (Report 2), Bull JSME, 16-92 (1973), 301/311.

[6]   Yoa-Ming, X. and Jun, S. Hydrostatic Bearing with Intermittent Oil Supply for the Distributor of Axial Plunger Pump, Proc. Intn. Conf. Fluid Power Trans. Conf., Hangzhou, China, 667/677 (1985).

[7]   Uehara, K., Akasaka, T. and Ishihara, T., Trust Load Carrying Capacity on Sliding Surfaces of Axial Piston Pumps (Report 2, in Japanese), Trans. JSME, 40-332 (1974), 1211/1221.

[8]   Foster, K., Hooke, C. J. and Modera, G., The Effect of Structural Deformation on the Performance of Port Plates in Axial Piston Pumps and Motors, Proc. Conf. Stress Anal. Group, Inst. Phys. Comput. Div., Exp. Numer. Stress Anal., Pap.9 (1976).

[9]   Matusmoto, K. and Ikeya, M., Friction Characteristics between the Piston and Cylinder for Low-Speed Conditions in a Swashplate-Type Axial Piston Motor (in Japanese), Tran. JSME, 57C-540 (1991), 2729/2733.

[10] Matusmoto, K. and Ikeya, M., Leakage Characteristics between the Valve Plate and Cylinder for Low-Speed Conditions in a Swashplate-Type Axial Piston Motor (in Japanese), Trans. JSME, 57c-541 (1991), 3008/3012.

[11] Yamaguchi, A., Sekine, H., Shimizu, S. and Ishida, S.,
      Bearing/Seal Characteristics of the Oil Film between a Valve
      Plate and a Cylinderblock of Axial Piston Pumps (Report 3, in
      Japanese), J. JHPS, 18-7 (1987), 543/550.
[12] Yamaguchi, A. Formation of a Fluid Film between a Valve Plate
      and a Cylinder Block of Piston Pumps and Motors (Report 1), Bull.
      JSME, 29-251 (1986), 1494/1498.
[13] Yamaguchi, A. Formation of a Fluid Film between a Valve Plate
      and a Cylinder Block of Piston Pumps and Motors (Report 2), Bull.
      JSME, 30-259 (1987), 87/92.
[14] Blackburn, J. F., Coarley, J. L., Ezekiel, F. D., and et al.,
      Fluid Power Control, John Wiley (1960), 278.
[15] Yamaguchi, A., Motion of Pistons in Piston-Type Hydraulic
      Machines (Report 3), Bull. JSME, 19-130 (1976), 413/419.
[16] Yamaguchi, A., Motion of the Piston in Piston Pumps and Motors
      (Report 2 in Japanese), Trans. JSME, 57B-537 (1991), 1689/1694.
[17] Yamaguchi, A., Motion of the Piston in Piston Pumps and Motors
      -Experiments and Theoretical Discussion, JSME Intern. J., 37B-1
      (1994), 83/88.
[18] Fang, Y. and Shirakashi, M. Mixed Lubrication Characteristics
      between the Piston and Cylinder in hydraulic Piston Pump-Motor,
      ASME, J. Tribo. 117 (1995), 80/85.
[19] Tanaka, K., Nakahara, T. and Kyogoku, K., Piston Rotation and
      Frictional Forces between Piston and Cylinder of piston Pump
      and Motor, JHPS Intern. Symp. Fluid Power, Tokyo (1993), 235/240.
[20] Iboshi, N. and Yamaguchi, A., Characteristics of a Slipper
      Bearing for Swash Plate Type Axial Piston Pumps and Motors
      (Report 1), Bull. JSME, 25-210 (1982), 1921/1930.
[21] Iboshi, N., Characteristics of a Slipper Bearing for Swash Plate
      Type Axial Piston Pumps and Motors (Report 3), Bull. JSME, 29-254
      (1986), 2529/2538.
[22] Matusmoto, K. and Ikeya, M., Friction and Leakage
      Characteristics between the Slipper and Swashplate for Starting
      and Low-Speed Conditions in a Swashplate-Type Axial Piston Motor
      (in Japanese), Trans. JSME, 57C-541 (1991), 3013/3018.
[23] Fang, Y., Matusmoto, K. and Ikeya, M., Experimental Analysis of
      Leakage Characteristics for Starting and Low-Speed Conditions of
      Hydrostatic Slipper Bearing in Swashplate-Type Axial Piston
      Motor (in Japanese), J. JHPS, 23-1 (1992), 107/112.
[24] Hooke, C, J. and Li, K. Y., The Lubrication of Overdamped Slipper
      in Axial Piston Pumps -Centrally Loaded Behaviour, Proc. IME, 202-
      C4 (1988), 84/88
[25] Yamaguchi, A. and Matsuoka, H., A Mixed Lubrication Model
      Applicable to Bearing/Seal Parts of Hydraulic Equipment, ASME, J.
      Tribo., 114 (1992), 116/121.
[26] Kazama, T. and Yamaguchi, A., Application of a Mixed Lubrication
      Model for Hydrostatic Thrust Bearings of Hydraulic Equipment,
      ASME, J. Tribo., 115 (1993), 686/691.
[27] Kazama, T. and Yamaguchi, A., Experiment on Mixed Lubrication of
      Hydrostatic Thrust Bearings for Hydraulic Equipment, ASME, J.
      Tribo., 117 (1995), 399/402.
[28] Kazama, T. and Yamaguchi, A., Optimum Design of Bearing and Seal
      Parts for Hydrostatic Equipment, Wear, 161 (1993), 161/171.
[29] Kazama, T., Iwasaki, N. and Yamaguchi, A., Effects of Elastic
      Deformation on Hydrostatic Bearings (in Japanese), J. JHPS, 25-3
      (1994), 433/438.

Pump Testing

Roland J. Bishop Jr.[1] and George E. Totten[1]

TRIBOLOGICAL TESTING WITH HYDRAULIC PUMPS:
A REVIEW AND CRITIQUE

REFERENCE: Bishop, R. J. Jr. and Totten, G. E., **"Tribological Testing With Hydraulic Pumps: A Review and Critique,"** *Tribology of Hydraulic Pump Testing, ASTM STP 1310,* George E. Totten, Gary H. Kling, and Donald J. Smolenski, Eds., American Society for Testing and Materials, 1996.

ABSTRACT: hydraulic fluids serve two functions in the hydraulic system; to transfer energy and to lubricate the system. The energy transfer role is relatively well understood while lubrication assessment has been hindered by inability to adequately model hydraulic systems in general by small-scale pump prototypes or by bench testing. In general, most conventional bench tests do not adequately model wear which may accompany hydraulic fluid use in various hydraulic pump designs. Therefore, this has necessitated the use of hydraulic pumps as tribological test stands. The objectives of this paper are to: 1) review common failure modes of different types of hydraulic pumps, 2) provide an overview of the deficiencies of bench testing of hydraulic fluids, 3) review the tribological criterion for successful lubrication of vane, piston and gear pump testing, 4) review the most common pump tests reported to date, and 5) critique the quality of these tests relative to their ability to provide the required data.

KEYWORDS: pump test, tribology, hydraulic, hydraulic fluids, lubrication

---

Fluids play two vital roles in hydraulic pump operation. Perhaps the best known and understood is their role in energy transfer. However, their role as lubricants is more poorly understood. This problem is compounded because there are very few standardized methods of evaluating and reporting the relative ability of a hydraulic fluid to adequately lubricate the various critical components in a hydraulic system.

Silva published a thorough review of the wear mechanisms in hydraulic pump operation [1]. The role of cavitation, adhesion, corrosion and abrasion wear was described. Also discussed was the role of fluid viscosity and the speed and loading at the wear contact as modeled by the classic Stribeck curve illustrated in (Fig. 1). However, discriminating methods of experimental modeling of pump wear was not discussed.

There are numerous references to the use of hydraulic pump tests to evaluate component durability [2,3] or to evaluate some aspect of component design features on either the efficiency or mechanism of energy transfer. However, there are fewer references on the development

---

[1]Union Carbide Corporation, 777 Old Saw Mill River Road, Tarrytown, NY 10591.

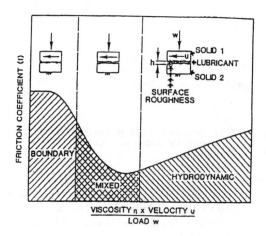

Fig. 1 -- Illustration of the effect of wear contact loading and speed
and fluid viscosity on wear (Stribeck-Hersey curve).

and use of hydraulic pumps as "tribological tests" to evaluate fluid
wear.

There have been some references describing the impact of
fundamental lubrication properties of hydraulic fluids, such as film
thickness, on pump wear using the hydraulic pump as a tribological test.
[4] However, there are a number of problems with such testing
procedures which include: cost of pumps, energy and components,
relatively long testing times, relatively poor manufacturing precision
of some of the components for use in reproducible tribological testing,
volumes of fluid required and subsequent disposal. Therefore, there has
been a longstanding effort to develop "bench test" alternatives to
evaluate hydraulic fluid lubricity [5].

Examples of bench tests include: Shell 4-ball, Falex pin-on-V-
block, and many others. Bench tests have a number of advantages
relative to pump tests including: relatively low fluid volumes are
required, numerous material variations and wear contact configurations
are possible, high-quality and reproducible test components are readily
available and the cost of testing is substantially lower.

Although it would be desirable to conduct bench tests as
alternatives to hydraulic pump testing, with few exceptions these tests
provide poor correlation with pump lubrication. For example, Renard and
Dalibert found no correlation between either the Shell 4-ball "wear"
(ASTM D-2596-T) or "extreme pressure" (ASTM D-2783) tests and the wear
results obtained with a Sperry-Vickers V-104 vane pump test (ASTM D-
2882) [6]. Knight reported that: "...Regrettably the data from these
tests not only failed to give quantitative correlation with data
determined from machines, but even the ranking between different fluids
failed to agree. ... ."[7] More recently, Lapotko, et.al. reported that
4-ball test results "... may deviate considerably from the data obtained
in actual service..."[8]

Recently, Tessmann and Hong [9] and Perez, et.al.[10] have
reported testing modifications of the Falex pin-on-V-block and Shell 4-
ball tests respectively, that provide acceptable correlation with vane
and ring wear in a Sperry-Vickers V-104 and 35VQ pumps.

Mizuhara and Tsuya extensively studied the wear correlation of a
broad range of hydraulic fluids evaluated in a piston, vane and gear
pump with a block-on-ring (Timken) test conducted according to ASTM D-
2741-68 [11]. From these studies, they concluded:

- "Load-carrying capacity" has nothing to do with hydraulic pump antiwear performance. An "antiwear test" must be used.
- Accelerated tests usually give the wrong results.
- Material pairs for the test pieces must be the same as the wear contact of interest in the hydraulic pump.
- It is necessary to evaluate hydraulic fluids under a wide range of test conditions. A single set of test conditions is usually inadequate.

One method of evaluating a hydraulic fluid under varying lubrication conditions is to construct a performance map such as that illustrated in (Fig. 2) to determine the mechanistic transitions between hydrodynamic, elastohydrodynamic (EHD), mixed EHD and boundary lubrication as a function of wear contact loading and rolling and sliding speed.[12]   The anti-friction characteristics within each transition can then be evaluated as shown in (Fig. 3). Of course, as Mizuhara and Tsuya showed, this work should be performed with the appropriate wear contact material pairs [11].

Although standardized bench tests have been generally shown to be unreliable, it is possible to modify these tests to achieve limited correlation with lubrication performance in a pump. Alternatively, pump lubrication can be examined fundamentally by modeling pump lubrication using a range of test conditions and appropriate material pairs [12]. However, regardless of what test is performed, it is always necessary to first calibrate the bench test to hydraulic pump wear and then to validate bench test conclusions by pump testing. The objective of this paper is to provide an overview of various pump testing procedures that have been used to study hydraulic fluid lubrication. This discussion will include vane, piston and gear pump testing procedures and also a summary of evaluation criteria.

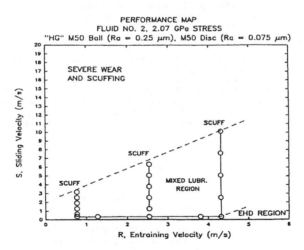

Fig. 2 -- Illustration of the mechanistic transitions of lubrication as a function of rolling and sliding speed.

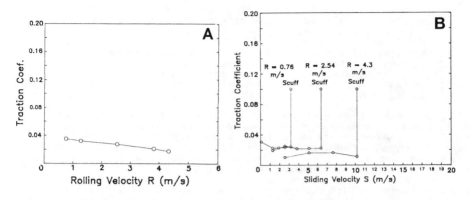

Fig. 3 -- Traction coefficient variation with varying rolling (A) and
         sliding (B) speeds.

DISCUSSION

A. Pump Failure Modes

There are numerous reasons for conducting laboratory hydraulic pump
tests. These include [13]:

1. To compare the differences in equipment performance.
2. To obtain controlled performance experience with new fluids.
3. To troubleshoot field performance problems.
4. To obtain comparative lubrication data on various hydraulic
   fluid; types and manufacturers.

Pump failures are typically accompanied by increased noise and loss
of volumetric efficiency. During the course of conducting pump tests, as
a minimum, observations must be made to determine the cause of pump
performance loss or failure. Preferably, more quantitative wear
determinations coupled with photographic record keeping and continuous
analysis of pump operation should be performed. A summary of some common
failure modes for gear pumps, axial piston pumps and vane pumps are
provided in (Table 1) [13,14,15].

Table 1 -- Common Pump Failure Modes

Pump Type	Failure Mode
Gear	Roller bearing fatigue, gear tooth surface pitting and fatigue, seal plate scoring and seizure.
Piston (Axial)	Roller bearing fatigue, valve plate scoring and erosion, slipper pad failures, piston and cylinder wear, cavitation.
Vane	Severe vane and ring, port plate, bearing and cavitation wear.

B. Fluid-Specific Properties

In addition to pump testing, and preferably prior to it, a number
of fluid-specific properties should be determined. These include metal
and non-metal compatibility. For example, vapor, liquid and dry film
corrosion properties should be examined [15]. Copper plate and other
soft metal corrosion should also be determined [16,17]. Propensity for
sludge formation, viscosity and viscosity stability, pH (for water-
containing fluids), and wetability are other important variables.
Feicht has recommended that, due to compressibility differences among
hydraulic fluids, valve housing stability be examined. Non-metal or
seal compatibility is vitally important [15].

C. Vane Pump Testing

The output pressure of a vane pump is directed to the back of the
vanes which holds them against the ring. The leading edge of the vane
forms a <u>line contact</u> with the ring and the rotation of the vane against
the ring generates a <u>sliding motion</u>. The challenge is to maintain
adequate lubrication between the vane tip and the cam ring with
increasing pressures and rotational speeds [4].

Ueno, et.al. have used the vane pump as a tribological test stand
to study the effects of pump delivery pressure, rotational speed,
eccentricity and hardness of the cam ring and vane tip curvature [4,18].
Their results showed that wear rate simply cannot be estimated from the
friction work due to these factors. However, the occurrence of severe
wear is proportional to $P_H V$ where, P_H is the maximum Hertzian contact
stress and V is the sliding velocity.

Hemeon reported the application of a "yardstick formula" to
quantitatively analyze and report vane pump wear. The critical part of
the analysis was weight loss of the ring, since dimensional changes due
to wear will significantly affect volumetric efficiency and leads to
noise and pulsations [19]. The yardstick formula is represented by
equation (1):

$$Y = K \times 1.482 \times mep \times gpm \qquad (1)$$

where: Y = Duty load on the pump in BTU/hr.
         K = A constant to correct for air entrainment, degraded or
             contaminated oil and fluid turbulence.
         mep = Mean Effective Pressure.
         gpm = Flow in gallons per minute.

Although K may be as high as 1.4, a value of 1.03 is typical. The
conversion constant 1.482 permits the use of pressures in psi and weight
loss in grams. Interestingly, in order to obtain reproducible and
reliable weight loss data, it was reported that the ring had to be
washed and baked at 200°F for 24 hours because the porous metal adsorbed
solvent and gave incorrect weight data.

Bosch (Racine Fluid Power) also utilized a cycled pressure vane
pump test. The pressure-time sequence and test circuit is illustrated
in (Fig.4) [20]. Based on their research, this is a more representative
test since it better incorporates pressure spikes that will invariably
be experienced by the system during circuit activation and deactivation.
At the conclusion of the test, the weight loss of the ring, vanes, port
and cover plate, and body and cover bearings were measured. The ring
and bearings were inspected for unusual wear patterns and for evidence
of corrosion, rusting and pitting.

Lapotko, et.al. have reported an alternative vane pump test
designated as the "MP-1 Test" [8,21]. A schematic of the MP-1 vane pump
is shown in (Fig. 5). Although the MP-1 may be run at pressures up to

10 Mpa (1450 psi), the reported test pressure is 7 Mpa with a total fluid volume of 0.7 liters. In addition to lower volume, the MP-1 test is conducted for only 50 hours (and in some cases for only 10 hours). The wear rate is based on the weight loss of the vanes only after the test is completed. In view of the relatively small size, this test comes as close to a "bench hydraulic pump test" as has been reported to date.

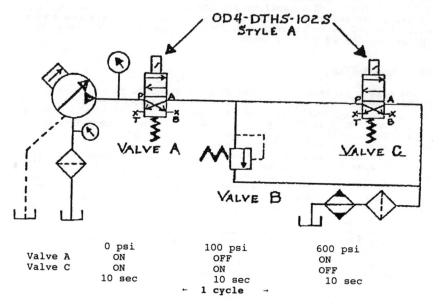

	0 psi	100 psi	600 psi
Valve A	ON	OFF	ON
Valve C	ON	ON	OFF
	10 sec	10 sec	10 sec
		← 1 cycle →	

Fig. 4 -- Test circuit for Racine cycled-pressure vane pump test utilizing a 7.5 hp electric motor, SV-10 vane pump and a 20 gallon reservoir.

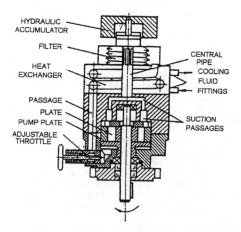

Fig. 5 -- A schematic illustration of the Russian MP-1 vane pump.

The Vickers V-104 vane pump shown in (Fig. 6) and (Fig. 7) continues to be the most commonly utilized hydraulic fluid pump test. [22]   There are at least three national standards based on the use of this pump; ASTM D-2882, DIN 51389, and BS 5096 (IP 281/77).   A comparison of the test conditions is provided in (Table 2).   The total weight loss of the vanes + ring at the conclusion of the test is the quantitative value of wear.

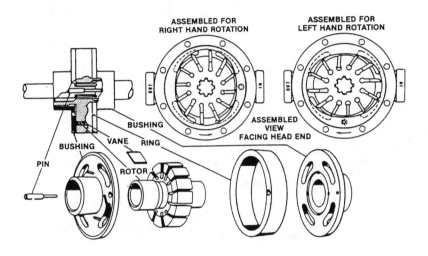

Fig. 6 -- Illustration of the Sperry-Vickers V-104 vane pump.

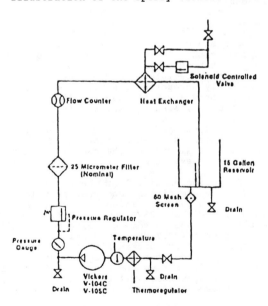

Fig.7 -- Test circuit for Vickers V-104 Vane Pump Test.

TABLE 2 -- Comparison of Sperry-Vickers V-104
Vane Pump Testing Procedures

Test Parameter	ASTM D-2882	DIN 51389	BS 5096 IP281/77
Pressure	14 MPa 2000 psi	10 MPa 1500 psi	14 MPa[1] 2088 psi
			11 MPa[2] 1540 psi
Rev/Min	1200	1500	1500
Time (Hr)	100	250	250
Fluid Volume (L)	56.8	55-70	55-70
Fluid Temperature	150°F	(3)	(3)

(1) For mineral oil type fluids.
(2) For HFA (oil-in-water emulsion), HFB (water-in-oil emulsion) and HFC (water-based) Fluids.
(3) The temperature is selected to give 46 cSt viscosity.

There is currently an effort underway within ASTM (D.02 N.07 committee) to replace the V-104 pump with a more current design; the Vickers 20VQ5 vane pump. Some differences between the two pumps are:

	V-104	20VQ5
Vane Load (lbs/1000 psi)	47	25
Max. Pressure (psi)	1000	3000
Max. Speed (rpm)	1500	2700
End Plates	Cast Bronze	Sintered Bronze

nitial round robin studiees involving the simple replacement of the V-104 vane pump in the ASTM D-2882 procedure gave mixed results. One shortcoming is, that while the correlation of the results is generally the same, the absolute agreement, as expected, is relatively poor. Also, previously unreported test results have shown that while there seems to be a correlation between results obtained by the new pumps, absolute agreement, also as expected is poor. This work is described in Ref. 23.

A test has been developed using a Chinese-built YB-6 dual function vane pump operating under ASTM D-2882 testing conditions [24]. In addition to measuring total wear of the cam ring and vanes, the temperature rise, noise level and the volumetric efficiency were measured.

One of the deficiencies of all of the previously described vane pump tests is that they are either conducted at relatively low pressure or rpm, compared to vane pumps in industrial use. Although not a national standard, the Vickers 35VQ pump test is often acknowledged to provide a more rigorous, and in some cases, a more realistic accelerated industrial pump wear test (20.7 MPa, 3000 psi, 2400 rpm) [25]. A schematic of this test is provided in Figure 8. The maximum weight loss of the vanes for each of the test cartridges must not exceed 15 mg and the total weight loss of the ring for each cartridge must not exceed 75 mg for the three cartridges used. If the weight loss of any of the cartridges exceeds these values, two more tests with new cartridges must be run.

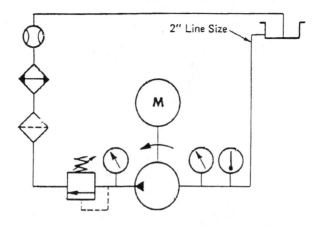

Fig. 8 -- Schematic of the Vickers 35VQ test stand.

Hagglunds-Denison has also developed a two-part recommended pump testing protocol (HFO). Although this test is used widely in the hydraulics industry, it is not a national standard test. The HFO procedure is composed of two parts, a vane pump test and a piston pump test. Only the vane pump test will be discussed at this time, the piston pump will be discussed later.

The test circuit for the Hagglunds-Denison vane pump wear test is based on their T5D-042 vane pump, which is located in a "bootstrap" circuit with the Denison series 46 piston pump. The T5D vane pump test circuit is shown in (Fig 9). The vane pump is operated at 2400 rpm and 2500 psi and is driven by a 136 hp electric motor. At the conclusion of the test, the procedure requires that the following be recorded:

- Cam-rotor and cam-vane clearance.
- Vane-and-slot clearance for all vanes.
- Tracings of lip contours of vanes 1,4 and 8.
- Visual appearance of cam ring interior surface.
- Visual appearance of both plate surfaces that are against rotating pumping elements.

Hagglund-Denison has also developed a SK-30320 vane pump test to evaluate vegetable oil durability. This test is also based on the Hagglunds-Denison T5D-042 vane pump and is a 600 hour, cycled pressure test (300 hours under "dry" conditions and then 300 hours after the addition of 1% of water). The pressure cycles from 10 to 240 bars each second and the fluid temperature is 70°C.

Most of the references described above utilize a gravimetric determination of wear by either measuring the weight loss of the vanes, ring or both. However, some hydraulic fluids exhibit non-newtonian viscosity behavior leading to internal leakage and loss of power transmission efficiency. These problems can be detected by experimental determination of pumping efficiency by flow rate measurement [26].

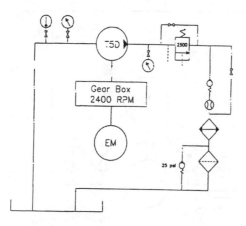

Fig. 9 -- Hagglunds-Denison T5D vane pump hydraulic test circuit.

D. Piston Pump Testing

There are numerous wear surfaces in a piston pump. In addition to sliding wear, such as pistons in cylinders, there is mixed rolling and sliding which would occur with rolling element bearings, corrosion and cavitation wear which might occur on the swash plate, etc. The relative amount of wear that would occur would be critically dependent on the material pairs of construction of the wear contacts. In view of the wide range of materials used for construction and design of piston pumps, this is one reason why most of the "standard" pump tests developed to date have been vane pump tests.

The objective of piston pump design is to minimize energy consumption while at the same time optimizing hydrodynamic lubrication to minimize wear and internal leakage. The performance parameters are fluid flow, speed, torque, pressure, viscosity, and inlet pressure. To minimize friction and internal leakage, wear contact loading (pressure), speed and viscosity must be optimized as shown in Figure 1 [1]. In a piston pump, the piston clearances may vary with eccentricity due to load and fluid viscosity. This may produce a change in the lubrication mechanism, e.g. hydrodynamic to boundary, resulting in increased wear and friction.

One piston pump test commonly used in the hydraulic fluid industry is the "Sundstrand Water Stability Test Procedure" [27]. The test circuit containing a Sundstrand Series 22 axial piston pump is shown in Figure 10. The test conditions are:

Input Speed	3000 - 3200 rpm
Load Pressure	5000 psi
Charge Pressure	180 - 220 psi
Case Pressure	40 psi max.
Stroke	1/2 of full
Reservoir Temperature	150 +/- 10°F
Loop Temperature	180 +/- 10°F
Maximum Inlet Vacuum	5 Inches Hg

The objective of this test is to determine the effect of water contamination on mineral oil hydraulic fluid performance. The duration of the test is 225 hours. In addition to disassembly and inspection for wear, corrosion and cavitation, the test criteria is "flow degradation". Flow degradation of 10% is considered a failure.

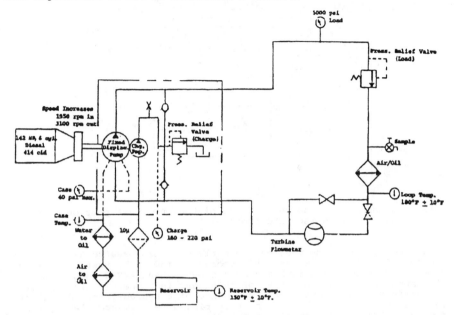

Fig. 10 -- Schematic of the Sundstrand Water Stability Test Circuit.

A cyclic loading variation of this test was recently reported to evaluate the performance of a water-glycol hydraulic fluid under these relatively high loading conditions [28,29]. In this test, a Sundstrand swash plate, axial piston pump was tested at 3175 rpm which was driven by a Sundstrand Series 20 swash plate piston motor at 890 rpm. The load was supplied by a Vickers vane pump at 975 rpm. The test sequence is summarized in Table 3. The rotating components are visually inspected for wear.

TABLE 3 -- 500 Hour Cycled Pressure Sundstrand Pump Test

Duration[1] (sec)	Vickers (MPa)	Sundstrand (MPa)
130	1.17	21.37
325	0.72	17.24
60	2.07	31.03
85	0.72	17.24

(1) The total time per cycle is 600 seconds.

The test circuit for the piston pump portion of the Hagglunds-Denison HFO piston pump is illustrated in (Fig. 11). This test, which utilizes a Dennison P46 piston pump was developed to evaluate the multi-metal compatibility of a fluid and its corrosiveness against soft metals in a severe hydraulic environment. This test is conducted for 100 hours and then the pump components are visually inspected for wear, corrosion and cavitation at the conclusion of the test.

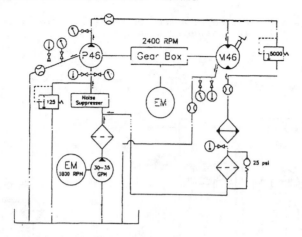

Fig. 11 -- Schematic of the Denison P46 piston pump test portion of the Hagglunds-Denison HFO protocol.

Another piston pump test has recently been proposed to ASTM Committee D.02 N.07 by The Rexroth Corporation.[30] The test circuit is provided in Figure 12 and is based on a Breuninghaus A4VSO swash plate, axial piston pump. The objective of this test is to better discriminate and classify hydraulic and wear performance of a hydraulic fluid. It is proposed that this would be done by prescribing performance levels to be achieved. Establishment of these performance levels has not been developed as of this time.

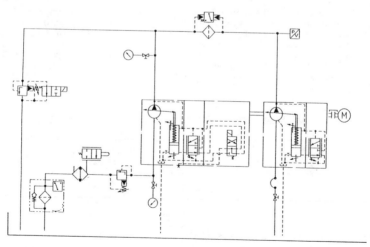

Fig. 12 -- Schematic of the proposed Rexroth Piston Pump Test.

less well known cycled pressure piston pump test is the Vickers AA 65560-1SC-4 piston pump test [31]. Previous work showed that if there is no cavitation resulting from inlet starvation and if the

hydraulic fluid is clean, then there are four principle modes of
failures: 1) spalling of the yoke and spindle bearing group, 2) fatigue
failure of the control piston assembly, 3) rotating group assembly
failure due to worn front and tail drive bearings and 4) static and
dynamic O-ring failures. In a sense, piston pump tests are excellent
bearing and cavitation tests.

Janko used high-speed piston pump tests conducted for 1250 hours
to supplement successful preliminary vane pump testing.[32]   This test
was conducted at constant 140 bar pressure for 1000 hours and then
completed by cycling the pressure between 70 and 140 bars at 0.1 Hz.
The pump was disassembled every 250 hours and visually inspected. The
stressed components including the pistons, piston slippers, cylinder
barrel, and reversing plate were inspected and measured for wear. In
addition, it was learned that the hydraulic fluids being studied caused
such severe bearing wear that they had to be replaced every 250 - 500
hours. The same fluids produced severe wear on the drive shaft with
spline ring and thrust pins. These results show the value of piston
pump testing to significantly increase the wear stress of hydraulic
fluids.

Edghill and Rubbery studied the correlation of laboratory testing
with the type and frequency of field failures [33]. There analysis
showed that the most critical areas for failure are:

For Strength:	1. Piston Necks
	2. Shafts
	3. Body Kidney Ports
For Bearing Surfaces:	1. Slipper/Cam Plate Interface
	2. Piston Skirt /Cylinder Bore
	3. Cam plate Trunion Liners

Recently, Ohkawa, et.al. developed a piston pump test to evaluate
biodegradable vegetable oil derived hydraulic fluids [34]. This test
stand (see Fig. 13) utilized a Komatsu HPV35+35 twin-piston pump under
the cycled-pressure test conditions summarized in (Fig. 14). The test
criteria included: pump efficiency change, wear and surface roughness,
formation of lacquer and varnish, and hydraulic oil deterioration.

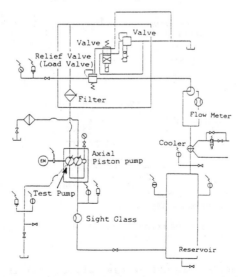

Fig. 13 -- Komatsu HPV35+35 piston pump test stand circuit.

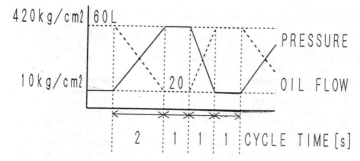

Pressure: 10-420 kg/cm² (140-6,000 psi)
Oil Flow: 16 gal (@140 psi), 5 gal (@6,000 psi)
Speed: 2100 rpm
Temperature: 80°C.
Tank Volume: 16 gallons
Duration: 500 hrs

Fig. 14 -- Summary of Komatsu test conditions.

Some piston pumps are used in very broad range temperature environment (-46 - 204°C). There are two tests that have been reported for these applications. Hopkins and Benzing utilized the test circuit shown in Figure 15, which uses a Manton Gaulin Model 500 HP-KL6-3PA, three-piston pump [35]. The modifications used for this pump did not facilitate analysis of pump wear surfaces. Instead, the tests, which were conducted at 3000 psi, 550°F for 100 hours were designed to evaluate fluid degradation, corrosion, lacquering and sludging tendencies.

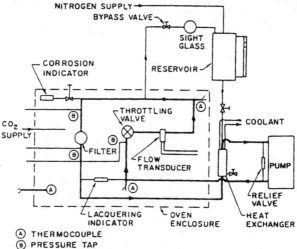

Fig. 15 -- Hopkins and Benzing high temperature modified Manton Gaulin piston pump test circuit.

Gschwender, et.al. used a test circuit utilizing a Vickers model PV3-075-15 pump to evaluate the wear of high temperature (122°C/250°F) poly(alpha-olefin) based hydraulic fluids [36]. The test circuit is illustrated in Figure 16. The tests were conducted at 20.4 MPa (2960 psi) and 5400 psi with the throttle valve closed and 5000 psi with the throttle valve open.

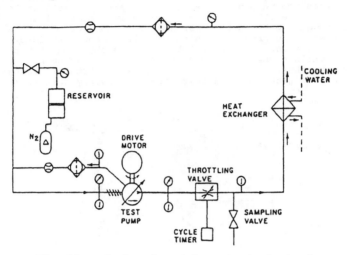

Fig. 16 -- Gschwender piston pump test stand.

E. Gear Pump Testing

In the above discussion regarding vane pumps, it was observed that the primary mode of wear, although certainly not the only one, was sliding wear of the vanes on the ring. For piston pump testing, failure analysis was more complex since there were numerous surfaces that must be inspected. Sliding with hydrodynamic wear was still the primary component. However, while sliding wear is still important in gear pumps, the wear mechanisms are even more variable. For example, in open gears, hydrodynamic, EHD, mixed EHD and boundary lubrication mechanisms may all occur simultaneously, depending on the position, and speed of the gear as seen in Figure 17 [37].

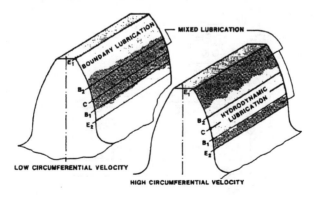

Fig. 17. -- Lubrication mechanisms on the gear tooth as a function of speed.

Frith and Scott performed a detailed theoretical analysis of gear pump wear. They noted that the primary areas of wear are the side plate near the suction port, the gear meshing zone, the gear tips and casing, especially near the suction port [38]. In addition, the imbalance of pressure across the pump may cause the gear shaft to deflect toward the inlet creating a reduction in gear tip clearance at the inlet. Also, a hydrodynamic wedge between the gear ends and the side plate, in combination with the pressure behind the plate, tends to force the side plate against the gears in the inlet region as shown in Figure 18. All of these conditions may affect the efficiency of gear pump operation. Interestingly, there have been relatively few reports of hydraulic fluid lubrication in a gear pump and there are no industry standard gear pump lubrication tests.

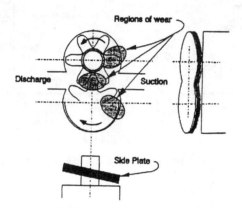

Fig. 18 -- Side plate action in a gear pump.

One study that has been reported was conducted by Knight who used the multiple gear pump test stand shown in Figure 19 [39]. Seven Hamworthy Hydraulics Ltd. Type PA 2113 gear pumps were run at 14.3 MPa (2075 psi). The test is run until pump failure, usually due to needle bearing fatigue. The condition of the roller bearings are monitored at least once every 24 hours.

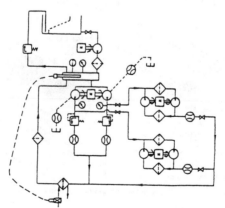

Fig. 19 -- Gear pump test circuit used at the British National Coal Board.

In another study, a cycled load over 500,000 cycles from zero to the maximum rated pressure, speed and temperature for the pump and the fluid was recommended [40].   In addition to the cycled loading test, an endurance test (maximum pressure, speed and temperature for the fluid for 250 hours, "proof" test where the pump is operated under extreme conditions at relatively short periods of time, e.g. 5 hours and an initial run-in test. Toogood stated that although degradation in performance would occur under extreme conditions, it was not clear if the damage was greatest under constant pressure or cycled pressure conditions [40].

Wanke conducted a study of the effects of fluid cleanliness with a multiple gear pump test stand shown in Figure 20 [41].   Two test sequences were used.   One was a cycled pressure test (1 cycle/10 seconds).   The other sequence was an endurance test.   As a result of this work, it was recommended that although flow was an adequate measure of the pump's integrity, monitoring the torque throughout the test would provide greater insight into overall pump integrity during the test.

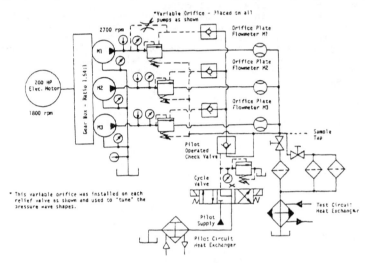

Fig. 20 -- Wanke gear pump test stand.

F. Evaluation of Pump Lubrication Results

The most commonly used analytical methods to monitor hydraulic pump operation have been reviewed in detail [42].   These methods include: flow analysis, volumetric and overall efficiency, cavitation pressure, minimum inlet pressure and others.   Johnson has provided a detailed review of pump testing methods to monitor pump durability and hydraulic operation [43].

Hunt has viewed pump lubrication analysis from the standpoint of failure correlation [44].   For example, hydraulic pumps are periodically inspected throughout the test for:

- Cavitation of port plate.
- Bearing wear and break-up.
- Piston slipper pad wear and blockage.
- Blockage of control devices.
- Piston and cylinder wear.
- Case seal defects.
- Gear teeth wear or fracture.

In addition to inspection, temperatures, pressures and flow throughout the test should be monitored since they are indicative of friction generation and fluid contamination. Also, wear analysis and vibration analysis as continuous monitors of wear should be performed [44,45,46]. Common vibrational and acoustical monitoring procedures and interpretive methods have been reviewed previously [47,48].

CONCLUSIONS

Although not exhaustive, an overview of vane, piston and gear pump testing circuits and procedures from the viewpoint of tribological testing have been provided here. With the exception of the various tests based on the Vickers V-104 vane pump and perhaps the Hagglunds-Denison HFO piston pump test, none of the pump tests described here have achieved broad industry acceptance. Most of the tests are pass/fail tests based either on inspection for cavitation and wear, fluid flow leakage, or weight loss of critical rotating components. Most of these tests do not employ continuous monitoring of hydraulic efficiencies and pressures, torque, vibrational analysis, etc. Therefore, although it may be desirable to either use or modify some of the tests reported here in the development of national standards, they should all be updated to reflect current engineering practice and possibilities to better model the tribological performance of the pump during use with a particular hydraulic fluid.

REFERENCES

1. Silva, S., SAE Trans., Vol.99, 1990, pp. 635-652.

2. Blanchard, R. and Hulls, L.R., RCA Engineer, Vol.20 No.6, 1975,pp. 36-39.

3. Hibi, A., Ichikawa, T., and Yamamura, M., Bull. of JSME, Vol.19, 1976, pp.179-186.

4. Ueno, H. and Tanaka, K., Junkatsu (J. Japn. Soc. Lub. Engrs.), Vol.33, 1988, pp.425-430.

5. Shrey, W.M., Lubr. Eng., Vol.15, 1959, pp. 64-67.

6. Renard, R. and Dalibert, A., J. Inst. Petr., Vol.55, 1969, pp.110-116.

7. Knight, G.C., "The Assessment of the Suitability of Hydrostatic Pumps and Motors for Use With Fire-Resistant Fluids", Rolling Contact Fatigue: Perform. Test. Lubr., Pap. Int. Symp., Eds. R. Tourret and E.P. Wright, 1977, pp.193-215.

8. Lapotko, O.P., Shkol'nikov, V.M., Bogdanov, Sh.K., Zagorodni, N.G., and Arsenov, V.V., Chem. and Tech. of Fuels and Oils, Vol. 17, 1981, pp.231-234.

9. Tessmann, R.K. and Hong, I.T., SAE Technical Paper Series, Paper Number 932438, 1993.

10. Perez, J.M., Hanson, R.C., and Klaus, E.E., Lubr. Eng., Vol.46, 1990, pp.248-255.

11. Mizuhara,K. and Tsuya, Y., Proc. of the JSLE Int. Tribol. Conf., 1985, pp. 853-858.

12. Wedeven, L.D., Totten, G.E., and Bishop Jr.,R.J., SAE Technical Paper Series, Paper No. 941752, 1994.

13. Platt, A., and Kelley, E.S., "Life Testing of Hydraulic Pumps and Motors on Fire Resistant Fluids", Proceed. Of The 1st Fluid Power Symposium, The British Hydromechanics Research Association, January, 1969, SP 982.

14. Turski, A.B., "Studies of Engineering Properties of Fire Resistant Hydraulic Fluids for Underground Use", Mining & Minerals Engineering, February, 1969, pp. 50-59.

15. Knight, G.C., SAE TECHNICAL PAPER SERIES, Paper Number, 810962, 1981.

16. Horiuchi, T., "Hydraulic Fluids and Trends in Oil Pumps and Motors", Nisseki Review, Vol. 21 No.3, 1979, pp 151-158.

17. Feicht, F., "Factors Influencing Service Life and Failure of Hydraulic Components", Oilhydraulik und Pneumatic, Vol. 20 No. 12, 1975, pp. 804-806.

18. Ueno, H., Tanaka, K. And Okajima, A., "Wear in the Cam Ring of a Vane Pump", Nippon Kikai Gakkai Ronbunsho Bhen, Vol 52 No. 480, 1985, pp. 2990-2997.

19. Hemeon, J.R., Appl. Hydraulics, August, 1955, pp.43-44.

20. This test procedure was communicated verbally by Mr. Paul Schacht as the standard cycled pressure vane pump test recommended by (Robert Bosch) Racine Fluid Power, Racine, WI.

21. Arsenov, V.V., Sedova, L.O., Lapotko, O.P., Zaretskaya, L.V., Kel'bas, V.I., and Ryaboshapka, V.M., Vestnik Mashinostroeniya, Vol.68, 1988, pp.32-33.

22. Thoenes, H.W., Bauer, K. and Herman, P., "Testing the Antiwear Characteristics of Hydraulic Fluids: Experience with Test Rigs Using a Vickers Pump", IP Int. Symposium, Performance Testing of Hydraulic Fluids, Oct. 1978, London, England, Ed. R. Tourret and E.P. Wright, Published by Heyden and Son Ltd.

23. Bishop, R.J. Jr. and Totten, G.E., "Comparison of Water-Glycol Hydraulic Fluids Using Vickers V-104 and 20VQ Vane Pumps", Ed. G.E. Totten, G.H. Kling and D.M. Smolenski, Tribology of Hydraulic Pump Testing STP ###, American Society for Testing and Materials, Philadelphia, PA, 1995.

24. Gengcheng, L. And Huanmou, M., "Test Stand Testing of Water-Glycol-Based, Non-Flammable Hydraulic Fluids", Runhua Yu Mifeng, Vol. 3, 1985, pp. 23-28.

25. "Pump Test Procedure for Evaluation of Antiwear Fluids for Mobil Systems", Vicker's Form No. M-2952-S.

26. Maxwell, J.F., Schwartz, S.E., and Viel, D.J., ASLE Preprint, No. 80-AM-713-2, 1980.

27. This is the so-called "Sundstrand Water Stability Test" described in Sundstrand Bulletin 9658. The test Protocol described was conducted by Southwest Research Institute in San Antonio, Texas.

28. Totten, G.E. and Webster,G.M., "High Performance Water-Glycol Hydraulic Fluids", Proceed. of the 46th Natl. Conf. on Fluid Power, March 23-24, 1994, p.185-194.

29. Lefebvre,S., "Evaluation of High Performance Water/Glycol Hydraulic Fluid in High Pressure Test Stand and Field Trial", Presented at STLE Annual Conference, May 1, 1993, Montreal, Canada.

30. This is the test procedure most recently recommended by Mr. Hans Melief (The Rexroth Corporation, Leheigh Valley, PA), as of 11/8/95. This test proposal is being studied by the ASTM D.02 N.07 committee.

31. "Conduct Test-To-Failure on Hydraulic Pumps", Vickers AA-65560-1SC-4), NTIS No. AD 602244, Sept. 1963.

32. Janko,K., _J. Synth. Lubr._, Vol.4, 1987, pp.99-114.

33. Edghill, C.M. and Rubery, A.M., "Hydraulic Pumps and Motors - Development Testing: Its Relationship With Field Failures", First European Fluid Power Conference, Paper No. 31, Sept. 10-12, 1973.

34. Ohkawa, S., Konishi, A., Hatano, H., Ishihama, K., Tanaka, K. And Iwamura, M., "Oxidation and Corrosion Characteristics of Vegetable-Base Biodegradable Hydraulic Oils", _SAE TECHNICAL PAPER SERIES_, Paper No. 951038, 1995.

35. Hopkins, V. and Benzing, R.J., _Ind. Eng. Chem. Prod. Res. Develop_, Vol.2, 1963, pp.77-78.

36. Gschwender, L.J., Snyder, C.E. Jr., and Sharma,S.K., _Lunr. Eng._, Vol.44, 1987, pp.324-329.

37. Paton, C.G., Maciejewski, W.B., and Melley, R.E., _Lubr. Eng._, Vol.46, 1990, pp.318-326.

38. Frith, R.H. and Scott,W., _Wear_, Vol. 172, 1994, pp.121-126.

39. Knight, G.C., "Experience with the Testing and Application of Fire-Resistant Fluids in the National Coal Board", _Trans. Sae_, Vol.90, 1981, pp. 2958-2969.

40. Toogood, G.J., "The Testing of Hydraulic Pumps and Motors", _Proc. Natl. Conf. Fluid Power, 37th_, Vol.35, 1981, pp.245-252.

41. Wanke, T., "A Comparative Study of Accelerated Life Tests Methods on Hydraulic Fluid Power Gear Pumps", _Proc. Natl. Conf. Fluid Power, 37th,_, Vol.35, 1985, pp.231-243.

42. American National Standard, "Hydraulic Fluid Power - Positive Displacement Pumps - Method of Testing and Presenting Basic Performance Data", ANSI/B93.27-1973.

43. Johnson,K.L., _Proc. Natl. Conf. Fluid Power, 30th_, Vol.28, 1974, p.331-370.

44. Hunt, T.M., _Technical Diagnostics_, Nov. 17-19, 1981, pp.89-99.

45. Bashta,T.M. and Babynin, I.M., _Soviet Engineering Research_, Vol.3 No.5, 1983, pp.3-5.

46. Avrunin, G.A. and Bakakin, G.N., _Soviet Engineering Research_, Vol.9 No.10, 1989, pp.37-39.

47. Maroney, G.E. and Fitch, E., _3rd International Fluid Power Symposium_, Paper C5, May 9-11, 1973, pp.C5-81-C5-96.

48. Dowdican, M., Silva, G. and Lowery, R.L., Oklahoma State Univ.-Fluid Power Research Center, Report No. OSU-FPRC-A5/84, 1984.

Jürgen Reichel[1]

Pump Testing Strategies and Associated Tribological Considerations—Vane Pump Testing Methods ASTM D 2882, IP281, and DIN 51389

REFERENCE: Reichel, J., **"Pump Testing Strategies and Associated Tribological Considerations–Vane Pump Testing Methods ASTM D 2882, IP281 and DIN 51389,"** *Tribology of Hydraulic Pump Testing, ASTM STP 1310,* George E. Totten, Gary H. Kling, and Donald J. Smolenski, Eds., American Society for Testing and Materials, 1996.

ABSTRACT:
Various test methods have been developed to determine the performance limits of various classes of hydraulic fluids. Lubrication capacity depends on various fundamental fluid parameters including viscosity and anti-wear properties. Critical elements of hydraulic pump and motor lubrication, which is characterized by sliding line-contact wear, will be discussed here. In vane pumps, the pressure loaded tips of the vanes are under Hertz-type load in contact with the surface of the cam ring, and rotate at a high speed creating a sliding line-contact. Due to this sliding line contact, the vane pump is the best-suited instrument for determination of the anti-wear performance of hydraulic fluids within acceptable time and at reasonable expense.

Alternatively, hydraulic pump and motor testing may require greater energy, 150 kW or more over a period of more than 1000 hours significantly increasing the cost of testing. Furthermore, tests on smaller versions of one type of pump or motor do not necessarily correlate with larger units of similar design. Therefore, it would be desirable to develop a laboratory test that: utilizes a rig with standard wear parts, provides a selective method for identifying various forms of lubricant failure, and that permits tests correlation with a wide variety of hydraulic pumps and motors used in the industry today.

KEYWORDS: Hydraulic, vane pump, tribology, lubrication, wear, hydraulic fluid

INTRODUCTION

A hydraulic fluid, which is used for energy transmission and control in a hydraulic system, should exhibit minimal pressure drops and losses due to mechanical friction over the broadest temperature range possible. Friction reduction is achieved by optimizing the anti-wear (or lubricity) properties of the fluid. The behavior of the fluid in the hydraulic system is dependent on fluid viscosity, which is usually expressed as dynamic (absolute)viscosity and is reported as Ns/m^2 (poise) or Pa s (1 Poise = mPa s). In practice, the viscosity/density ratio (kinematic viscosity) expressed in terms of cSt or mm^2/s is used. Generally, an operational viscosity range from 100 to 13 mm^2/s is required over the time the hydraulic system is in operation. Fluid characteristics are referenced to the viscosity at 40 °C (ISO-VG 46) which is determined according to ISO 3448.

[1]DMT-Gesellschaft für Forschung und Prüfung mbH, D-45307 Essen, Germany

Hydraulic fluids for power transmission systems are classified according to ISO 6743/4 standard which is shown in Table 1 and Table 2. Fluids may be either mineral oil or synthetic fluid derived.

They may, or may not, contain additives for improving special properties, such as corrosion protection, anti-wear properties, or viscosity as a function of temperature. Hydraulic fluids may also be classified as fire-resistant, aqueous or non aqueous. Some hydraulic fluids may be classified as rapidly biodegradable, a classification which is currently being developed into an international standard, which is largely being driven by Germany. Table 2 is showing the classification of biologically degradable fluids which either may be soluble in water or not.

The performance limits of hydraulic fluids can be improved by the addition of additives. Since fluid performance may than be additive-dependent, hydrostatic system manufacturers are interested in specifying appropriate fluid testing and qualification methods which adequately reflect the expected performance of the hydraulic fluid in their pump and motors. These testing methods should be repeatable, run at reasonable cost, and allow assessment of the in-use performance of the various classes of hydraulic fluids shown on Tables 1 and 2.

Table 1--Hydraulic fluids (mineral oil - ISO 6743/4)

CATEGORY	COMPOSITION Typical Properties	APPLICATIONS Operating Temperatures
HH	Non-inhibited refined mineral oils	- 10 to 90 °C
HL	Refined mineral oils with improved anti rust andanti oxidation properties	- 10 to 90 °C
HM	Oils of HL type with improved anti-wear properties	Generell hydraulic systems which included highly loaded components - 20 to 90 °C
HR	Oils of HL type with improved viscosity/ temperature properties	-35 to 120 °C
HV	Oils of HM type with improved viscosity/ temperature properties	Mobile applications - 35 to 120 °C
HS	Synthetic fluids with no specific fire resistant properties	- 35 to 120 °C

Fluid anti-wear performance is difficult to model quantitatively and be valid for the numerous types of wear contacts and material pairs found in the large variety of pump designs used in the fluid power industry today. The antiwear results obtained from laboratory testing apparatus including Four Ball Wear Test, Pin-on-disk Test, Timken Wear

Test and Cyclic Stress Vane Test can vary depending on load application, contact geometry and material pairs. Such test results may be valid only to a limited extent due to the multifunctional conditions prevailing on the wear surfaces in modern fluid power systems. These tests may not correlate well with actual field performance of hydraulic fluids. For this reason, the Hydraulic Pump Test is currently widely used to determine the wear property of fluids.

Table 2-- Hydraulic fluids (other than mineral oil - ISO 6743/4)

CATEGORY	COMPOSITION	APPLICATIONS
	Typical Properties	Operating Temperatures
	WATER BASED FLUIDS	
HFA E	Oil-in-Water -Emulsions mineral oil/ synthetic ester Concentration by vol. < 20%	Power transmissons - powered roof supports in mining -high working pressures
HFA S	Mineral oil free aqueous synthetic solutions Concentration by vol. < 20 %	Hydrostatic drives -low working pressures 5 to < 55°C
HFB	Water-in-Oil-Emulsions mineral oil portion < 60 %	Hydrostatic drives 5 to < 55°C
HFC	Aqueous polymer solutions water content > 35 %	Hydrostatic drives - 20 to < 60 °C
	Water free synthetic fluids	
HFD R	Phosphate ester base not soluble in water	Turbine governor fluids and hydrostatic drives < 120 °C
HFD S	Based on chlorinated hydrocarbons	Hydrodynamic couplings < 150 °C
HFD T	Blends of chlorinated hydrocarbons and phosphat ester	Hydrostatic drives 10 to 70 °C
HFD U	Anhydrous fluids based on other components	Hydrostatic drives -35 to < 90 °C
	Water free environmentally acceptable fluids (substitutes to mineral oil - ISO 6743/4, Draft)	
HEPG	Polyalkylene glycols soluble in water	Hydrostatic drives -30 to < 90 °C
HETG	Vegetable oils (Triglycerides, Rape seed oils) not soluble in water	Hydrostatic drives -20 to < 80 °C
HEES	Synthetic esters (Polyolester) not soluble in water	Hydrostatic drives - 35 to < 90 °C

Currently, a combination of two wear tests are often used o determine the performance of a hydraulic fluid under mixed-friction conditions. One is a vane pump test run according to either DIN 51 389 or ASTM D 2882 testing methods. The second test is to evaluate the fluid in an FZG gear test rig according to DIN 51 354 part 2 or ASTM D 1947.These two tests are used in hydraulic fluid performance specifications. The standards DIN 51524 and ISO/DIS 11158 "Specifications for Hydraulic fluids" specify the minimum requirements for mineral oil based hydraulic fluids of the types HM, HV and HG (see Table 1).

The FZG gear test rig is the predominant test used in Europe for evaluation of gear oils. The FZG-A/8,3/90 test provides performance limits of a lubricant utilizing a test gear at a circumferential speed of 8.3 m/s and a starting fluid temperature of 90 °C, with stage-wise increase of load induction through the spur gear over 12 testing stages. When larger damages occur, the test is stopped. The result is stated in terms of the load stage reached when such damage occurs. Type HM hydraulic oils must achieve at least stage 10 before damage occurs. Although, the anti wear properties of axial piston pumps are usually determined by the FZG test, specifically designed piston pump tests may also be used.

However, these tests have usually been limited to specific types of fluids. In this report, the development of   more general vane pump tests will be described to evaluate the antiwear properties of hydraulic fluids.

DISCUSSION

1. Constant-Volume Vane Pump Testing

Hydraulic fluids tested according to the FZG gear test [1] provide a reliable forecast of wear occuring in axial-piston pumps [2]. A similar correlation for vane pumps and gear pumps has not been shown. The anti-wear properties of a hydraulic fluid used in vane pumps may, however, be modelled when the fluid is tested in a Vicker's V 104 vane pump according to ASTM D 2882, CETOP RP 67 H, DIN 51389, part 2, and part 3 or IP 281. This standards apply to mineral oils other non aqueous fluids and aqueous fire-resistant fluids. The results of ASTM D 2882 DIN 51389 are not comparable since ASTM D 2882 is not run at a specified hydraulic fluid viscosity, but at specified temperature. Also the rotational speeds and testing times are not equivalent. However CETOP RP 67H, the British Standard IP 281 and the German Standard DIN 51389 are technically equivalent. The testing conditions for all of these standards are summarized in Table 3.

Pump Test Procedures vane pump V 104 C						
Procedure	Fluid Volume L	Time h	Pump RPM	Outlet Pressure bar	Inlet Temp./V. °C/cSt	Power Requirements kW
ASTM D 2882	57	100	1200	140	79°C	7
DIN 51389/2	70	250	1500	140	13 cSt	7
IP 281	70	250	1500	140	13 cSt	7
CETOP RP 67	70	250	1500	140	13 cSt	7

These vane pump tests are applicable to hydraulic fluids of ISO-viscosity grades VG 22 to VG 68. The total mass loss due to wear on the vanes and on the ring at the conclusion of the test is reported as ''total mg of wear''. The appearance of the cartridge must be recorded prior to the test. As three Round Robbin tests according to DIN 51389/2 conducted within Germany have shown [3], an assessment of the precision of the test is only possible if the mass loss is greater than 120 mg for the ring and greater than 30 mg for the vanes and if the total wear is greater than 150 mg. Since weight losses are frequently less than these values, it is often stated that the differences in wear figures for ring and vanes is attached too much importance.

The significance of the results decrease with increasing operating temperatures when the test is conducted at increasing temperatures resulting in increasingly lower fluid viscosities. In these cases, the anti-wear properties of the hydraulic fluid have not shown any correlation to in-field use. Also, there have been no reliable extrapolations reported to related vane pump wear obtained at the test temperature and viscosity to wear that would be expected to result at higher temperatures.

2. Wear Analysis of Vane Pump Cartridges

A schematic of the V-105C vane and ring assembly cartridge used in Vickers V-105C (and V-104C) is shown in Figure 1.

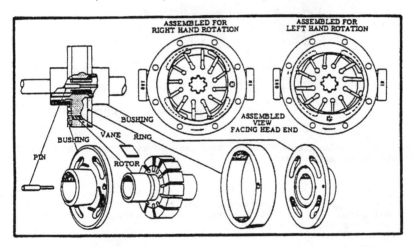

Fig. 1 Illustration of the constant volume vane pump cartridge for the Vicker's V-105C pump

Recently, there have been an increasing number of complaints about pump downtime due to rotor and segment failure and worn or seizing brass bushings DIN 51389 vane pump testing which is conducted with V-104 C or V-105 C type vane pumps and HM 46 hydraulic oils. An illustration of this wear is provided in Figure 2.

Bushing failure was accompanied be increasingly abnormal noise, strong pressure gauge pulsation, decreasing delivery flow rates and a decrease in the adjusted test pressure. Rotor failure was accompanied by sudden pumping noises sudden collapse of the test pressure, and in exceptional cases, a failure of the pumps shaft.

In order to address these problems, pump manufacturers funded a program to be conducted at the Technical University of Darmstadt and other testing organizations within Germany to evaluate this pump failures with respect to:

- Material analyses
- Dimensional accuracy of the pump components
- Geometry of the parts sliding past each other
- Load on the parts sliding past each other
- Assessment of the standard testing procedure

Figure 2 --Illustration of common bushing and rotor failure modes for V-105C vane pump cartridges

In addition, experimental investigations on highly stressed linear slide contacts in vane pumps were run at the Aachen Technical University and tribological analyses of the wear contacts were conducted [4]. The results of this studies showed:

1. Bushings failure were reproducible. The 2016 (cover-side) exhibited chatter marks and grooving with strong localized wear. The several hundredths of a mm was mainly caused by the vanes.

2. Rotor wear accounted for only a relatively a small portion of total wear. Rotor failures did sometimes occur during the first hours of the tests (Temperature > 93°C, pressure > 135 bar).

3. The material analyses on bushings of various manufacturing dates yielded no quantitative material chemistry correlation with wear.

4. Deviations were found in dimensional accuracy, however, no correlation of failure with these variations was observed.

5. No improvement in premature failure by bushing wear was observed with increasing fluid viscosity from 13 to 30 mm²/s (lower testing temperature).

6. Further test were run with a water-cooled pump cover. With such a cover, bushing wear, which primarily occurs during the first hour of testing, could be avoided.

7. A further important finding was that the type HM 46 hydraulic fluid exhibited an overall beneficial effect on wear reduction. Within the first hours, failure due to tarnishing of the bushings occurred in several testing laboratories. In other laboratories, where the tests were run on an oil HM 46 supplied by another manufacturer, no failure occurred and the total material loss for the ring and vanes was less than the recommended minimum requirement of 150 mg total wear. In the Timken tests run with load of 30 lbs, this tested hydraulic fluid resulted in less than 4 mg of abrasive wear [5].

Tests with a modified pin-on-disc tribometer were conducted at Darmstadt Technical University. In this test, the pin was replaced by a "vane" [6]. Interestingly the anti wear properties of various hydraulic fluids tested in this way correlated not in every case with wear rates obtained in the vane pump test. It would appear that the vane-on-disc tribometer may provide a cost saving alternative for developing or formulating new hydraulic fluids. Vane pump wear testing times, however, may be reduced by improvement of the measuring technology on the vane-and-cam-ring contact.

3. Further Vane Pump Tests Development

ASTM D 2882 and DIN 51389 are generally recognized worldwide as reasonable modelling the anti wear properties of AW (anti-wear) hydraulic fluids required for in-field use. The high sliding velocity and the highly stressed sliding contacts of the vane tips are difficult testing conditions to model in any other way. However, the standard testing pressure of 140 bar no longer correspond to the present state of vane pump technology. Also, it is expected that the production of the test cartridges will soon be no longer produced by the current manufacturer. A comparison of test results obtained with the vane pump V 104C as per ASTM D 2882 and a modern type 20VQ5 vane pump is reported in the papers submitted by E. Broszeit [7] and L. Honary at all [8]. Also pumping tests with higher rated vane pumps, such as the Type 35VQ25 pump, will be reported by H.T. Johnson [9]. These test proposals utilize 210 bar test pressure, and are advantageous because this alternative pumps are available worldwide. Therefore, it is proposed that international standards be developed using test procedures utilizing this pumps. Good repeatability and comparability of the results and standardization of a reference fluid will assure that testing failures may be clearly assigned to the appropriate cause. Furthermore, worldwide availability of test cartridges will be assured.

4. Pump Testing Strategies and Tribological Considerations.

Although significant advances in the evaluation of non-aqueous hydraulic fluids, especially antiwear mineral oil fluids, by bench testing and pump testing have been made, these success have not been equaled for aqueous fire-resistant hydraulic fluids, especially HFC fluids (see Table 2). Thus considerable work is still to be done to develop meaningful and reliable antiwear tests for this class of fluids.

Thus far, no small-scale pump test or simplified laboratory test that approximately models the tribological contacts encountered in a hydraulic pump has been identified which adequately models the actual wear obtained in hydraulic pumps in industrial operation. Therefore, it is still generally necessary to avaluate HFC hydraulic fluids in a variety of hydraulic pumps that are representative of those in use in the industry. In order to address this significant testing deficiency, it is first necessary to adequately differentiate and model sliding wear and rolling wear modes that will be encountered.

a. Sliding wear

The testing method "Mechanical testing in the vane pump, method B, testing of water based hydraulic fluids", DIN 51389, Part 3, has permitted the development of HFC hydraulic fluids with significantly improved sliding anti-wear properties. HFC 46 hydraulic fluids with a viscosity of 30 mm²/s are tested at 105 bar pressure and total wear rates on the vanes and the cam ring are typically less than 200 mg after the 250 hour test. This should be contrasted with the total wear rates typically obtained with earlier generations of these fluids was greater than 20,000 mg.

In trials run at DMT in Essen, Germany,   has found that an anti-wear layer builds up on the wear surface within 24 h of operation even with 140 bar operation pressure, as specified for mineral oil, no further increase in wear is observed. These results were confirmed in tests with internally pressurized radial-piston pumps and internal gear pumps with sliding bearings and pressure-controlled axial and radial gap compensation, under alternating load at 300 bar operation pressure. Progress in the development of HFC fluids recorded over the last 6 years is shown on Fig. 3.

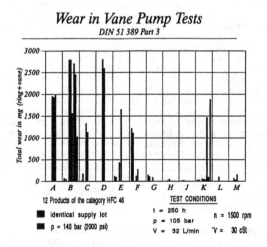

Fig. 3 --Test results conducted according to DIN 51 389 - Part 3 recorded for HFC hydraulic fluids over the past 6 years at DMT in Essen, Germany.

More recently developed HFC hydraulic fluids are capable of achieving the failure stage of 10 when tested in the FZG gear test rig at 50 °C according to method A 8,3 according to DIN 51354 part 2. It should be noted, that this result does not indicate that equivalent success is assured when these fluids are tested in the vane pump according to DIN 51389.

b. Rolling wear

The rolling anti-wear behavior of various roller bearing designs was evaluated using low-speed and high-speed positive-displacement pumps and HFC fluids was evaluated using the FZG test rig. Depending on the design, it was found that these bearings achieved lifetimes of 10 to 90 % of those obtained when using mineral oils. These results are illustrated in Figure 4.

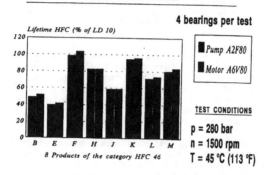

Fig. 4 - Lifetime of beveled ball bearings in pump tests conducted with HFC hydraulic fluids.

Failure of bearings made from the standard 100 CR 6 steel quality is almost always due to material fatigue (fatigue wear). A remedy is the use of Cronidur 30 high-nitrogen steel (HNS) qualities for roller bearings [10]. These novel stainless nitrogen-treated steel qualities are now successfully used by the bearing manufacturer FAG. In test runs at DMT, in Essen, Germany inclined ball bearings in bent-shaft axial-piston pumps recorded a longevity of 2,305 h, without recognizable damage, when the operation temperature increased from 48 to 65 °C. Even though this corresponds to 164 % of the 1,402 h of life time obtained with HFC 46 tested, the useful life of the bearings was not determined since the pump was still successfully running, even when the test was terminated. The trials was stopped due to increasing leakage on the radial shaft seals. The combination of high fatigue life and corrosion resistance assures a wide application range for high nitrogen steel qualities. Combined with ceramic balls, high nitrogen steel could frequently constitute optimized solutions to bearing problems. [11]

Summary

In hydrostatic drives, ISO 6743/4 fluids are used predominantly. Type HM and HV hydraulic oils, and automatic-transmission fluids (ATF) set the standards for antiwear performance in axial piston pumps, vane pumps, or gear pumps in modern heavy-duty power transmissions which are to be used pressures of up to 420 bar (6,000 psi) and temperatures up to 120 °C (250 °F). The anti-wear properties of these hydraulic fluids may be further improved by the addition of additives. The effectiveness of the anti-wear additives is difficult to quantitatively assess and to determine there effectiveness at all wear contacts of the various pump designs available today.

Over the past 20 years, vane pump testing methods have been developed that provide excellent models of the actual wear conditions encountered in industrial use. The wear characteristics of hydraulic fluids, are determined with constant-volume vane pump tests which have been widely adopted into national and international standards. For verification of the effectiveness of anti-wear additives, two standard testing methods based on the Vicker's V-104 vane pump are now well proven for the determination of hydraulic fluid antiwear performance.

A new testing method using a type 20VQ5 vane pump is being developed at present in the United States which provides the opportunity for the development of a global standard that is a cost-effective testing method to be run on existing test rigs with pressures of 210 bar.

A second testing method is to evaluate hydraulic fluids in the FZG gear tester as per DIN 51354, Part 2. In Europe, this testing unit is the most widespread procedure used for testing gear oils. For type HM and HV hydraulic oils the failure stage ten is specified as a minimum requirement. Also in case of axial-piston pumps, the pressure/wear behavior is normally determined by FZG testing. However, more recently piston pump tests have been increasingly used.

Additional pump testing methods must be developed for hydraulic fluids such as biodegradable fluids, fire-resistant fluids, or synthetic fluids. In these cases, conducting a pump test on load-test rigs is expensive, but very useful alternative, if expensive industrial equipment failures are to be eleminated.

References:
[1] Seitzinger, Kurt: Grenzen und Möglichkeiten der FZG Zahnrad-Verspannungs-Prüfmaschine.
Schmierungstechnik + Tribologie, (18) 1971, Seite 223

[2] Thoenes, H.W.: Anforderungen an Druckflüssigkeiten
ingenieur digest B. 11 (1972), Heft 12, Seiten 2-4

[3] n.n.: Workinggroup DIN 51389, Protokoll dated Dec. 1975

[4] Ortwig, Harald : Berechnung der physikalischen Belastungsgrößen im Flügel/Hubring-Gleitkontakt, einer Flügelzellenpumpe,
o+p Ölhydraulik und Pneumatik 34(1990) Nr. 11, Seiten 786-791

[5] n.n: Workinggroup DIN 51389, Protokoll dated Sept. 1989

[6] Kloos, K.H., E. Broszeit and F. Schmidt: Tribologische Prüfungen von Hydraulikflüssigkeiten,
Tribologie + Schmierungstechnik, 32 (1985) Heft 3 , Seiten 136-144

[7] Broszeit E. and A. Kunz: Comparison of several Vickers Vane Pump Test Procedures; Suggestions for an Improved Testing Procedure. Tribology of Hydraulic Pump Testing ASTM STP   , George E. Totten, Gary H. Kling, and Donald M. Smolenski, Eds., American Society for Testing and Materials, Philadelphia, 1995

[8] Honary L. ,T.A.Smith and E. Walles: Comparison of ASTM D 2882 and ASTM D 2271 Conducted Using Vickers V-104 and 20VQ5 Vane Pumps. Tribology of Hydraulic Pump Testing ASTM STP   , George E. Totten, Gary H. Kling, and Donald M. Smolenski, Eds., American Society for Testing and Materials, Philadelphia, 1995

[9] Johnson H.T. and T.I. Lewis: The Vickers 35VQ25 Pump Test. Tribology of Hydraulic Pump Testing ASTM STP   , George E. Totten, Gary H. Kling, and Donald M. Smolenski, Eds., American Society for Testing and Materials, Philadephia, 1995

[10] Lösche T., M. Weigand: Quantifying and Reducing the Effect of Con-tamination on Rolling Bearing Fatigue Life(FAG OEM und Handel AG). ISSN 0148-7191 SAE Technical Paper Series 952123, 1995, pp 41-51

[11] Reichel, J.: Fluid Power Engineering With Fire Resistant Hydraulic Fluids, Experences with Water -Containing Hydraulic Fluids. Lubrication Engineering, Vol. 50, No. 12 Dec. 1994, PP 947-952

G. Michael Gent[1]

REVIEW OF ASTM D 2882 AND CURRENT POSSIBILITIES

REFERENCE: Gent, G. M., **"Review of ASTM D 2882 and Current Possibilities,"** *Tribology of Hydraulic Pump Testing, ASTM STP 1310,* George E. Totten, Gary H. Kling, and Donald J. Smolenski, Eds., American Society for Testing and Materials, 1996.

ABSTRACT: In April of 1995, various laboratories in the United States and U.K. were contacted regarding, problems encountered, solutions and their current practice using the ASTM test method D 2882 (Standard Test Method for Indicating the Wear Characteristics of Petroleum and Non-Petroleum Hydraulic Fluids in a Constant Volume Vane Pump). Participants were sent the collected comments and techniques which had been shared by ASTM D-2 submittee N07 members over the last fifteen years. Participants were asked to review the document and add their own comments. The purpose was to collect and compare all available information about the test in an effort to resolve the operational problems associated with this test method. The objective of this paper is to present the results of this survey and to describe the engineering research performed by Conestoga to identify and remedy the operational problems associated with this method.

KEYWORDS: hydraulic fluids, wear test, vane pump

1. Conestoga USA, Inc., P.O. Box 3052, Pottstown, PA 19464
   (610) 327-2882

INTRODUCTION

In April, 1995, various laboratories in the United States were contacted regarding problems encountered, solutions and their current practice using the ASTM test method D 2882 (Standard Test Method for Indicating the Wear Characteristics of Petroleum and Non-Petroleum Hydraulic
Fluids in a Constant Volume Vane Pump). The purpose was to
collect and compare all available information about the test in an effort to resolve the operational problems associated with this test method. This survey was conducted by sending, to each participant, a copy of the actual test method with various comments and additional information added after the appropriate section in the method. The respondents were asked to add additional information and other comments regarding their experience in running the test.

The objective of this paper is to summarize the results of this survey. A total of six (6) complete operating procedures and over 150 comments were received. Only those subjects which were addressed by most of the respondents or those comments which are known to significantly affect the vane pump test will be discussed here. In addition, an overview of recent engineering research work conducted by Conestoga USA, Inc on various aspects of this test will also be provided.

DISCUSSION

A. Survey Results

The objective of this section is to summarize the most common comments received from multiple laboratories who conduct the ASTM D 2882 test method. This will be done by citing the subject of the comment followed by a brief summary of the comments received.

1. Parts Preparation: Pump parts require extensive inspection and hand preparation. Typically there is a high rejection rate of vanes and bushings due to their being out of tolerance. Other items which may be rejected if they can not be brought into tolerance by parts preparation procedures include the rotor, shaft, and housing. Although interlaboratory preparation techniques varied, finished part tolerances fell within a reasonably narrow band. General comments concerning this subject rated parts preparation and inspection as the most important factor in being able to run successful tests.

2. Rotor Failure: Rotor failure occurred regardless of the inspection and preparation techniques utilized.

3. Bushing Failure: Bushing failure has been reduced or eliminated by careful attention to parts preparation, maintaining adequate clearances in the cartridge and various torquing techniques.

4. Wear Results: There are indications that existing variations of the material condition of the wear parts and/or the geometry of the vanes may influence wear results.

5. Reuse Of Parts: It is common practice to reuse rotors for multiple tests. Bushings are reused if new acceptable bushings are not available.

6. Torquing: Assorted torquing techniques were reported. A summary of these methods is provided in Table 1.   The ASTM D 2882 method advises that 10 in-lb increments be used, up to a usual final level of 100-140 in-lb. Most laboratories which quoted values do not use 10 in-lb increments and none of the laboratories use the recommended final torque values.   Feeler gages and shim stock are used in torquing to ensure that the head has not cocked and has been drawn in sufficiently to seal evenly against the pump cartridge.

Table 1 -- Torquing Schedule

Laboratory Number	Torque Increment	Final Torque Value	Feeler Gauges Used (Yes/No)
D 2882 Requirement	10in-lb	100-140in-lb	no
1	5	50-90	yes
2	unknown	75	unknown
3	10	unknown	yes
4	5	65	no
5	5	40-70	yes
6	unknown	unknown	yes
7	5	40-60	unknown

7. Pump Maintenance: Reported pump maintenance routines varied and are summarized in Table 2. Several laboratories inspect the housing for an oversized bore which may contribute to ring cracking. These laboratories either replace the housing or "pack-out" the ring by wrapping a 0.001 inch shim around it. Several laboratories reported that bushing failures may be associated with a change of the housing. If this occurred, they simply replaced the housing again.

Table 2 --Pump Maintenance Schedule

Laboratory Number	Replace Seals	Replace Shaft & Bearings
D 2882 Requirement	each test	5 tests
2	10 tests	3-5 tests
3	3-5 tests	3-5 tests
5	each test	bearings each test, shaft as required
7	when fluid type changed	15,000 hours

8. Test Conditions: Test conditions have been modified by some laboratories in an effort to reduce rotor failures. Reducing the test pressure from 2000 psi to 1900-1950 psi appears to have reduced the incidence of rotor failure.

9. Fluid Volume: The fluid volume used for the test varied widely among the different laboratories.   Some laboratories reported operational difficulties with "small", 3-5 gallon, systems. Most laboratories use baffles in their reservoirs.

10. Filtration: While the D 2882 method prescribes 20 micron filtration, 10 micron filtration was used by all laboratories who addressed this point.

11. Flushing: No reporting laboratory, as shown in Table 3, followed the D 2882 prescribed flushing procedure.

B. Recent Engineering Research Results

Over the past ten (10) years, various engineering research projects have been undertaken at Conestoga USA, Inc. to address the operational problems that have been encountered with the D 2882 vane pump test. The experimental strategy has been to address these problems individually in order to identify and remove these variables by improving the engineering of the test componentry.

Table 3 -- Flushing Procedures

Laboratory Number	PSI	°F	Fluid	Time. Min.
D 2882 Req., 1st flush	1,000	100-120	Stoddard	30
D 2882 Req., 2nd flush	1,000	100-120	test fluid	30
1, 1st flush	50-150	100	Stoddard	30
1, 2nd flush	50-150	100	test fluid	30
2	500	160	base stock	120
3	Does not flush each run			
4, 1st flush	102	86	Stoddard	30
4, 2nd flush	102	86	Mineral oil	30
5	0	100	Test oil	60
6	100-120	unknown	Stoddard	unknown
8	100	100-120	Stoddard	unknown
9	Disassemble and wash system			

To prevent bushing and rotor failure, proper vane selection and preparation are essential. Vane wear of the bushings can cause bushing failure by opening a flow path directly between the suction and pressure ports. To prevent this wearing action, the vanes must properly track the ring and must not be subjected to unbalanced forces which will drive the vanes toward either bushing. Also, as shown in Fig. 1, a vane is in contact with the rotor and ring along three lines. For the vane to be in balance, the following three lines must be parallel:

A.   The line defined where the rotor slot trailing face meets with the vane surface.

B.   The line defined where the rotor slot leading face meets the vane bottom edge.

C.   The line defined where the vane chamfer meets with the ring surface.

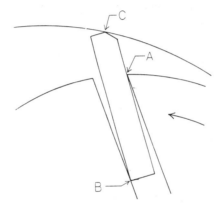

Fig. 1 -- Illustration of the three lines of vane contact with the rotor and ring.

To determine if the depth of the vane champfer had any affect on total weight lost during the test, duplicate tests using the same oil were performed in 1984. In the first test vanes with deeply cut chamfers (approximately .020" deep) were used and total weight loss of ring and vanes was 116.0 mg. In the second test vanes with lightly cut chamfers (approximately .007" deep) were used and total weight loss of ring and vanes was 59.5 mg. This potential source of inconsistent wear prompted Conestoga USA, Inc. to produce vanes with chamfers of consistent depth. In addition, Conestoga specified that the chamfer be

perpendicular to the sides and parallel to the rear edge (as shown in Fig. 2).

Standardization of the shape of the vanes would ensure that any bushing wear produced by these vanes could be traced to other components in the pump. However, continued bushing and rotor failures when utilizing these vanes showed that all pump failures could not be attributed to inconsistencies of the vanes.

Since standardization of the vanes did not prevent pump failure, the pump housing (illustrated in Fig. 3) was carefully examined. Several problems were identified including: bushing bores, cartridge bore and bearing seats were not necessarily concentric, nor were the body and head faces perpendicular to the bores. In a housing with these deficiencies, it may be impossible to get the bushings to seal properly against the housing faces while keeping the rotor true to the axis of the pump. Initially these problems were corrected by remachining all of the bores and faces. It proved to be more practical for Conestoga to design and machine its own pump housing which also permitted upsizing of the bearings, the addition of external case drains, and the inclusion of mounting feet as part of the casting. An added advantage of having a housing with true faces was to permit examination for any cocking of the head with the use of feeler gages to check the amount of the gap between the head and the body.

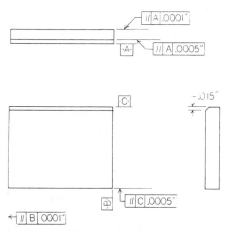

Fig. 2 -- Illustration of custom prepared vanes with precise chamfers for wear testing.

By careful measurement of the pump housing and cartridge, it is possible to determine exactly what gap should be present around the periphery of the head when the cartridge has seated properly.

The shaft was also custom machined to ensure that the splines were true to the axis and had smooth, straight sides so that the rotor was being driven evenly with no tendency to wear into a bushing face. So that pump alignment can be quickly and accurately accomplished, the outer diameter of the pump-motor coupling pair was ground concentric to its bore, so that a manual alignment check (with straight edge and feeler gage) can hold shaft to shaft alignment within .002".

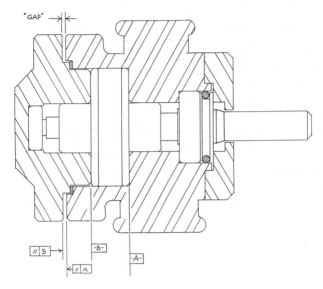

Fig. 3 -- Schematic of the vane pump housing.

Having removed the performance variables from the vanes, housing and drive components, potential bushings problems were next addressed. The bushing is illustrated in Fig. 4. By custom machining the bushings, it was possible to make the bushing shank with a larger outside diameter in order to fit more snugly with the housing bore. Since the bushing incorporates the journal bearing which supports the rotor, the bushings should not be able to shift relative to the housing. The suction ports were also eliminated from the outer bushing since they serve no purpose in that position and can only be a source of internal leakage.

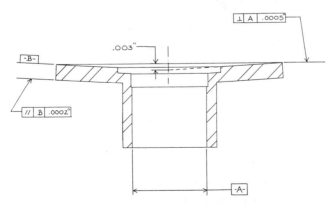

Fig. 4 -- Schematic of the vane pump bushing.

On the basis of this work, it was determined that there was very little at fault with the ring and nothing which would cause the persistent bushing and rotor failures being experienced was identified. However, rings were custom machined to fit a standard housing without having to wrap the OD with shim stock and which would have consistent weight, geometry and surface finish.

After eliminating all perceived variables from the components mentioned above, the pump failures continued and we concluded that the rotor, shown in Fig. 5, being the last variable component, must be a major factor in bushing failure as well as being an obvious one in rotor failure. Prior to this point, we had been inspecting the rotors and regrinding the outer diameter true to the journals to eliminate any taper which might influence vane movement. Other deficiencies inspected for included machining marks on the slot faces (which could interfere with vane motion), ragged broaching marks down the length of the terminal holes and slots cut off center of the terminal holes (which weaken the rotor segment). It was also discovered that the rotor slots might not be cut parallel to the rotor journals (which will definitely influence vane motion). Other than grinding the outer diameter, these conditions could not be corrected; therefore, only those rotors which passed inspection could be used. Since the number of rotors which passed inspection were prohibitively few, work was begun to custom produce the rotor.

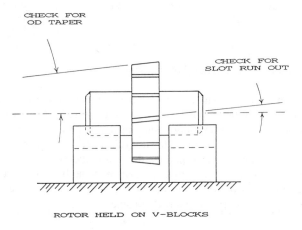

Fig.5 -- Rotor held on V-blocks.

The first rotor was produced from AISI 4340 alloy steel with a titanium nitride (TiN) coating to provide a wear surface for the vanes. Of two trials performed, one rotor survived 2000 psi for 100 hours and one failed within 12 hours. It was concluded that the fatigue endurance limit of the AISI 4340 material was too low relative to the expected stresses at the rotor segment root. It was also concluded that TiN coating was not a good choice since it did not cover the internal portions of the slots uniformly enough to present a good wear surface for the vanes.

As an alternative, rotors were prepared from Teledyne Vasco Company's VascoMax C-300, a maraging steel of 18% nickel, 9% cobalt and 5% molybdenum. Its predicted fatigue endurance limit (125,000 psi at 10^8 cycle) far exceeded the calculated stress at the rotor segment root. The rotor was nitrided after aging to create a hard surface for the vanes to ride against. This rotor was used in a pump run at 2,000 psi, 175°F with a premium polyol ester using the custom manufactured pump housing with the custom machined 8 gallon per minute ring, vanes, bushing and shaft. There was no decay of the flow rate during the test. At 40 hours the pressure was increased to 2,100 psi. After 2 hours of running at 2,100 psi, the rotor failed. A definitive cause of the failure has not yet been established but the condition of the components and the steady flow rate indicate that the failure was probably due to inadequacies of the rotor itself. It is possible that the nitriding cycle caused the material to overage and weaken. Currently the plan is to continue experiments with this and other materials until a rotor capable of extended performance at 2,000 psi is produced.

CONCLUSIONS

In this report, an overview of the survey results showed that none of the laboratories conducted their tests exactly according to the test method. Each of these test modifications were made to address pump performance and test reproducibility problems encountered in assessing the antiwear performance of both petroleum oil and non-petroleum oil derived hydraulic fluids. It is hoped that the survey results can serve as a guide for further improvement of ASTM D-2882.

Also in this report, the results of an engineering research program conducted to examine the effect of pump component construction variables on pump performance was discussed. As a result of this work, a number of possible pump component alterations were identified, with the intention that these suggestions may be helpful tc users of ASTM D-2882

ACKNOWLEDGEMENTS: Sincere appreciation to those labs so gracious as to contribute their knowledge to the investigations described herein. Particular appreciation to the ASTM D.02 N07 committee for their cooperation and guidance.

William M. Nahumck[1] and Thelma Marougy[2]

SECTION D02.N.7 STATUS REPORT ON VICKERS 20VQ5 VANE PUMP TEST DEVELOPMENT

REFERENCE: Nahumck, W. M. and Marougy, T., **"Section D.02.N.7 Status Report on Vickers 20VQ5 Test Development,"** *Tribology of Hydraulic Pump Testing, ASTM STP 1310,* George E. Totten, Gary H. Kling, and Donald J. Smolenski, Eds., American Society for Testing and Materials, 1996.

ABSTRACT: Concerns about operational difficulties with the ASTM D2882 Test Method and whether the Vickers 104C vane pump will be continued in production has prompted the need to develop a replacement test for the ASTM D2882 Test Method. ASTM Section D02.N.7 has been investigating for the last five years such a potential replacement test using the Vickers 20VQ5 vane pump. The goal is to develop a replacement test that uses a pump of current and expected future manufacture that will provide results consistent with results obtained previously using the D2882 Test Method. The initial study that was the basis for test development was originally published in May 1992 in the STLE publication, "Lubrication Engineering". Since that time there has been an ongoing effort to complete a comparison using four (4) reference oils in nine (9) laboratories, including European participants. Difficulties and problems related to test operation and test cartridge preparation have occurred and are reviewed in this paper.

KEYWORDS: antiwear, hydraulic fluid, vane pump, pressure, galling, rippling, vane tip contours, vanes, cam ring, internal leakage, flow rate

For many years, ASTM Test Method D2882 [1] has been used by the producers of industrial hydraulic fluids as one of the primary "industry standard" tests to evaluate the antiwear performance characteristic of various hydraulic fluid formulations. The D2882 test utilizes the conditions outlined in Table 1 using the Vickers 104C vane pump. The test circuit diagram is shown in Figure 1, operating at a discharge pressure of 138 bar, gauge (2000 psig). This pressure is

---

[1]Project engineer, Mechanical Engineering Department, The Lubrizol Corp., 29400 Lakeland Blvd., Wickliffe, OH 44092.

[2]Manager, Fluids Laboratory, Vickers, Inc., 2730 Research Dr., Rochester Hills, MI 48309-3570.

100% over the rated capacity for this pump at the test speed of 1200 r/min. Many of the testing laboratories in the industry have experienced difficulties conducting consistent fluid evaluations under the specified test conditions of ASTM D2882. These difficulties are related to the test pressure and to variations in manufacturing.

TABLE 1--**ASTM D2882 test conditions**

Fluid Volume, Liters	Test Length, Hours	Pump Speed, r/min.	Discharge Pressure, bar (gauge)	Inlet Fluid Temp., °C	Power Requirement, kW
57	100	1200	138	79.5	6.9

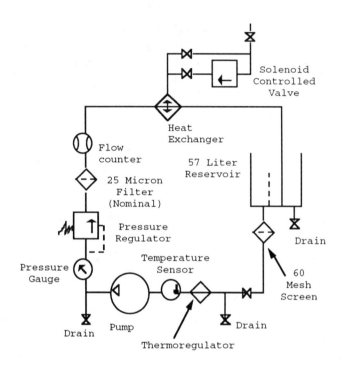

FIG. 1--Schematic of hydraulic fluid wear test

INITIAL INDEPENDENT STUDY

The initial independent study [2] selected the Vickers 20VQ5 pump and used the existing ASTM D2882 test conditions as a starting point for test development. The objective of the initial study was to develop a low cost replacement test using a modern high pressure vane pump, yet duplicate results obtained in D2882 using hydraulic fluids of known quality (Table 2) [3][4]. An automatic transmission fluid (ATF) was also evaluated because several manufacturers use a modification of the D2882 which is used to evaluate pump performance of an ATF. Also ATF's are regularly used for some hydraulic system applications.

The Vickers 20VQ5 pump was chosen because minimal hardware changes are required to convert the test stand from the D2882 configuration. Only hose and fitting changes would be needed in most cases. If successful, there would be value to the hydraulic fluid industry by providing easy conversions of the numerous D2882 test stands that currently exist for evaluating hydraulic fluids.

The 20VQ5 vane pump has been designed to operate at pressures in excess of 200 bar (2900 psig) at speeds of up to 2700 r/min. By contrast, the Vickers 104C is rated at 70 bar (1000 psig) at 1200 r/min. This new test hardware would also provide capabilities to the industry that are not possible with the pump used for the D2882 test. More stressful test conditions, such as higher discharge pressure and inlet temperature, can now be considered for future test development to improve discrimination of hydraulic fluid performance.

TABLE 2--**Commercial hydraulic fluids**

Code	Description (Zinc Dithiophosphate Type)	Specification Approvals	ISO Grade
R & O	None	Denison HF-1	46
AW-1	Stabilized Primary	Vickers M-2952-S ASTM D2882* Denison HF-0	46
AW-2	Primary	Vickers M-2952-S ASTM D2882* Denison HF-0	46
AW-3	Secondary	Vickers M-2952-S ASTM D2882* Denison HF-2	46
ATF	Secondary	Vickers M-2952-S ASTM D2882* Denison HF-2	32

* D2882 does not "specify" a limit for the ring and vane wear.
However, many specifications exist which limit the allowable wear in this test.

This initial non-ASTM study [2] with the Vickers 20VQ5 pump utilized the demonstrated performance of the ASTM D2882 test conditions (Table 1) as a reasonable point to begin the investigation. In addition, other test parameters (pressure and temperature) were modified as shown in Table 3 by Procedures A, B and C. A summary of multiple test runs from each of the Procedures A, B and C is shown graphically in Figure 2. These results show the expected separation of known antiwear (AW) versus rust and oxidation (R & O) formulations. Also, comparing procedure "B" to other recognized industry vane pump tests confirmed that the ranking of the test fluid results has not been changed (Figure 3). The authors of the original procedure study [2]

TABLE 3--**20VQ5 alternative pump test procedures**

Procedure	Fluid Volume, Liters	Test Length, Hours	Pump Speed, r/min.	Discharge Pressure, bar (gauge)	Inlet Fluid Temp., °C	Power, kW
A	57	100	1200	138	79.5	4.3
B	57	100	1200	138	93	4.3
C	57	100	1200	207	79.5	6.5

suggested that Procedure B gave the degree of separation desired
without excessive catastrophic wear that could damage the test stand
components and not be as discriminatory.

In the D2882 Test Method, the only performance criteria listed is
the ring and vane weight loss. Although this is an important parameter
that can be measured, most original equipment manufacturers (OEM) also
visually evaluate the condition of the vane and cam ring. The
condition of the various parts give a more complete understanding of
the fluid performance related to such areas as air entrainment,
cavitation and type of wear. The authors wanted to include the OEM's
concerns as guidelines for determining overall performance of a
hydraulic fluid under demanding test conditions. They were able to
develop a test method that had the following performance requirements.

- Primary reported measurement is ring and vane weight loss.
- For good performing oils, vane tip contours remain unchanged.
- No visible rippling evident on the cam ring surface for good
  performing oils.
- Side plates should remain essentially unchanged.
- No evidence of metal smearing or galling observed between the
  vanes and rotor slots for good performing oils.
- No visible deposit formation or discoloration shown on the
  vanes or cam ring.

With this background having been presented to ASTM, the proposed
procedure was reviewed by ASTM Section D02.N.7 in 1990. The membership
agreed that this study provided good potential for beginning the
development of a replacement test for the aging ASTM D2882 procedure
and associated Vickers 104C hardware. Section D02.N.7 decided to use a
modification of Procedure A instead of Procedure B because the test
conditions of Procedure A more closely matched those of D2882 Test
Method.

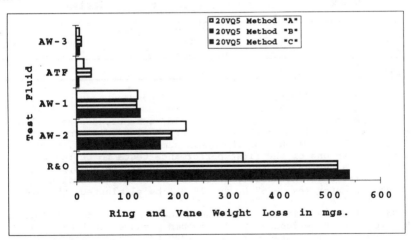

FIG. 2--Results obtained with the 20VQ5 pump and alternative procedures

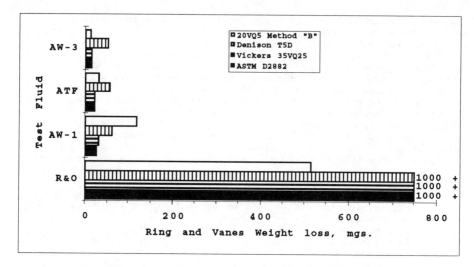

FIG. 3--Summary of vane pump wear results

INITIAL ASTM ROUND ROBIN TEST METHOD

Plans were prepared within ASTM Section D02.N.7 to begin a Round Robin using the agreed upon test conditions summarized in Table 4, incorporating two modifications to Procedure A of the initial study. The test method was written in a draft format to allow the laboratories to begin testing using a specific test procedure[5].

TABLE 4--**First round robin test conditions**

Fluid Volume, Liters	Test Length, Hours	Pump Speed, r/min.	Discharge Pressure, bar (gauge)	Inlet Fluid Temp., °C	Power Requirement, kW
19	100	1200	138	79.5	6.9

The first modification done of the independent test conditions was to standardize on a reduced reservoir volume from 57 L. to 19 L. This was done to reduce the amount of fluid to be blended as well as hazardous waste handling at the end of test. Also this mimics the D2882 Test Method. Because of the small sump and low cost, this type of test is often one of the first performance tests to be run in the evaluation of a typical hydraulic fluid. Standard methods, by their nature, require that the ability to choose between two versions of the same test parameter (ex. - 57 L. vs. 19 L.) be eliminated, wherever possible.

Secondly, the ASTM Section changed the test pump to the Vickers 20VQ8 pump. The 8 designates that the pump produces an 8 gal/min (30.3 L/min) flow rate at a rated pump speed 1200 r/min and pressure of 6.9 bar, gauge. It was anticipated that this combination would provide a similar flow rate to the 7.5 gal/min (28.4 L/min) and a vane tip speed

similar to the Vickers 104C pump used in the D2882 Test Method. The
initial independent study did not evaluate these test conditions using
a 20VQ8 cartridge. ASTM Section N.7 was felt that the higher flow rate
(20VQ8 pump vs. 20VQ5 pump) in combination with the expected higher
vane tip speed would give an opportunity to increase test severity at
the lower pump speed that was carried over from the ASTM D2882 Test
Method. This pump would also provide for expanded future capabilities,
such as running at 206 bar, gauge, if desired. Other stand
modifications, such as a larger electric motor, would be required to
most stands to accommodate higher pressures and greater speeds.

INITIAL ASTM ROUND ROBIN

Four (4) different test oils were chosen that would be similar to
the types used in the initial independent study. They were identified
as Oils A, B, C and D (Table 5). An ATF fluid was not used because the
Section wanted to concentrate their efforts on mineral based hydraulic
fluids. The ASTM test matrix was started with thirteen (13)
laboratories volunteering to run duplicate tests using the conditions
outlined in Table 4 with the modifications of the 20VQ8 cartridge and a
19 liter reservoir. All laboratories were asked to run Oil A first to
verify that they were capable of obtaining the level of severity that
was required for Oil A.

TABLE 5--Commercial hydraulic fluids used for ASTM round robin

Code	Description (Zinc Dithiophosphate Type)	ISO Grade	Typical D2882 Wear Results, TWL (mgs.)
A	None	46	1000+
B	Stabilized Primary	46	25
C	Secondary	46	15
D	Primary	32	15

Table 6 shows the results of the first ASTM Round Robin testing
using the conditions shown in Table 4 using a 19 L. sump and the
Vickers 20VQ8 pump. The wear results are described by the Total Weight
Loss (TWL) in mgs. which is the combined weight loss from the ring and
vanes. Test data were developed by ten different laboratories, two
thirds of which ran all four reference oils. There was only partial
agreement by some laboratories as shown in Table 6. The results
obtained with Oil A were not consistent from laboratory to laboratory.
Some labs obtained low wear with Oil A whereas several participants
were able to get the expected high level of wear. In some cases, the
trend was reversed compared to the original data, showing Oil B with
higher wear than Oil A. With Oil B being an antiwear oil with known
performance, there should have been an improvement in wear performance
compared to Oil A.

Testing in this matrix was terminated prior to the cooperators
completing all of their runs. As a result of the inconsistent results
of the runs using Oil A, the Section agreed that further testing would
waste valuable resources that were volunteered by the cooperators.
Investigations were begun to review several different aspects that

TABLE 6--First ASTM 20VQ round robin wear results

(Table 4 test conditions with 19 L. sump and 20VQ8 pump)
Total ring and vane weight loss, TWL (mg.)

Laboratory	Run #	Oil A	Oil B	Oil C	Oil D
1	1	5	104	1	0
	2	1	106	4	0
2	1	0	150	20	10
	2	30	110	50	10
3	1	54	250	1	3
	2	38	159	0	3
4	1	280	156	8	8
	2	249	119	5	2
	3		147		
5*	1	363	173	5	3
	2	535	115	6	0
6	1	66	25	0	4
	2	94	64	15	22
7	1	4	89	0	3
	2	11	84	6	2
8	1	31			
	2	4			
9**	1	218			
	2				
10	1	168			
	2				

* Tests conducted at 1440 r/min.
** Tests conducted at 1750 r/min.

could relate to test severity, including the physical stand configurations and cleaning methods used by the various laboratories. After reviewing the preliminary data from the first round robin, the Section decided that it was necessary to have the participants supply information for review in the following areas.

1. Stand configurations
2. Rebuild methods
3. Operating procedure
4. Verification of proper oil shipment
5. Flushing procedures

Stand descriptions and photographs were compiled. There were no visible significant differences that the committee felt would be related to severity. As most stands have been in service for many years, some laboratories needed conduct maintenance, such as hose repair or replacement, on some stands. Rerouting of some hoses were recommended. Full flow shut off valves were not present on all stands and were installed. As a precaution, all laboratories were requested to verify their instrumentation calibration before continuing any further testing.

Some of the participants had indicated that they had difficulty with proper cartridge assembly prior to installation in the test pump housing. This could have affected internal leakage, resulting in lower flow rates. A pump cartridge alignment tool, an automotive piston ring compressor of the correct diameter, was recommended in the procedure, but apparently was not used by all. This tool provides quicker and more accurate alignment of the cartridge components.

One laboratory had noted that for one of their runs they were able to obtain the 138 bar discharge pressure, but were unable to develop the minimum acceptable flow rate (greater than 13.2 L/min at the specified discharge pressure). Upon investigating the problem, they discovered that the pump cartridge was assembled with some of the vanes installed backwards. Another laboratory reported that with one cartridge, one vane was considerably shorter than the others, giving a vane/rotor relief of 35.6 μm. This caused the flow rate to be at 6.43 L/min, instead of the desired 13.2 L/min. Replacement of the short vane restored the flow rate to the desired level.

Some of the procedural instructions were not being followed. The inlet temperature was not the same for all runs. A few tests were conducted at 66°C (150°F) instead of 79°C (175°F). This parameter must be controlled for a consistent test procedure. It was stressed that for all future testing that all runs must be conducted at 79°C (175°F).

One of the procedural parameters that became a variable was pump speed. The procedure specifies that the nominal "North American" motor speed of 1200 r/min be used. However, in Europe, the closest standard motor speed used for electric motor is 1440 r/min. This speed has also been required in the pump test specified by the IP 281 or DIN 51389 specification, both of which are similar to D2882 [6][7]. The laboratory that participated from Europe, Laboratory 5, gave expected typical results for reference Oil A with the higher pump speed. (See Table 6.) Also, Laboratory 9 conducted their single test at a standard fixed motor speed of 1750 r/min as this was all they had available. The level of wear for Oil A in this lab was on the low side of the acceptable range. (See Table 6.)

All participating laboratories were asked to analyze a fresh sample of the Reference Oil A that was used in their laboratory. All the results that were reported indicated that there was no antiwear additive chemistry in any of the Reference Oil A runs. This verified that Oil A was to an antiwear formulated fluid. Reference Oil A has no zinc dithiophosphate (ZN DTP) present compared to the other oils that are being evaluated in the round robin. The conclusion was that Reference Oil A was properly shipped to the participating laboratories.

Flushing techniques were reviewed and were found to be somewhat variable based on the past practices of that laboratory. Prior to starting the next mini Round Robin, the participants were reminded that the flushing procedure was specified in the draft test procedure.

MINI ROUND ROBIN TO INVESTIGATE SEVERITY

After reviewing the data that were available from the first Round Robin, the Section prepared for the next phase of evaluating appropriate test conditions to increase test severity before requiring significant changes to the test stand configuration. Table 8 shows the conditions that were agreed upon for conducting a mini Round Robin using only Reference Oil A. The primary change was the higher discharge pressure. If this effort was not successful, then it would become necessary to consider significant changes in test conditions that would require increasing the size of the electric motor to drive the pump used in most laboratories. This could be costly and would likely reduce participation of some volunteers in future round robin testing.

TABLE 7--Mini round robin test conditions

Fluid Volume, Liters	Test Length, Hours	Pump Speed, r/min.	Discharge Pressure, bar (gauge)	Inlet Fluid Temp., °C	Power Requirement, kW
19	100	1200	206	79.5	10.3

These conditions also required the following caveats.

1. All tests are to be run using 20VQ5 pump cartridges.
2. All 20VQ5 pump cartridges are to be from a single assembly batch.
3. Inlet suction pressure was to be checked and adjusted, if necessary, to obtain 0.0 to 0.2 bar, gauge.
4. Flow rate data (before and end of test) is to be reported for all tests. If the flow rate of the pump falls below 13.2 L/min once the pump has reached test conditions, the cartridge should be inspected and either replaced or rebuilt.

It was felt by Section N.7 that if the participants of the mini Round Robin were able to obtain the appropriate level of wear with Oil A using these new test conditions, we would have a sound basis for conducting a larger Round Robin, similar to the one previously described using Reference Oils A, B, C and D.

Seven laboratories volunteered to conduct testing using the new conditions. The numbering sequence for identifying the laboratories from the first Round Robin has been retained for this Round Robin. Table 8 shows the wear results obtained in this investigation with the conditions shown above. The wear results are described by the Total Weight Loss (TWL) in mgs. which is the combined weight loss from the ring and vanes. Also, the flow rates from the 0 hours and the end of test are listed, where available.

TABLE 8--ASTM 20VQ mini round robin test results
(Table 7 test conditions with 19 L. sump and 20VQ8 pump)
REFERENCE OIL A

Laboratory	Run #	Weight loss, mgs.			Flow Rate, L/min	
		Ring	Vanes	Total	0 hrs.	EOT
1	1	Scheduling				
	2	Problems				
3	1	593	76	669		
	2	658	81	739		
4	1	283	5	288	14.4	14.0
	2	298	10	308		
	3	337	12	349		
6	1	Running				
	2	Problems				
9*	1	200	8	208	28.0	25.8
	2	300	17	317	26.5	23.5
10	1	50	4	54	14.4	14.0
	2					
11	1	530	30	560		
	2	311	84	398		

* Tests conducted at 1750 r/min.

Upon reviewing the test data (Table 8) from some of the laboratories, the flow rates reported were found to be not repeatable, either at an individual laboratory and among different laboratories. Laboratory 6 questioned the rebuild instructions that were provided in the 2nd draft procedure distributed by the Section. This happened after their technician invested extensive time into the pump cartridge rebuild and was not able to obtain the new test conditions (Table 7) with that cartridge. The vane/rotor relief was originally stated as 25.4 µm to 22.9 µm in the procedure. By following the procedure exactly, higher internal leakage would result. This will contribute to low flow rates and difficult pressure control.

Rebuild instructions were corrected to show in the 3rd revision of the draft procedure that "... the top to bottom vane widths must be 0.0 µm to 22.9 µm less than the average rotor width". This change was issued prior to completing the Mini Round Robin with modified conditions. As will be shown later, Laboratory 6 was able to get the expected performance with Oil A using the proper vane/rotor relief specification.

The wear data suggest that the increase in pressure was sufficient in most cases to increase the severity level. This is particularly evident in the results with Reference Oil A for Laboratory 3. Previous conditions had produced an average wear result for Oil A of 46 mgs. The revised conditions produced average wear results of 704 mgs.

SECOND ASTM ROUND ROBIN

ASTM Section D02.N.7 reviewed the data in Table 8 during the December 1994 meeting and concluded that there was significant improvement in the test severity to continue using these test conditions to develop future test data. It was requested that testing continue with Reference Oil A for those laboratories who had not completed their runs. The membership also authorized completing the test matrix shown in Table 9 with Reference Oils B, C and D. This matrix is referred to as the Second Round Robin.

The second 20VQ5 Round Robin for the ASTM Section D02.N.7 is currently in progress. This test procedure has been developed over a long period of time with the latest draft being the 4th version. The limited data that have been developed suggest that there is a basis for a new constant volume vane pump test to replace the ASTM D2882 Test method. The performance criteria that were used as guidelines by the original procedure authors has been maintained during the ASTM procedure development. Reference Oils B, C and D were shipped to the laboratories shown in Table 9 so that they may complete the required runs for the Second Round Robin with the new conditions. Please note that Table 9 includes the runs shown in Table 8.

Additional cooperators have since volunteered. They are planning on conducting duplicate runs with all four reference oils. This brings the current number of cooperators to nine with the four different reference oils being evaluated. There have been indications of additional volunteers from North America and Europe. They are willing to conduct tests on the reference oils to add to the database. The European connection could provide data using the higher pump speed of 1440 r/min and its influence of a higher pump speed on hydraulic fluid evaluations.

Testing with the other reference oils commenced in 1995 with the conditions outlined in Table 7, using the 4th revision to the draft procedure. Data available at the time of this writing is given in

Table 9. The scheduling problems for Laboratory 1 have been partially resolved.

Reproducibility issues needed to be addressed and precision data need to be developed. Section D02.N.7 will assist those laboratories that have produced low wear results with Reference Oil A in determining the probable causes. Once these issues have been addressed, the

TABLE 9--ASTM 20VQ second round robin test results
(Table 7 test conditions with 19 L. sump and 20VQ5 pump)

Lab #	Run #	Oil A TWL, mgs	Oil A Flow Rate, L/min 0 hrs.	Oil A EOT	Oil B TWL, mgs	Oil B Flow Rate, L/min 0 hrs.	Oil B EOT	Oil C TWL, mgs	Oil C Flow Rate, L/min 0 hrs.	Oil C EOT	Oil D TWL, mgs	Oil D Flow Rate, L/min 0 hrs.	Oil D EOT
1	1	190											
	2	216											
3	1	669											
	2	739											
4	1	288	14.4	14.0				12	12.9	13.3	6	10.6	10.6
	2	308	14.4	14.4				8	11.4	12.5	5	13.7	14.4
	3	349	11.0	11.4				11	14.8	14.8	8	12.1	12.9
6	1	346	15.2	15.2	54	14.8	14.0	7	13.3	12.5	10		
	2				37			20			7		
8	1	23	9.5	9.5									
	2												
9*	1	208	28.0	25.8									
	2	317	26.5	23.5									
10	1	54	14.4	14.0									
	2												
11	1	560											
	2	395											
12	1	131	21.6	19.3									
	2	160	22.7	20.5									

* Tests conducted at 1750 r/min.

remaining test work should be performed as expected, based on known performance of the reference fluids. Severity issues at Laboratory 8 and 10 are under investigation and hoped to be resolved soon.

The running problems for Laboratory 6 were resolved by correcting the rebuild clearances that were stated in the procedure and some minor stand maintenance related to hoses. Not only were the running difficulties overcome there, but the expected severity level for Reference Oil A has been obtained.

CONCLUSIONS AND FUTURE WORK

Although it is too early to draw final conclusions, it does appear that the Section has a reasonable chance at producing a new constant volume vane pump test using the Second Round Robin test conditions. This new vane pump test has been developed to reflect higher pressures seen in actual field conditions.

If the additional Round Robin work shows that this procedure has correlation to D2882 and known field performance, this will be the first hydraulic pump test in many years to be considered for an ASTM standard test method. Reproducibility issues need to be resolved

before the test method can be considered for an ASTM standard.   The
cost of converting a test stand from D2882 to the Vickers 20VQ5 pump
should be small (less than $1000) and manageable.   This will permit a
quick conversion should a laboratory decide to convert an existing
D2882 test stand.

This test stand configuration also allows for expanded testing
capabilities and ability to vary pressures and speeds, if desired, to
demonstrate other performance criteria.   This was not possible with the
previous hardware used to support the D2882 Test Method.

With other types of hydraulic fluids available in the market,
such as water based or vegetable oil based fluids, this test method
needs to be evaluated to determine if it is applicable to these fluids.
Many of the existing tests have not addressed the concerns of the
producers and users of these alternate fluids to the satisfaction of
that part of the industry.   There may be other parameters that may be
significant to demonstrate their performance in the field.   Using this
basic test format and stand configuration, changes to these parameters
may need to be explored to reflect either the running conditions, stand
hardware or performance requirements to develop meaningful test methods
for these alternate hydraulic fluids.

ASTM Section D02.N.7 is looking forward to soon presenting to the
main body of Committee D02 a new, improved test method for indicating
the wear prevention characteristics of hydraulic fluids.

References

[1]   "D2882 - 90, Standard Test Method for Indicating the Wear
      Characteristics of Petroleum and Non-Petroleum Hydraulic Fluids
      in a Constant Volume Vane Pump", Annual Book of ASTM Standards,
      Section 5, 05.02, Petroleum Products and Lubricants, American
      Society of Testing And Materials, Philadelphia, 1991.

[2]   Perez, R. J. and Brenner, M. S., "Development of a New Constant
      Volume Vane Pump Test for Measuring Wear Characteristics of
      Fluids," Journal of the Society of Tribologists and Lubrication
      Engineers. Lubrication Engineering, May, 1992, pp 354-359.

[3]   Denison HF-0 Fluid Test Procedure, Denison Hydraulics, Inc.,
      Marysville, Ohio, 1995.

[4]   Form M-2952-S (Rev. 8/88), "Vickers Pump Test Procedure for
      Evaluation of Antiwear Fluids for Mobile Applications", Vickers,
      Inc., Troy, Michigan, 1988.

[5]   Draft of "Proposed Standard Test Method for Indicating the Wear
      Characteristics of Petroleum and Non-Petroleum Hydraulic Fluids
      in a Constant Volume Vane Pump", Section N.7 of the ASTM
      Committee D02., October, 1990.

[6]   "IP281/80 (Reapproved 1988), "Anti-Wear Properties of Hydraulic
      Fluids - Vane Pump Test," Standard Methods for Analysis and
      Testing of Petroleum and Related Products Institutes of Petroleum
      1992, Volume 2, Methods IP276-392, The Institute of Petroleum,
      London, 1992.

[7]   "DIN 51389, Part 1, "Mechanical test of hydraulic fluids in vane
      pumps, General working principles", Deutsches Institut fur
      Normung e.V., Berlin, 1982.

George E. Totten[1] and Roland J. Bishop Jr.[1]

EVALUATION OF VICKERS V-104 AND 20VQ5 VANE PUMPS FOR ASTM D-2882
WEAR TESTS USING WATER-GLYCOL HYDRAULIC FLUIDS

REFERENCE: Totten, G. E. and Bishop, R. J. Jr., "Evaluation of Vickers V-104 and 20VQ5 Vane Pumps for ASTM D 2882 Wear Tests Using Water-Glycol Hydraulic Fluids," *Tribology of Hydraulic Pump Testing, ASTM STP 1310,* George E. Totten, Gary H. Kling, and Donald J. Smolenski, Eds., American Society for Testing and Materials, 1996.

ABSTRACT: Work is currently underway within the ASTM D.02 N.07 "Hydraulic Fluid Testing" Committee to evaluate the potential of replacing the Vickers V-104 vane pump with a newer, more current model, 20VQ vane pump for use in an updated ASTM D-2882 and other national standards. All of the round robin work conducted within the committee thus far has involved the use of non-aqueous hydraulic oils. Although there are some significant inter-laboratory reproducibility problems, it appears that the overall ranking of the hydraulic oils by most laboratories appears to be consistent.

To further evaluate the potential utility of replacing the Vickers V-104 pump with the 20VQ pump, a series of different water-glycol hydraulic fluid formulations with significantly different wear rates were evaluated using a "modified" ASTM D-2882 testing procedure. The results showed that expected catastrophic pump failures, which occurred with the V-104 pump, not only did not occur with the 20VQ pump, but the relative orders of wear rates for some of the fluids were also different.

In this paper, a brief history of the development of the ASTM D-2882 test, as currently conducted, will be presented, a comparison of the features of the V-104 and 20VQ pump and cartridge assemblies will be given and a comparative tabulation of the pump test using the V-104 pump and the 20VQ pump using different testing conditions will be provided. Finally, recommendations for future work will be made.

KEYWORDS: hydraulic, pump test, vane pump, water-glycol, hydraulic fluid

---

INTRODUCTION

There are two methods of evaluating the potential lubrication properties of a hydraulic fluid. One method is to conduct a pump test either as part of a laboratory test or as part of a field trial. However, either method may be costly and is further complicated by the numerous pump manufacturers, types, models and especially varying material friction pairings used in the pump design. Over the years, only pump tests based on the Vickers V-104 vane pump such as ASTM D-2882 and DIN 51389 have achieved wide acceptability in the fluid power industry.

In view of the cost and time to conduct these tests and of the fluid volumes typically required, it is desirable to perform at least

---

[1]Union Carbide Corporation, 777 Old Saw Mill River Road, Tarrytown, NY 10591.

preliminary screening tests using standardized laboratory bench tests such as the Shell 4-ball, Timken, FZG, and other customized tests [1]. Bench tests have a number of advantages relative to pump testing including: relatively small fluid volumes are required, numerous material variations and wear contact configurations are possible, high-quality and reproducible test components are readily available and substantially lower cost of testing.

With the exception of previously reported work by Tessmann and Hong [2], and Perez, et.al. [3], currently available single bench tests have not successfully replaced pump tests. In fact, it has been reported that there is no correlation at all with currently available standard bench tests and wear results obtained with various hydraulic fluids [1]. However, Mizuhara and Tsuya have extensively studied the wear correlation of a broad range of hydraulic fluids tested according to ASTM D-2741-68 and wear obtained at various critical points in a gear, piston and vane pump [4]. As a result of this work, it was concluded that:

- "Load carrying capacity" has nothing to do with hydraulic pump antiwear performance. An "antiwear" test must be used.

- Accelerated tests usually give the wrong results.

- Material pairs for the test pieces must be the same as the wear contact of interest in the hydraulic pump.

- It is necessary to evaluate hydraulic fluids over a wide range of test conditions; loads, speeds, material pairs, and where necessary, contact geometry. A single set of test conditions is usually inadequate.

Although Mizuhara and Tsuya did achieve success, it was still necessary to calibrate bench wear test results against a pump test. Presumably the same testing strategy could be applied to other bench tests. However, regardless of which approach is taken, it is always first necessary to calibrate the bench test to hydraulic pump wear and then to validate conclusions by pump testing. Therefore, it is always critically necessary to evaluate hydraulic fluid lubrication by hydraulic pump testing.

One of the most commonly encountered wear modes, particularly in vane and certain parts of a piston pump, is sliding wear. In a vane pump, the outlet pressure is directed to the back of the vanes which holds them against the ring as shown in Fig.1. The leading edge of the vane forms a "line contact" with the ring and the rotation of the vane against the ring generates a "sliding motion". This is the tribological condition being modeled by a vane pump and is shown schematically in Fig.1 [5].

Numerous pump tests have been proposed, however in view of their relative simplicity vane pump tests have achieved the greatest acceptance as tribological tests for hydraulic fluid evaluation. For example, excellent tests have been introduced by Bosch (Racine Fluid Power) [6] and Denison [7]. Recently, an interesting Russian vane pump test [8,9] has been proposed which utilizes only 0.7 liters of fluid compared to >50 liters of fluid required for the Vickers V-104 and other vane pumps.

Of all of these tests, the use of the Vickers V-104 vane pump has achieved the greatest acceptance in the fluid power industry. Currently, there are three national standards based on the Vickers V-104 pump; ASTM D-2882, DIN 51389 and IP 281. However, Vickers has suggested that these national standards be updated to incorporate a more current 20VQ vane pump. Therefore, in view of the ongoing work to accommodate Vicker's request, a project was undertaken to determine the correlation of wear results that would be obtained using these two pumps, if any, using various water-glycol hydraulic fluids with significantly varying compositions.

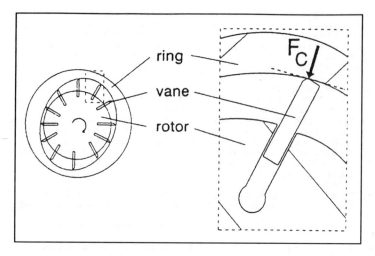

Fig. 1 -- Illustration of the sliding line contact with vane on ring
         assembly.

     The objectives of this paper are to provide a brief history of the
development of ASTM D-2882, to provide a description of the various
design differences between the V-104 and 20VQ pump, to summarize
comparative wear data obtained using varying testing conditions with the
20VQ pump and to provide recommendations for future work.

DISCUSSION

A.   Historical Overview of ASTM D-2882 Test Development

     In 1959, Shrey used a V-104 vane pump test to evaluate the
antiwear properties of both aqueous and non-aqueous fluids using a
hydraulic circuit similar to that shown in Fig. 2 [10]. Most of his
tests were conducted at 1000 psi, 1000 hr, 1.9 gal/min and 1120 rpm. The
total weight loss of the vane and ring were determined and reported.
     A variation of this test was also reported where the pressure of
the pump was cycled for 20 seconds on then 20 seconds off while the
pressure relief valve was set at 1000 psi to set up a "shock running
load" from 0-1000 psi. No other comparisons or conclusions were drawn
other than to state that no damage to the moving parts was observed.
     Reiland used essentially the same constant pressure test circuit
reported by Shrey to evaluate antiwear hydraulic oils [11]. However,
another variant in this now-classic test was evaluated. In this case,
the V-104 vane pump was run for 100 hours (versus 1000 hrs used
previously) at 2000 psi. (The pump is only recommended for use at 1000
psi.) These are the test conditions currently used in ASTM D-2882. The
conclusion was that "the more severe the test, the better its ability to
discriminate between oils" [11]. However, these running conditions
resulted in frequent breakage of the rotor which still occurs today
[12].
     The relative lubricity of a hydraulic oil was determined by visual
comparison of wear curves that were constructed by plotting total weight
loss of the vanes + ring versus the total time as shown in Fig. 3 [11].

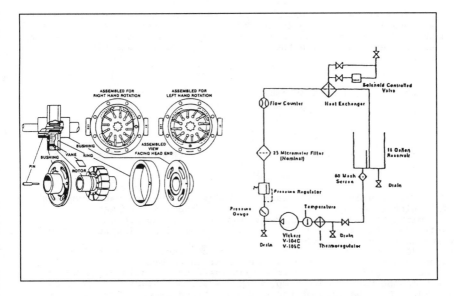

Fig. 2 -- Illustration of the Sperry-Vickers V-104 vane pump and
Schematic of test circuit used for V-104 vane pump tests.

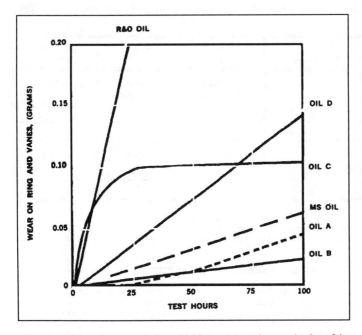

Fig. 3 -- Wear curves obtained for different antiwear hydraulic oils
using the Vickers V-104 vane pump at 2000 psi and 100 hours.

Alternatively, it was also proposed that the antiwear properties of a hydraulic oil are reflected by the "unit compressive pressure" which is determined by photomicrographing the leading edge of the vanes after the test, measuring the wear scar and solving equation (1):

$$P_a = P_b \times [A_b/A_a] \qquad (1)$$

where: P_a = unit pressure on the leading edge of the vane,
P_b = unit pressure on the back of the vane,
A_a = area on the leading edge of the vane,
A_b = area on the back of the vane.

For 2000 psi pump operating pressure, equation 1 becomes:

$$P_a = 2000 \times [A_b/A_a] \qquad (2)$$

Weaver used a radio-tracer technique to measure vane pump wear [13]. Either the vanes or the ring were irradiated and wear was determined by measuring the gamma radiation of the radioactive particles. While interesting, this procedure is of little practical value as an industry-wide wear test.

Currently, the Vickers V-104 pump continues to be the most commonly used pump for the evaluation of the lubrication properties of hydraulic fluids [12]. There are at least three national standards based on the use of this pump; ASTM D-2882, DIN 51389, and BS 2000 (IP 281/80). A comparison of the test conditions is provided in Table 1. The total weight loss of the vanes + ring at the conclusion of the test is the quantitative value of wear.

TABLE 1 -- Comparison of Sperry-Vickers V-104 Vane Pump Testing
Procedures

Test Parameter	ASTM D-2882	DIN 51389	BS 2000 IP281/80
Pressure	14 MPa (2000 psi)	10 MPa (1500 psi)	14 MPa[1] (2000 psi) 10.5 MPa[2] (1520 psi)
Speed (r/min)	1200	1500	1500
Time (h)	100	250	250
Fluid Vol.(L)	56.8	56.8	55-70
Fluid Temp.	150°F	(3)	(3)

(1)   For mineral oil type fluids.
(2)   For HFA, HFB and HFC fluids.
(3)   Temperature selected to give 46 cSt viscosity.

B.   Comparison of the Vickers V-104 and 20VQ5 Vane Pumps

Within the ASTM D.02 N.07 working group there is a round robin underway to study the potential replacement of the V-104 pump with the 20VQ5 pump. Although the objective is simple replacement of the V-104 pump, it is important to acknowledge that the tribological conditions are significantly different for both pumps. A comparison of the recommended test conditions are summarized in Table 2.

TABLE 2 -- Comparison of the Test Conditions of the Sperry-Vickers
V-104 and 20VQ5 Pumps

PARAMETER	V-104	20VQ5
VANE LOAD (lbs/1000psi)	47	25
MAX. PRESSURE (psi)	1000	3000
MAX. SPEED (rpm)	1500	2700
END PLATES (Bronze)	CAST	SINTERED

In addition, it is also important to recognize that the vane
design is fundamentally different for the two pumps. The 20VQ5 pump has
an "intravane" configuration, whereas the V-104 pump has a single vane
arrangement. These vane arrangements and the configuration of the vane
cartridge in the two pumps is compared in Fig's. 4-6.
In addition, to these physical differences in the two pump
arrangements, it should also be noted that under the recommended test
conditions the current V-104 pump is being operated at 2000 psi which is
outside of the recommended pressure limit of 1000 psi for the pump. The
recommended pressure for the test using the 20VQ5 pump is 3000 psi,
which is also the recommended pressure limit of the pump [14]. This is
particularly important since every pump has its own characteristic wear
regions as a function of speed and load [15]. Almost certainly, under
these conditions, the V-104 pump would have a significantly greater
hydraulic film load requirement to provide adequate lubrication than
would the 20VQ5 pump. This is especially true in view of the different
vane design between the two pumps. Therefore, it is not likely that
there is a direct wear correlation between the two pumps.

C.   Summary of Comparative Wear Data

Experimental water-glycol hydraulic fluids were selected which
would provide significant differences in vane pump total wear (vanes +
ring) after completion of the test. Table 3 shows the operating
parameters used for the V-104 and 20VQ5 pumps in this study.
A comparison of the wear results obtained for these fluids using a
variety of test conditions is provided in Table 4.

The results show that:

• The worst formulations tested in the V-104 pump, fluids B and
  E, gave very low wear in the 20VQ5 pump.

• Generally for a given fluid, the 20VQ5 pump gives lower wear
  than the V-104 pump, as illustrated in Fig. 7.

Fig. 5 shows a graphical comparison of the wear data obtained with
the V-104 pump using a "modified" ASTM D-2882 protocol (2000 psi, 1200
rpm, 1 gal. reservoir) with that obtained using the 20VQ5 pump at 3000
psi and 1200 rpm. This data clearly illustrates the 20VQ5 to lack
sufficient wear latitude to be useful in evaluating these fluids.
Future work using the 20VQ8 pump is planned.

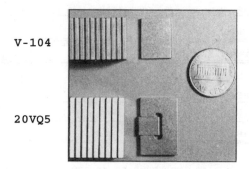

V-104

20VQ5

Fig. 4 -- Comparison of the vanes in the Vickers V-104 and 20VQ5 pumps.

V-104   20VQ5
Fig. 5 -- The internal view of the Vickers V-104 and 20VQ5 pumps.

Fig. 6 -- Comparison of the assembled cartridges of the Vickers V-104
and 20VQ5 pumps, respectively.

TABLE 3 -- Comparison of Pump Operating Parameters

PARAMETER	V-104	V-104	20VQ5	20VQ5	20VQ5
Flow (gpm)	8	10	5	10	11.3
Pressure (psi)	2000	1500	3000	3000	3000
Speed (rpm)	1200	1500	1200	2400	2700[1]
Power (hp)	9.3	8.8	8.8	17.5	19.7
Torque (in-lbs)	490	370	460	460	460

(1)   Maximum rated speed at 3000 psi using a 5 gpm cartridge.

TABLE 4 -- Results of Pump Tests

FLUID	FLOW (gpm)	PRESSURE (psi)	SPEED (rpm)	RING/VANE Wt. Loss (mg)	END PLATE Wt. Loss (mg)	V-104 PUMP	20VQ PUMP
A	8	2000	1200	21		X	
	10	1500	1500	31	23	X	
	5	3000	1200	13			X
	11.3	"	2700	34			X
B	8	2000	1200	205	7	X	
	10	1500	1500	199	38	X	
	5	3000	1200	12	60		X
	10	"	2400	10	482		X
C	8	2000	1200	22	20	X	
	10	1500	1500	116	14	X	
	5	3000	1200	18	40		X
	10	"	2400	7.3	130		X
D	8	2000	1200	37		X	
	5	3000	"	37			X
E	8	2000	1200	5110		X	
	5	3000	"	16			X
F	8	2000	1200	83		X	
	5	3000	"	93			X
G	8	2000	1200	40		X	
	5	3000	"	9.5			X
H	8	2000	1200	34		X	
	5	3000	"	20			X
I	8	2000	1200	25		X	
	5	3000	"	22			X

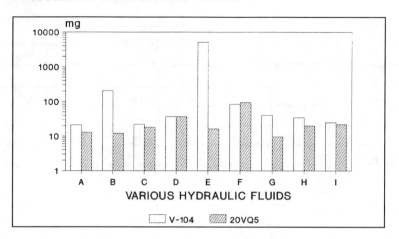

Fig. 7 -- Chart of ring + vane wear (in milligrams) from V-104 and 20VQ5
          pumps running at 1200 r/min with 9 fluids at a pressure of
          2000 and 3000 psi for the V-104 and 20VQ5 pumps, respectively.

        In another test, a fluid formulation was tested at two different
speeds (1200 and 2400 rpm) using the 20VQ5 pump. Although the fluid
exhibited very low total wear of the vanes + ring, fluid "B" exhibited
dramatic cavitation of the bushing at 2400 rpm as shown in Fig. 8. This
is also important since neither the ASTM D-2882 or the proposed 20VQ5
pump testing protocols require the reporting of any wear except on the
vanes and ring. Clearly, this should be addressed in any new revisions
of these testing procedures.

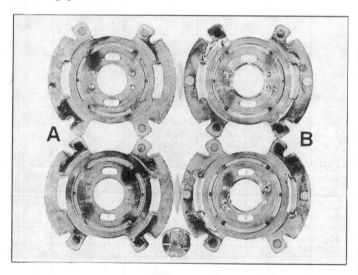

Fig. 8 -- Comparison of 20VQ5 bushing wear (by cavitation) obtained at
          two speeds, A) 1200 r/min and B) 2400 r/min, at 3000 psi using
          a fluid that exhibited very low wear on the ring and vanes,
          (12 and 10 mg total wear after 100 hours).

CONCLUSIONS AND RECOMMENDATIONS

The results reported here show that there is little, if any direct correlation between the wear obtained by the current ASTM D-2882 using the Sperry-Vickers V-104 vane pump and the proposed testing procedure using the 20VQ5 pump. These differences are most likely due to the differences in the vane design of the two pumps and to the expected differences in lubrication requirements resulting from differences in vane loading and pump speed. It is recommended that these problems be systematically examined in order to better model at least nearly equivalent tribological testing conditions with both pumps and perhaps more importantly, to create realistic testing regimes that better model lubrication properties of the hydraulic fluids and minimizing unavoidable fatigue failure of the pump components.

In addition, it is recommended that the pump testing protocol better utilize currently available technology to optimize the quality of data being obtained from the test. Some possibilities include:

1. Measurement of pump leakage and torque over the duration of the test in order to obtain a better measure of the break-in properties of the fluid in the pump and to determine if wear stabilization has been achieved at the conclusion of the test.

2. Pump wear performance should be continuously monitored and a written record of any changes in the pump performance, and other events such as unexpected shut-downs, should be reported at the conclusion of the test.

3. Wear measurement should not, indeed must not, be limited to the vanes and the ring. Measurement, or at least visual analysis, must be made of other components such as bushings and bearings, metal compatibility, and failure modes. These should be photographed and be required as part of the final report.

4. If the pump is to be operated in excess of its design limits, some assurance must be made that the wear data obtained is actually due to the lubrication properties of the hydraulic fluid and not be dominated by the fatigue limits of the pump components. That is, the data obtained should make tribological sense.

REFERENCES

1. Totten, G.E., Bishop, R.J. Jr., and Kling, G.H., "Evaluation of Hydraulic Fluid Performance: Correlation of Water-Glycol Fluid Performance by ASTM D-2882 Vane Pump and Various Bench Tests", SAE TECHNICAL PAPER SERIES, Paper Number 952156, 1995.

2. Tessmann, R.K. and Hong, I.T., SAE TECHNICAL PAPER SERIES, Paper Number 932438, 1993.

3. Perez, J.M., Hanson, R.C., and Klaus, E.E., Lubrication Engineering, Vol. 46, 1990, pp 248-255.

4. Mizuhara, K. and Tsuya, Y., Proc. of the JSLE Int. Tribol. Conf., 1985, pp. 853-858.

5. Gellrich,R., Kunz, A., Beckmann, G. and Broszeit, E., "Theoretical and Practical Aspects of the Wear of Vane Pumps Part A: Adaptation of a Model for Predictive Wear Calculation", Wear, Vol. 181-183, 1995, pp. 862-867.

6. This test was communicated verbally by Mr. Paul Schacht as the "standard" cyclic pressure vane pump test recommended by (Robert Bosch) Racine Fluid Power, Inc. in Racine, WI.

7. The Hagglunds-Denison vane pump wear test is contained in their recommended two-part HFO pump testing protocol. The vane pump test circuit contains a Denison T5D-042 vane pump.

8. Lapotko, O.P., Shkol'nikov, V.M., Bogdanov, Sh., K., Zagordni, N.G. and Arsenov, V.V., Chem. and Tech. of Fuels and Oils, Vol. 17, 1981, pp. 231-234.

9. Arsenov, V.V., Sedova, L.O., Lapatko, O.P., Zaretskaya, L.V., Kel'Bas, V.I., and Ryaboshapka, V.M., Vestnik Mashinostroeniya, Vol. 68, 1988, pp. 32-33.

10. W.M. Shrey, Lubrication Engineering, Vol. 15, 1959, pp. 64-67.

11. Reiland, W.H., Hydraulics & Pneumatics, March, 1968, pp. 96-98.

12. Thoenes, H.W., Bauer, K. and Herman, P., "Testing the Antiwear Characteristics of Hydraulic Fluids: Experience with Test Rigs Using a Vickers Pump", IP Int. Symposium, Performance Testing of Hydraulic Fluids, Oct. 1978, London, England, Tourret, r. and Wright, E.P., Eds., Publ. Heydon and Son Ltd.

13. Weaver, J. J., Lubr. Eng., Vol. 21, 1965, pp. 12-15.

14. Vickers, Inc.,"Vane Pump & Motor Design Guide For Mobile Equipment", No. 353, revised 11-1-92, p. 10.

15. Runhua, T. and Caiyun, Y., "The Vane Profile Improvement for a Variable Displacement Vane Pump", Proceed. of the International Fluid Power Applications Conf., March 24-26, 1992, National Fluid Power Assoc., Milwaukee, WI.

16. Wedeven, L.D. and Totten, G.E., "Testing Within the Continuum of Multiple Lubrication and Failure Mechanisms", TRIBOLOGY OF HYDRAULIC PUMP TESTING, George E. Totten, Gary H. Kling and Donald M. Smolenski, Ed., American Society for Testing and Materials, Philadelphia, 1995.

Harry T. Johnson[1] and Thomas I. Lewis[2]

VICKERS' 35VQ25 PUMP TEST

REFERENCE: Johnson, H. T. and Lewis, T. I., **"Vickers' 35VQ25 Pump Test,"** *Tribology of Hydraulic Pump Testing, ASTM STP 1310*, George E. Totten, Gary H. Kling, and Donald J. Smolenski, Eds., American Society for Testing and Materials, 1996.

ABSTRACT: In the late 1970's, Vickers, Inc., recognized that their methods of establishing petroleum based fluid recommendations had become inadequate. They provided poor correlation with the field results in the higher pressure and higher speed pumps that had been developed for the mobile hydraulics market. A test procedure based on the high performance VQ product line that had the most severe fluid antiwear requirements was therefore developed.

This paper discusses the resulting 35VQ25 vane pump test; summarizes the rationale used to select test parameters; and presents an interpretation of results with various fluids. The influence of test parameters and variations in production pump components on the overall results are outlined.

Although not developed as a replacement for existing industry standards, the test procedure has become widely used within our industry over the last 15 years. As a result, this paper will attempt to compare the pros and cons of this procedure to other alternatives such as ASTM D-2882.

KEYWORDS: vane pump, petroleum, hydraulic fluids, antiwear, test procedure, ASTM D-2882

[1]Section Chief (Retired), Vane Product Engineering, Vickers, Inc., 6600 North 72nd Street, Omaha, NE 68122-1798

[2]Senior Project Engineer, Vane Product Engineering, Vickers, Inc., 6600 North 72nd Street, Omaha, NE 68122-1798

INTRODUCTION

In December 1979, Vickers, Inc., completed the development of a pump test procedure for evaluation of antiwear fluids for mobile systems. This development was the result of a recognition that their method of establishing petroleum based fluid recommendations, by specifying the amount of zinc present in the fluid, was not an adequate measure of fluid antiwear characteristics. The utilization of other test procedures, such as the ASTM Standard Method for Indicating the Wear Characteristics of Petroleum and Non-Petroleum Hydraulic Fluids in a Constant Volume Vane Pump (D-2882), provided unsatisfactory correlation with mobile system field results. The fluid antiwear performance demands imposed by the higher performance pumps that had been developed for the mobile hydraulics market were not sufficiently duplicated by such tests.

The primary objective of this development program was to establish a petroleum-based fluid evaluation test based on a Vickers' product that was supplied to the mobile equipment market and had the most severe fluid antiwear requirements. The selection of a VQ vane pump resulted from field experience that had identified that petroleum-based fluids which were satisfactory for use in vane pumps also provided acceptable results in piston and gear pumps. An accept/reject criterion was established that was capable of defining the acceptability of a particular fluid without reference to the nature of the given antiwear package.

It should be noted that it was not an objective of the development program to establish a test procedure that could also serve as a universal industry standard test for hydraulic fluids. No comparative testing was conducted on alternative pump designs.

TEST PROCEDURE DESCRIPTION

A Vickers 35VQ25 fixed displacement (81 cm³/rev.) vane pump is utilized as the test pump. This is a replaceable cartridge, intravane design that is rated at 2500 r/min. and 207 bar. The pump is installed in the test circuit described in Figure 1:

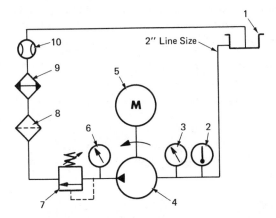

DESCRIPTION OF COMPONENTS

1. Reservoir (50 gallons
   minimum; elevated above
   pump centerline to provide gravity feed)
2. Temperature gage or
   thermocouple
3. Inlet pressure gage
4. Pump: 35VQ25A-11*20
   (cartridge kit
   P/N 413421)

5. Electric motor (125 HP)
6. Outlet pressure gage
7. Pressure relief valve
8. Filter (10 micrometer
   nominal)
9. Cooler
10. Flow meter

Figure 1: Test Circuit

The evaluation requires that a minimum of three vane pump cartridges be tested for 50 hours each. The 50 hour duration includes a one hour plus break-in period during which the inlet fluid temperature and pump outlet pressure are increased to the final test levels in stages. The remainder of the test is conducted in a steady-state operating mode at the following conditions:

Speed	:	2400 r/min.
Outlet Pressure	:	207 bar gage
Inlet Pressure	:	0 - 0.15 bar gage with flooded inlet
Inlet Temperature	:	93°C

The accept/reject guidelines for each fluid basically follow the approach of ASTM D-2882. They are based on the quantitative vane tip/cam ring wear characteristics as defined by component weight loss. In addition, a visual inspection of components, particularly the cam ring, is required to cover instances where unsatisfactory performance is displayed at acceptable weight loss levels.

The reader is referred to Vickers' Form M-2952-S for a more complete description of this test procedure and the fluid acceptability guidelines. The latter also includes color photographs of acceptable or unacceptable cam ring wear characteristics.

TEST DEVELOPMENT SUMMARY

Test Procedure Rationale

As with most development programs, the development of this test procedure required a number of compromises to meet the objectives in a realistic manner. The following discussion outlines the reasoning used for this procedure.

Test Pump--A vane pump requires a high level of fluid antiwear performance when the combination of vane tip surface speed and loading against the cam ring surface is maximized. The latter occurs in pumps with large cam ring diameters and at high rotational speeds and outlet pressures.

The physical size and ratings of the 35VQ25 pump provided these "worst case" wear conditions. The use of the smallest displacement cam ring (25 gpm at 1200 r/min.) in the 35VQ frame size also permitted the required pump input power to be limited to 75 kW.

Test Parameters--In the design of this procedure, the following factors were considered:

1.   The 35VQ25 is a production (high volume/low cost) pump and not a specially controlled test specimen.

2.   The objective was to evaluate fluids and not the pump cartridges.

3.   The procedure would be unacceptable if the results were not repeatable.

With the above in mind, the test conditions were limited to the ratings of the 35VQ25 pump. A positive pressure/flooded inlet condition was required and the test speed specified at 100 r/min. below rated speed to ensure that cavitation was not encountered.

Only the fluid inlet temperature can be considered to be in excess of rated, since the test was originally intended for SAE viscosity Grade 10W fluids (Reference SAE J300). The pump minimum rated viscosity is 10 cSt; but at the test temperature, and with an SAE 10W fluid, a viscosity of 6 to 7 cSt would be expected. The latter is consistent, however, with field experience and the Vickers' product qualification test procedure.

To address production subtleties, multiple test specimen (cartridges) and a controlled test start-up/break-in were utilized. Subsequent tests results have proven that this approach is of particular value when making acceptability decisions relative to marginal fluids. The possibility of a defective cartridge is addressed by requiring that two additional cartridges be tested if one of the first three fails for any reason.

Experience has identified that the vane tip/cam ring wear characteristics can be most rapidly established by a steady-state pressure operation mode. The Vickers' product qualification test is a cyclic pressure test that is an order of magnitude longer in duration, but also addresses component fatigue capabilities. Although a shorter duration test was desirable, the 50 hours specified provides results with a "worst case" acceptable fluid that are consistent with acceptable product qualification test results.

Accept/Reject Criterion--As noted above in the discussion of the test duration, the 15 mg. vane set and 75 mg. cam ring maximum weight loss were also established to be consistent with the historical data obtained form the product qualification testing on the 35VQ pump. This quantitative measure is normally sufficient, but in the cases of marginal fluids a visual inspection was found to be necessary.

Pump Flow Rate--There are two circumstances for which the test should be terminated, but not considered a failure of the fluid being tested. Both relate to the 35VQ25 flow rate when operated a fluid viscosities below 10 cSt in combination with maximum speed and outlet pressure. The reduced viscosity increases internal pump leakage. As a result, there is a tendency for pump sideplate (bronze-faced flex plate) erosion and subsequent flow degradation. The result is not a fair test of the fluid. Test termination conditions where the initial flow rate is less than 136 L/min. or flow degradation exceeds 7.5 L/min. were therefore incorporated.

Development Baseline Fluid

Currently and for many years prior to the development of this procedure, Vickers has used the same petroleum-based antiwear fluid to conduct development and product qualification testing for the general market application of all its pump product lines (piston, vane, etc.). This fluid is a Newtonian, SAE 10W motor oil. It contains a 0.07% minimum zinc, as secondary alkyl zinc dithiophosphate, antiwear additive package. Compared to many newer technology fluids, it is considered a marginally acceptable product. It ensures that proper fluid tolerance is designed into the pump products as they are developed. It was also the fluid used to establish accept/reject guidelines during the development of this procedure.

Summary of Development Testing

As part of the development and validation of the procedure, a total of nine fluids (including the baseline) were tested in the Vickers Development Laboratory. These fluids represented technology that was either available or being developed in the late 1970's. Four of the fluids were found to be equal to or better than the baseline fluid. The remaining four displayed varying degrees of unacceptability, including one fluid that repeatedly permitted excessive vane tip/cam ring galling and subsequent cartridge failure during the break-in period.

In addition to the Vickers testing, the test laboratories of three major fluid and/or additive manufacturers participated in the test procedure validation process. They conducted testing on the baseline fluid and at least one of the other fluids. Appreciable correlation existed in terms of weight loss results and cam ring appearance between Vickers and the outside laboratories. The correlation was obtained using independent facilities and cartridges that were randomly obtained from the Vickers' distribution system.

The cam ring weight loss from 14 cartridges tested using the baseline fluid ranged from 15 to 60 mg. and averaged 36 mg. For comparison, the best fluid that proved very effective in minimizing pitting in highly loaded zones of the cam ring surface had a cam ring weight loss that ranged from 9 to 16 mg. and averaged 13 mg. One the unacceptable side, a non-zinc additive fluid had a cam ring weight loss that ranged from 82 to 283 mg. and averaged 171 mg.

It should be noted that the better fluids had minimal variation in the results. As the fluids became marginal, the variation from cartridge to cartridge became greater.

TEST RESULTS: OBSERVATIONS AND INTERPRETATIONS

This procedure uses a balanced, two-lobe, vane pump to conduct a fluid evaluation in a sliding, Hertzian contact, wear interface with a clean and dry hydraulic fluid. More specifically, the interface is between a hardened tool steel vane tip and the inner cam contour of an equal hardness, 52100 steel cam ring. Since the pump is balanced, the bearings are basically unloaded. As long as the pump is not subjected to an outlet pressure in excess of ratings, one should not expect sideplate wear with any reasonable lubricant.

The procedure qualifies the fluid at worst case conditions and it is reasonable to expect that some fluids that fail would perform acceptably at higher viscosities and lower rotational speeds and outlet pressures. Since lower speeds and higher viscosities are common in industrial market systems, two of the fluids that failed during the procedure development were re-tested at lesser conditions.

The speed was reduced to 1800 r/min. and the inlet temperature to 74°C. The latter placed the viscosity in the 10 to 11 cSt range. When retested both fluids displayed significant reductions in component weight loss and were well within the established maximums. More interestingly, the quality level ranking of the fluids reversed. One fluid that previously failed during the break-in period had an average cam ring weight loss of only 20 mg. The average cam ring weight loss in the second fluid dropped from 171 mg. to 50 mg., but was also still considered a failure since the appearance of the cam surface wear pattern did not improve to a great extent.

The above example is one case where it was clear that weight loss alone was not sufficient to judge fluid capabilities, particularly when using the results of a test that was limited to 50 hours of duration. Unfortunately, visual inspections require an experienced, educated eye.

The cam ring surface is most likely to show incipient failure. The cam ring should not show evidence of vane tip/cam ring surface galling as a result of the frictional drag. It should also have a minimal amount of the pitting that is the result of subsurface fatigue from the sliding Hertzian contact. Fluid design and viscosity were shown to influence these wear conditions during the development of the test procedure.

Although the test was originally intended to be based on fluids with an SAE 10W viscosity or equivalent, tests were conducted with an SAE 30 viscosity version of the baseline fluid. This essentially doubled the test fluid viscosity. This change in viscosity reduced the average cam ring weight loss by 45% and virtually eliminated all evidence of pitting. Results like this have led to the conclusion that positive results with a higher viscosity base stock do not validate an additive package with a lower viscosity base stock oil.

Although the test procedure break-in was incorporated to address cartridge variables and improve repeatability, its affect on fluid performance should be noted. These observations are the result of comparison testing conducting using the Vickers 35VQ25 procedure and a Caterpillar, Inc., procedure titled VC3. Both procedures are based on the 35VQ** vane pump and relatively similar operating conditions. The major differences are that the VC3 procedure has no break-in and is only 30 minutes maximum in duration. The objective of the VC3 procedure was to establish qualitatively the vane tip/cam ring wear characteristics over a range of temperatures and pump outlet pressures up to 200 bar with the pump at 2200 r/min. The results were plotted as a failure curve on a pressure vs. temperature plot for the test fluid.

When a known good fluid was compared with both procedures, the answer was the same: The fluid is acceptable. When a marginal fluid was compared, however, the results did not correlate. Complete cartridge failures were encountered within five minutes of VC3 testing at much reduced operating conditions compared to the 35VQ25 procedure. Although this fluid was established as marginally unacceptable by the 35VQ25 procedure, maximum ring weight loss did not exceed 130 mg.

The introduction of the 35VQ25 break-in prior to testing a cartridge per VC3 resulted in completely successful cartridge performance. The complete 30 minute maximum duration could be completed with no evidence of vane tip or cam ring distress. The major difference in results was striking.

Summary on Meaning of Results

The results of the 35VQ25 pump test basically establish the ability of an antiwear fluid to provide the necessary antiwear protection for a Vickers' intravane vane pump. The results probably apply to other vane pump designs with similar vane and cam ring metallurgy, particularly relative to the ranking order of the fluids. The difficult decision would be where to draw the line between acceptable and unacceptable fluids for a given pump design.

Certainly the test does not address some of the more unique material combinations and wear interfaces found in piston equipment or the question of highly loaded gear pump journal bearings. One must always remember that it is a test of a sliding, Hertzian contact, wear interface.

COMPARISON TO OTHER TEST METHODS

ASTM D-2882

This is an industry standard vane pump test procedure for indicating the wear characteristics of petroleum and non-petroleum hydraulic fluids. It uses a Vickers V104-C-vane pump as the test specimen at 138 bar outlet pressure. The ASTM is currently reviewing the use of Vickers' 20VQ5 vane pump at 207 bar outlet pressure in its place.

The vane tip loading in the above pumps and the 35VQ25 are comparable and approach a maximum of 30 N/mm at their respective test pressures. A major difference is noted, however, when one compares the vane tip surface speed relative to the cam rings as noted below:

Pump	Rotation Speed (r/min.)	Surface Speed (m/s)
V104-C	1200	4.2
20VQ5	1200	3.9
35VQ25	2400	11.4

Although both the V104-C and 20VQ5 have smaller cam ring minor diameters, the major reason for the large difference in surface speed is due to the test rotational speed. Based on this comparison it is reasonable to believe that the fluid antiwear performance demands imposed by the 35VQ25 test are significantly greater. This, of course, has been validated by test.

In addition, the fluid temperatures specified by ASTM D-2882 are not sufficiently elevated to provide comparable fluid test viscosities. In this case the pumps also provide a limitation relative to operating at reduced viscosities at 1200 rpm. The resulting internal leakage at the test pressures would probably be excessive relative to the pumped fluid in these low displacement pumps (V104-C or 20VQ5).

The above discussion suggests that ASTM D-2882 is inadequate, but this is only true if one is interested in addressing the requirements of mobile hydraulic systems where the demands on the petroleum-based fluids are typically more severe. It is a question of duplicating the operating conditions. The latter comes at a price, however. The 35VQ25 pump test requires roughly 10 times more power to drive the pump and a larger test facility in general to handle the greater pumped flow rates.

Bench Tests

There are numerous lubricant evaluation, load-carrying bench tests that are available. Vickers has not attempted to correlate the 35VQ25 pump test results with any of these bench tests. The development of a vane test rig was undertaken in an attempt to significantly reduce the test facility and power requirements inherent in pump tests like the 35VQ25. This work was not completed, but did result in a number of interesting observations.

The vane test rig design essentially eliminated the pumping action of a vane pump by utilizing a circular cam ring. Two vanes were hydraulically loaded against the cam ring by an external pressure source. Test fluid was circulated axially through the zone between the cam ring and the rotor. It seemed easy. A device that was low in required power, etc., and that could simulate the vane tip loading and surface speeds with relatively inexpensive test components.

Initial test results were totally surprising. Instead of the cam ring showing the initial wear, the vane tips wore. Subsequent review of the conditions led to the conclusion that although the loads and speeds were simulated accurately, the duty cycle was not. The vane tips were loaded continuously in the test rig. In a pump they are alternately loaded and unloaded. With only two vanes, the loading cycle rate at given point on the cam ring was much reduced. A test rig redesign added additional vanes and the vane loading was cycled in a synchronized manner so the same points on the cam surface were always loaded.

When the tests were repeated, the correlation with the results obtained in pump tests was much improved. Situations like the above tend to point out why bench test correlation is so difficult, particularly when one wants to know if a given fluid will work in a unique pump design.

CONCLUSIONS

The 35VQ25 pump test is a good basic test at what should be considered "worst case" operating conditions. It has a good track record relative to answering the question that Vickers wanted to know: "Does this fluid work in our product?"

It does a better job than ASTM D-2882 of identifying "bad" and "marginal" fluids at severe operating conditions simply because it subjects the fluid to higher performance demands. But like all high displacement, high pressure pump tests, this information comes at a price.

If one just wanted to rank the relative antiwear performance quality level of fluids, then there are probably many relatively inexpensive tests that are capable. It is far more difficult to establish if a given fluid will operate in a particular hydraulic system. One must closely simulate the field conditions. Fluid ranking is important, but probably not sufficient.

Certainly the 35VQ25 pump test is not a perfect test that provides the proper accept/reject decisions relative to all pump types, loads, speeds, and material combinations. That test will probably never exist. It should also be recognized, however, that most pump tests like the 35VQ25 are typically 20 years old. We as an industry should be able to do a better job today.

Axel J. Kunz,[1] *Erhard Broszeit*[2]

COMPARISON OF VANE PUMP TESTS USING DIFFERENT VICKERS VANE PUMPS

REFERENCE: Kunz, A. J. and Broszeit, E. **"Comparison of Vane Pump Tests Using Different Vickers Vane Pumps,"** *Tribology of Hydraulic Pump Testing, ASTM STP 1310,* George E. Totten, Gary H. Kling, and Donald J. Smolenski, Eds., American Society for Testing and Materials, 1996.

Abstract: Several test procedures with the Vickers V 104 C are compared with those of the Vickers VQ-series to show their different technological properties and tribological options for testing. Based on these investigations, the wear results achieved on different test rigs can be interpreted to form a statement on the actual scope of hydraulic pump testing with Vickers vane pumps. Starting from this and including the requirements of industry together with stimulations coming up from new developed fluid analysing devices, the demands for a new pump test method can be derived. Therefore, a comprehensive structure and stress analysis and the documentation of basic wear characteristics in the vane pump V 104 C using a pure mineral oil without additives will be introduced. Thus wear results achieved with this pump and the 35 VQ 25 using identical commercial fluids can be compared. Wear maps derived from wear calculations and verified with wear results obtained with the pure mineral oil can be used to estimate the test performance of the 20 VQ 5 as a substitute for the V 104 C, and will allow a theoretical explanation of the empirically derived upper wear limits used in all vane pump test.

Keywords: vane pumps, structure and stress analysis, wear calculation, wear maps

---

At the Institute of Material Sciences (IfW), Technical University of Darmstadt, vane pump tests with several types of Vickers vane pumps have long been used, both to investigate the tribological background of the test procedures and for commercial test activities for the lubrication industry. Based on these experiences, the paper intends to give an overview of the actual scope of hydraulic pump testing with Vickers vane pumps. Looking at different test procedures, different kinds of test intentions are shown. Results achieved on different test rigs and different pumps, are compared and similarities and the particular need of change are pointed out. To understand the wear phenomena a structure [1] and stress analysis of all Vickers vane pumps

---

[1] Dr.-Ing., Test Department, Borg-Warner Automotive, Kurpfalzring, 69123 Heidelberg, Germany

[2] Dr.-Ing. Dep. of Tribology, Institute of Material Sciences (IfW), Technical University Darmstadt, Grafenstr. 2, 64283 Darmstadt, Germany

will be introduced to show the different abilities of each test method. An attempt will be made to explain these effects, as well as the upper wear limits used for rating the fluids anti-wear properties theoretically with wear maps derived from predictive calculations based on the shear energy hypothesis. In the end, the combination of all (the structure and stress analysis, the predictive calculation, the requirements of the hydraulic industry and a new fluid analysing devices) allows to discuss the demands of a new pump test method.

Test procedures

Vane pump test runs are used to test the anti wear properties of hydraulic fluids at the final stage of the development of a new formulation. Therefore the weight loss of pump components, achieved after a certain period of time under constant high load conditions, is evaluated with empirically derived upper wear limits. If the wear of ring and vanes remains within the range marked by zero and these limits wear, the formulation fulfils the fluid specifications.

Two different test rigs are running at the IfW in Darmstadt. FIG. 1a and 1b show the rig and its hydraulic circuit for the smaller Vickers vane pump V 104 C (according to ASTM D 2882 or DIN 51 389 / IP 281). With regard to this category of fluids, German and British standard are technologically equivalent and seems to represent the European conception of vane pump tests.

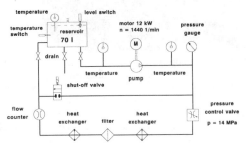

FIG. 1a--Front of rig for tests   FIG. 1b--Hydraulic circuit of rig for
according to ASTM / DIN (V 104 C)   according to ASTM / DIN

FIG. 2a and 2b depict the equivalent devices for the larger Vickers vane pump 35 VQ 25 (according to the Vickers standard M 2952-S), which has obtained a great acceptance in industry involved in mobile hydraulic systems.

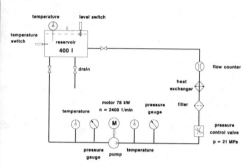

FIG. 2a--Front of rig according to Vickers M 2952-S standard (35 VQ 25)

FIG. 2b--Hydraulic circuit of rig according to Vickers M 2952-S standard

The suitability of the test with the small vane pump is in doubt since problems with so called typical damages have been reported [2]. However, a method to avoid such damages has been reported previously [3]. In addition, the testing conditions do not reflect the actual requirements of industrial hydraulic systems because of the obsolete design of the V 104 C, which only allows a maximum pressure of 14 MPa. As a substitute, the Vickers vane pump 20 VQ 5 is in discussion [4]. This model enables system pressures up to 21 MPa and can be easily adapted to the test rig shown in FIG. 1 without the need for large modifications. First experiences with such a modified test rig have already been made and documented [4] at the IfW as well. Basic testing conditions for all pumps are given in TABLE 1.

Obviously there are interesting differences among the test procedures using the same pump. All following statements concerning the V 104 C refer to the test conditions according to DIN 51 389. In addition to the number of revolutions and in duration, the test procedures with the small Vickers vane pump are different in considering fluids with different viscosities. According to German standard, the hydraulic fluids are tested with one fixed viscosity, whereas the American standard prescribes a fixed temperature in dependence of the viscosity class of the fluid to be tested. Looking at the details of all test procedures reveals that there are clear instructions on how to start a test run, but guidelines regarding the acceptable degree of fluid contamination before a test can be started are missing.

TABLE 1--Basic Vickers vane pump testing conditions

	35 VQ 25	V 104 C / ASTM	V 104 C / DIN	20 VQ 5 (IfW)
pressure [MPa]	20.5 - 21.0	13.79 ± 0.28	13.8 - 14.0	20.5 - 21.0
revolutions [1/min]	2350 - 2400	1200 ± 60	1440 ± 30	1440 ± 30
temperature [°C] at pump inlet respectively	90 - 96	65.6 ± 3 (≤ ISO VG 46) 79.4 ± 3 (> ISO VG 46)	-	-
cinemat. viscosity [mm²/s]	-	-	13	13
test duration [h]	50	100	250	250
filter width [μm]	10	25	10	10

In TABLE 2 the upper wear limits for the Vickers vane pump tests are given. If the weight loss of ring or vanes or both together exceeds these limits, the test must be considered as failed. The separate view on ring and vane wear can reveal irregularities which might have occurred in a test run much earlier than a rating of a summarised weight loss. More restrictive pass criteria are prescribed in the Vickers standard M 2952-S. Here the damage type of the worn inner ring surfaces is considered qualitvely as well. Some possible wear phenomena are not accepted even if the weight loss does not exceed the upper wear limits of ring and vanes.

TABLE 2--Upper wear limits for HM fluids

	35 VQ 25	V 104 C / DIN	20 VQ 5
ring [mg]	75	120	-
vanes [mg]	15	30	-

Due to lack of upper wear limits for the 20 VQ 5, tests with this pump cannot yet be run. In addition to explaining the large differences of the upper wear limits of the obsolete small pump and the modern and larger 35 VQ 25, the following passages will try to show where those limits can be fixed.

Analysis of the tribo system and wear results

A comprehensive structure and stress analysis as part of a systems approach according to [1], to investigate wear phenomena in Vickers vane pumps, was done for all pump models mentioned. In a quasi-static model (model of all internal forces, velocities, geometry and other input data derived from measurements) the contact force F_c (FIG. 3) as a parameter of huge tribological interest in the ring vane contact could be calculated [5].

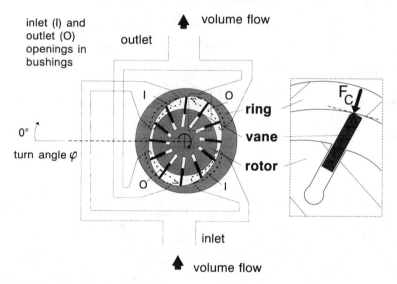

FIG.3--Constant volume flow vane pump (V 104 C) and contact force F_c

Due to the symmetry of constant volume flow vane pumps, all considerations can be reduced to half a rotor turn. The contact force F_c in the line contact of each vane and ring is given for all pump models (see TABLE 1) in FIG. 4. The prevailing parts of contact forces are caused by different pressures of the fluid, which surrounds each vane. For the V 104 C, the pressure distribution was measured with pressure transducers installed in the rotor [5], whereas the other progressions of pressure needed were calculated based on the steering geometry of each pump. Although the level of the curves in FIG. 4 are strongly determined by the working areas for the different pressures on each vane, this parameter is very important for the following considerations. The contact force can be assumed as constant for the whole testing period. For a direct comparison of the loading conditions in each pump, the Hertzian pressure can be calculated from the contact force. Therefore the contacting vane tips (see FIG. 3) are assumed as Hertzian cylinders sliding along a plain ring surface, disregarding (for simplyfication) any kind of lubrication. Due to wear, the radius of this cylinder increases. Thus, a decrease of Hertzian pressure over time is caused and therefore FIG. 5 reflects the starting conditions.

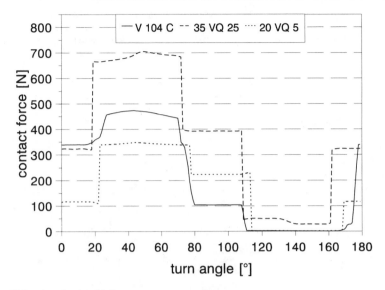

FIG. 4--Contact forces

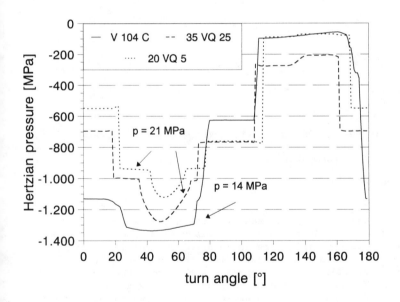

FIG. 5--Hertzian pressure for unworn components and full system pressure

It should be noted that, although the system pressure p for the two pumps of the VQ-series are 50 % higher, the maximum Hertzian pressure is still reached with the 60 years old V 104 C. However, the form of stress which occurs in every pump is very similar. The 20 VQ 5 and the 35 VQ 25 shows a increase of hertzian pressure (betweenn 30°-70°) because of the reduction of the contact area (additional inlet bore).

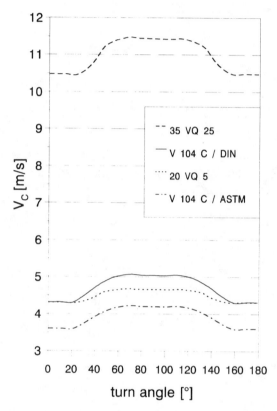

FIG. 6--Contact velocities

In FIG. 6 considerable differences between the various testing methods, with possible effects on the lubricant film conditions between ring and vanes, become visible. The increase of the rolling components on vane motion due to wear was measured in a test series with a pure mineral oil without additives and represents therefore the maximum change (FIG. 7). How the differences in load, velocity and other testing conditions are reflected by wear is shown in FIG. 8, where test results with the 35 VQ 25 and the V 104 C / DIN with identical fluids can be directly compared due to the normalization of single wear data with the upper wear limits (see TABLE 2) of each test mode.

Also dissimilar are the various contact velocities V_C (FIG. 6) of the vanes along the inner ring surface. V_C means the velocity in sliding direction with a superimposed component of rolling, which is caused by a swivel motion of the vanes relative to the actual tangent to the inner ring surface (see FIG. 3). Although this component is small, it can not be neglected because it increases with wear (FIG. 7).

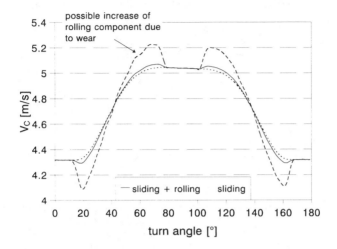

FIG. 7:Rolling component of contact velocity in dependence of wear V 104 C / DIN)

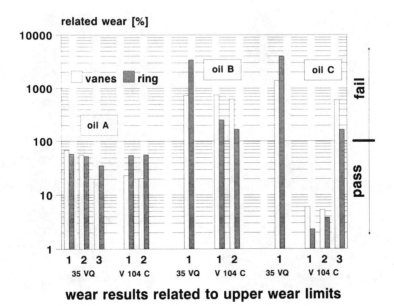

FIG. 8:Wear results achieved with identical fluids and different vane pumps

For fluid A the result is homogeneous. On both test rigs this fluid caused clear passes. The expectation was fulfilled, that testing on the 35 VQ 25 rig with 21 MPa system pressure leads to higher wear results. Fluid B represents a clear result as well. In all runs the fluid failed, but the results obtained in the 35 VQ 25 run exceeded the upper wear limits more than proportionally in comparison to those of the V 104 C runs. Finally, looking at the results with fluid C seems to remain unsatisfying, expecially by looking at test run #3 (V 104 C). The fulfilment of the pass criteria seems to depend on the choice of the test mode in this case. The reason for this can be found in the different fluid compositions which allows to interpret the remaining differences concerning temperature, viscosity and pressure between the testing methods to produce opposite results.

As another example of inhomogenous results among other possibilities, FIG. 9a and 9b show totally different wear mechanisms, which were all obtained in 35 VQ 25 pumps under equivalent test conditions when different fluids were tested. The prevailing wear mechanism can be found on the ring as well as on the vanes . In [5] it is reported that the mechanisms of adhesion, abrasion and fatigue can be found on these parts if a lubricant without any additive is used. Therefore, the obvious domination of one mechanism as shown in FIG. 9a or 9b can be interpreted as an insufficient attempt to surpress all possible wear mechanisms by additives.

FIG 9a--Abrasive wear on ring surface,   FIG 9b--Pittings on ring
        35 VQ 25   surface, 35 VQ 25

Wear maps for Vickers vane pumps

The upper wear limits given in TABLE 2 were derived empirically. A way to derive theoretically these limits in a qualitative manner would allow us to predict the capabilities of coming pump models for testing the fluid properties. Therefore, the main precondition is to allow only general pump parameters as input for those conclusional attempts. In a common research project of the University for Technology, Social Sciences and Economy in Zittau/Görlitz and the Technical University of Darmstadt a synthesis of energetic calculations on wear of vane pumps [5] and experimental investigations was performed. One result achieved was an approximate solution for this problem. Therefore a predictive model for wear calculation concerning the V 104 C / DIN and based on the shear energy hypothesis [6] was developed, where the wear debris is assumed to be proportional to shear forces in the surface regions and that there is a material specific wear constant for each friction partner and friction pair forming a tribo system according to [1]. The calculated results were improved by experiments [7, 8, 9] and these reinfluenced the further developing process of the mathematical model. Finally, a satisfying correspondence between theory and experiment, concerning the temporal wear progress, the wear masses of ring and vanes and the progression of local linear wear along the inner ring surface was achieved for a particular fluid. This lubricant was a mineral based oil without any additive which was chosen as a reference oil for test purposes from the German Research Association for Transmission Techniques. Thus, the disturbing influences of additives could be excluded.

Nevertheless, basic reactions of the tribo system Vickers vane pump could be retraced respectively predicted in general, firstly to the good correlations between experimental and calculated wear results for this fluid and secondly to the similarities of the different Vickers vane pumps with regard to specific stress parameters as shown in FIG. 4 - 6. To obtain an approximate calculated wear result for the pumps of the VQ-series, a variation of the input parameters load and speed (contact force and contact velocity, FIG. 4 and FIG. 6) was used, within a factor range of .5 to 1.5 of the input data for the V 104 C. The calculated wear results were normalised by the results achieved with the original V 104 C data. In this way three dimensional diagrams describing the wear behaviour in Vickers vane pumps as a function of characteristic inertial load and speed conditions were derived for test durations of 50, 100 and 250 hours (FIG. 10a-c). They are identical for ring and vanes. Strictly speaking, these wear maps were obtained only for a particular fluid without additives. For investigations in a more qualitative manner the maps can keep their meaning even if fluids enriched with additives are used, as long as the discussion of different pump models is concerning the same fluid.

All wear maps show the expected square root dependence of the relative contact force due to the Hertzian formulas. The dependence of relative velocity was found to be linear. This is in agreement with the linear rise of the lubrication gap in EHL contacts for contact velocity increase.

The calculated and experimentally improved results for the V 104 C can be found in FIG. 10c (250 h) on the diagram surface for a relative load and a relative velocity equal to 1. Similarly, the wear results for the 35 VQ 25 can be found in the 50 h wear map (FIG. 10 a), which is normalised as well as the 100 h diagram with the results of the V 104 C after 250 h. For the 35 VQ 25 the simplifying assumption was made, that for a contact force factor of 1.5 the contact velocity factor was 1.5 as well. A factor of about 2 - which can be derived from FIG. 6 - would have enlarged the uncertainties of this variation in an unacceptable manner. (The proposed substitute 20 VQ 5 in the following discussion concerning the V 104 C can be found for load and speed factors of 0.75.)

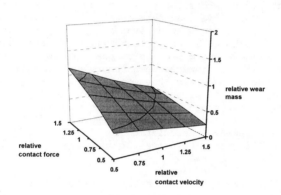

FIG. 10a--Wear map for 50 h of test duration

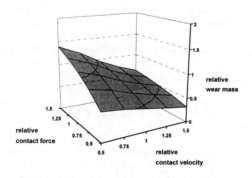

FIG. 10b--Wear map for 100 h of test duration

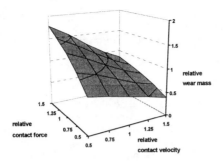

FIG. 10c--Wear map for 250 h of test duration

Keeping this in mind, it can be deduced that in theory only a small amount of wear - even after prolonging the test to 250 h - can be achieved on the 35 VQ 25 test rig in comparison to the V 104 C / DIN test. The theoretical results for the V 104 C test according to ASTM can be found for a contact velocity factor of 0.84 and a contact force factor of 1 in the diagram for a 100 h test duration. Due to this stress increase, wear values on a comparable level to those achieved in a DIN test after 250 h can be expected earlier. It should be noted, that for this comparison of 35 VQ 25 and various V 104 C tests, identical fluid viscosities for each test are expected. Therefore, these deductions need to be applied to fluids with higher basic viscosity values. In this case, the differences of prescribed viscosities for different procedures become smaller and the basis for those comparative investigations is expanded.

Independent of questions of viscosity, the wear values for the 20 VQ 5 can be found in FIG 10c (point 0.75 - 0.75 - 1), if the test procedure will be related to the intentions of the DIN 51 389 / IP 281 (see TABLE 1). They can to be expected to be on the same level with those of the V 104 C / DIN after 250 h.

Discussion

Looking at different test procedures revealed some differences with regard to testing conditions according to American and European standards. Therefore, it is necessary to mention exactly for each test result the standard to which it is related. The strategies of temperature control show the different intentions of the methods described in the standards. A fixed inlet temperature (35 VQ 25, ASTM) points to the relationship between the test procedure and industrial application problems. There the occurrence of a problem with a fluid is often related to a critical oil temperature determined by the application with only a small dependence to different viscosity groups of other potential hydraulic fluids. On the other hand, a prescribed

cinematic viscosity allows to have a choice between good or better
fluids, because it is impossible for the test fluid to hide behind its
lubrication abilities simply given by high viscosity values. But, such a
close connection to particular practical problems can not be achieved
with these tests as long as the prescribed viscosity outranges the
requirements of real hydraulic systems.

However, one similarity between all procedures seems to include
more relevance to allow an improvement to further vane pump test
methods. In ASTM D 2882, DIN 51 389 / IP 281 and M 2952-S no remarks
with regard to the necessity of a fluid cleaning process can be found,
although it is no secret that test fluids delivered in drums often show
a significant contamination and therefore need to be filtered before
testing [10, 11, 12]. This circumstance shows the need for reforming all
test procedures to increase the reproducibility and the repeatability of
wear results in any vane pump test.

During the development of a new test procedure fluid contamination
limits must be defined, which must fall short of the contamination
values before the test can be started. Without a need for time intensive
investigations this can be done by filtration together with newly
developed devices for online monitoring of fluid contamination. In the
meantime different systems are available, which can be used to count
particles in a particular fluid volume. The partial volume should be
extracted from the fluid flow as close as possible to the pump outlet.
In ISO 4406 [13] a method for coding the level of fluid contamination by
solid particles is described. The contamination level can be derived
directly from the particle count. Before a test can be started, the
filtration of the fluid in the hydraulic circuit has to be continued
until a defined contamination level is achieved. The severe disadvantage
of common vane pump test procedures namely that they do not specify the
initial starting condition of the object to be investigated, can be
overcome. This test reformation may help to achieve better
reproducibility and repeatability of the wear results obtained in
several tests on the same test rig as well as between wear results
coming from different laboratories.

Furthermore, measuring the contamination level periodically
without interrupting the test run would give the actual degree of fluid
contamination which is nothing but an indicator of wear. If a
correlation between the contamination level and corresponding wear is
found, a test run which will lead to results above the upper wear limits
would be indicated at an early phase during the test run. Those runs
would not need to be continued over the complete period, they could be
aborted after severe wear is confirmed in this way. Together with an
increase in reproducibility with a corresponding decrease in the amount
of necessary runs, a reduction of test duration will reduce costs
considerably. However, a lot of development must be done to work out the
capabilities of such a measurement, to find correlations between
contamination and wear with regard to a particular pump and to assure
the statistical reliability of such a process control.

In case of testing according to the Vickers standard M 2952-S with
the 35 VQ 25, an attempt was made to take the different wear mechanisms

(see FIG. 9a and 9b) into account. This can be interpreted as one step to extend the validation criteria to cover more important items related to wear with relevance for industrial applications than those simply expressed by wear masses. For a coming test procedure, this idea needs to be picked up. With a definition concerning the wear phenomena and/or the additional damage on the bushings (possible in 35 VQ 25 tests) to be tolerated, the validation possibilities of fluids will be enlarged considerably. Therefore, investigations have to be made for precisely specifying the additional criteria with regard to the pump to be chosen for this purpose.

The review of loading and stress conditions of pump models already in use for test purposes and those whose potential use is now being discussed reveals a great number of similarities. The shapes of load and stress parameter curves are comparable and in most cases situated within the same range.. Nevertheless, it is worth noting that the highest mechanical stress in combination with the lowest loading conditions occurs in the obsolete V 104 C. This reflects the advantages in vane pump design which have been achieved in the meantime. However, the level of mechanical stress in pump tests using up to date high loading conditions (21 MPa) does not exceed the Hertzian pressure achieved with a much lower system pressure (14 MPa) in the tests according to ASTM or DIN. Independent of these differences, all Hertzian pressure curves show the mentioned similarity as well as the shape of the different contact forces and contact velocities if their different levels are disregarded.

Therefore, a comparison of different pump models can be made, based on these parameters as shown in the wear maps. These maps (FIG. 10a - 10c) show that the empirically derived upper wear limits for the 35 VQ 25 test procedure must be lower than the corresponding pair of limits for the V 104 C because it can be deducted analogically from calculated wear results to corresponding wear limits. Therefore, the calculation itself is again qualitatively confirmed by the already existing upper wear limits for different test procedures. The upper wear limits can be estimated for a coming test procedure with a Vickers vane pump. If the test procedure is to reflect the requirements of the industry, a system pressure of 21 MPa as state of the art for vane pumps must be achieved. In the case of testing with the 20 VQ 5 and conditions as given in TABLE 1 following this request, the relative wear mass (point 0.75-0.75-1, FIG. 10c) is on the same level as the relative wear mass of the V 104 C (point 1-1-1). Therefore, the upper wear limits for a test procedure with the 20 VQ 5 at 21 MPa over 250 h can be assumed to be on the same level too. A system pressure of 14 MPa in the 20 VQ 5 will lead to much lower wear results. However, these theoretical deductions have to be verified by experiments.

Conclusion

Vickers vane pump tests have in common that
- they are capable of indicating the individual anti wear properties of different hydraulic fluids.
- the test procedures are basically similar, although there are differences concerning the test parameters.

- the test procedures need to be reformed because of the insufficient reproducibility of most test results and the small range of validation criteria: In case of the V 104 C a substitute must be found due to the obsolete pump design, which does not meet the requirements of hydraulic industry concerning a fundamental test parameter and which causes test handling problems.

A new test procedure with the Vickers vane pump 20 VQ 5 can overcome the disadvantages mentioned if
- the fluid contamination is controlled before and during the test runs. A filtration process to reach a specified contamination level of free solid particles will enlarge the test reproducibility.
- the experiences with the 35 VQ 25 test prescription concerning the optic validation of additional pump parts are used to develop more precisely a specification for wear phenomena on those parts to be tolerated.
- a system pressure of 21 MPa is chosen. This pressure level is not only related to industrial applications. As shown with the wear maps, wear masses comparable to those with the V 104 C can be expected. A large scale of possible wear results is necessary to indicate good and better fluids.
- these test procedure enlargements are supported by corresponding experimental investigations.

References

[1]   Czichos, H., "Tribology - A systems approach to the science and technology of friction, lubrication and wear," Tribology Series 1, Elsevier, Amsterdam 1978

[2]   Kunz, A. J., Matthes, B. , Broszeit, E., and Kloos, K.-H., "Application of TiN-coated bushings in the standard method for indicating the wear characteristics of hydraulic fluids in vane pumps," wear 162 -164 (1993), pp 966 - 970

[3]   Perez, R. J., and Brenner, M. S., "Development of a new constant volume vane pump test for measuring wear characteristics of fluids," Lubrication Engineering, May 1992, pp 354 - 359

[4]   Thoenes, H. W., Bauer, K., and Hermann, P., "Erfahrungen mit der Vickers-Flügelzellenpumpe," Schmiertechnik + Tribologie 26 (1979) 4, pp 124 - 127

[5]   Kunz, A. J., "Synthese experimenteller Untersuchungen und energetischer Berechnungen zum Verschleißverhalten des Tribosystems Flügelzellenpumpe," Dissertation, Technical University Darmstadt, Darmstadt 1995

[6]   Beckmann, G., "A theory of wear based on shear effects in metal surfaces," wear   59 (1980), pp 421 - 432

[7]   Gellrich, R., Kunz, A. J., Broszeit, E., and Beckmann, G., "Combination of theoretical and experimental investigations on wear of vane pumps," Vol. II, Proceedings of the 6th Nordic Symposium on Tribology, Uppsala 1994

[8]   Gellrich, R., Kunz, A. J., Beckmann, G., and Broszeit, E.,
       "Theoretical and practical aspects on wear of vane pumps. Part A:
       Adaption of a model for predictive wear calculation," wear 181 -
       183 (1995), pp 862 - 867

[9]   Kunz, A. J., Gellrich, R., Beckmann, G., and Broszeit, E.,
       "Theoretical and practical aspects on wear of vane pumps. Part B:
       Analysis of wear behaviour in the Vickers vane pump test," wear
       181 - 183 (1995), pp 868 - 875

[10]   Broszeit, E., and Steindorf, H., "Prüfung von Hydraulikflüssig-
       keiten," Tribologie und Schmierungstechnik 35 (1988) 4, pp 193-198

[11]   Broszeit, E., Steindorf, H., and Kunz, A. J., "Prüfung von
       Hydraulikflüssigkeiten mit Flügelzellenpumpen," Tribologie und
       Schmierungstechnik, 37 (1990) 4, pp   202 - 209

[12]   Kunz, A. J., and Broszeit, E., "Technologieprüfung von
       Hydraulikflüssigkeiten," in Bartz, W. J. (Editor):
       "Hydraulikflüssigkeiten," Band 475 Kontakt & Studium, expert
       Verlag, Renningen-Malmsheim 1995

[13]   ISO 4406: Hydraulic fluid power; Fluids; "Method for coding level
       of contamination by solid particles," ISO International
       Organization for Standardization, Beuth Verlag, Berlin 1987

Ken J. Young[1]

HYDRAULIC FLUID WEAR TEST DESIGN AND DEVELOPMENT

REFERENCE: Young, K. J., **"Hydraulic Fluid Wear Test Design and Development,"** *Tribology of Hydraulic Pump Testing, ASTM STP 1310,* George E. Totten, Gary H. Kling, and Donald J. Smolenski, Eds., American Society for Testing and Materials, 1996.

ABSTRACT: The current widely accepted "industry standard" pump test employs a Vickers V104C (or V105C) vane pump to assess the antiwear performance of hydraulic fluids. The two main versions of the test are documented as ASTM and IP methods. This pump test method, which was first developed in the 1960s, assesses only the steel-on-steel antiwear performance of hydraulic fluids, and this under constant speed, constant volume, and constant high pressure conditions. Constant conditions are not experienced in practical applications and it can be shown that hydraulic fluids which perform well at high pressure can fail to provide adequate antiwear performance at low pressure. Although other pump types may have different materials in contact, standard universally accepted procedures have not yet emerged to cover these. It can be shown that hydraulic fluids which have good steel-on-steel antiwear performance can have poor performance in steel-on-yellow-metal contacts. There is therefore a need for wear test procedures to be developed which cover the full range of operating conditions and material types found in practical applications. This paper will review experience gained in the author's laboratory, with different hydraulic fluid types, during the development of wear test procedures which more adequately cover the range of operating conditions and material contacts experienced in practical operating environments.

KEYWORDS: hydraulic fluid, pump, lubrication

Hydraulic fluids are used to transmit power. While practically any liquid could be used to perform this function, it is important that the fluid should lubricate and protect the hydraulic system components, and maintain its ability to function effectively over a long time period. Often a single fluid is selected by an operator for all of his systems,

[1]Senior Scientist, Industrial Lubricants Department, Shell Research Ltd., Thornton Research Centre, P.O. Box 1, Chester, CH1 3SH, England.

and it is expected to operate in hydraulic machinery having a wide variety of different designs, under a wide range of operating conditions, and in varying operating environments. It is important that the fluid exhibits the performance characteristics required by all of these systems.

In an effort to control hydraulic fluid quality, various specifying bodies, such as DIN (the Deutsches Institut fur Normung e.V), ASTM and ISO have issued, or are developing, standard specifications for hydraulic oils. The test method prescribed by these specifications to assess the degree of pump wear protection offered by a hydraulic fluid employs a constant volume Vickers V104C vane pump. There are two different versions of the test. The US ASTM D2882 'Test Method for Indicating the Wear Characteristics of Petroleum and Non-Petroleum Hydraulic Fluids in a Constant Volume Vane Pump' requires a 100 hour test to be carried out at a pressure of 14 MPa, a speed of 1200 rpm and a temperature of either 65.6°C or 79.4°C depending on the fluid viscosity. The European IP 281 method, 'Determination of Antiwear Properties of Hydraulic Fluids - Vane Pump Method', although specifying the same test pressure, requires a 250 hour test at a speed of 1440 rpm, at a temperature to give a bulk fluid viscosity of 13 mm^2s^{-1} at the pump inlet. The higher speed and longer duration makes this latter version more severe. Temperature differences between the two methods are most marked for ISO 46 fluid tests, but these are unlikely to have a major influence on severity differences.

The V104C vane pump test was first developed in the 1960s and pump designs have progressed significantly since then. It is run to assess steel-on-steel antiwear performance only, and this under constant speed, constant volume, and constant high pressure conditions. Although other pump types, such as axial piston pumps, may have different materials in contact, standard, universally accepted procedures have not yet emerged to cover these. Some hydraulic equipment manufacturers have issued their own independent specifications and/or test methods, however there is a need for universally accepted wear test procedures to be developed which cover the full range of operating conditions and material types found in practical applications.

In order to develop hydraulic fluids which will be capable of operating over a wide range of conditions and with a wide range of materials in contact, it is essential to have appropriate laboratory test methods in place. Since there is only one universally specified pump wear test, run using constant operating parameters (albeit with differences between the US and European versions of the method), it has been necessary, in the author's laboratory, to derive modifications to the standard V104C steel-on-steel wear test, and to develop a steel-on-yellow-metal wear piston pump test and screening test. Performance data with different fluid types, obtained using these methods, will be presented in this paper to illustrate the need for significant advances in the understanding of the tribology of pump testing and the development of more appropriate test methods.

EXPERIMENTAL METHODS

Steel-on-Steel Wear Tests

The standard IP 281 method requires the V104C test to be run at a constant pressure of 14 MPa and at a temperature to give a bulk fluid viscosity of 13 mm^2s^{-1} at the pump inlet. For an ISO 32 fluid the required temperature is between 65 and 67°C, depending on the VI of the fluid. Some practical hydraulic systems are run at lower pressures, and it is not uncommon for bulk fluid temperatures to exceed 65°C. A high pressure, constant temperature test therefore may not be sufficient to establish antiwear performance over a wider temperature or pressure range. V104C tests in the author's laboratory are therefore run on ISO 32 fluids at 14 MPa, and at 5 MPa, with additional tests being run at 105°C using an ISO 150 fluid. In order to continue to run V104C tests, it has been necessary to replace the port plates supplied with V104C cartridges with port plates manufactured from an alternative material, as from time to time there have been excessive port plate wear problems associated with differences in chemical composition and micro structure of the phosphor bronze port plates normally supplied. This change has not affected steel-on-steel wear results.

Vickers 20VQ5 and 20VQ8 pumps are being considered by the ASTM as replacements for the Vickers V104C pump. The test pressure selected for these later pumps has been set at 21 MPa. Tests in the author's laboratory have been run with these pumps at this pressure, and the significantly lower pressure of 5 MPa, for 250 hours at 1440 rpm with bulk fluid temperatures set to give $13mm^2s^{-1}$ fluid viscosity.

Steel-on-Yellow-Metal Wear Tests

A common feature in piston pump design is the use of yellow metal (bronze or brass) slipper heads. A fluid having good steel-on-steel antiwear performance can not be assumed to have good steel-on-yellow-metal antiwear performance.

An axial piston pump test has been developed in co-operation with a pump manufacturer to evaluate steel-on-yellow-metal wear performance. The test rig is shown in Plate 1. Test conditions, established by the pump manufacturer, are an inlet pressure of 21 MPa, a bulk fluid temperature of 80°C, and a speed of 3000 rpm.

Piston pump testing is expensive and time consuming, however, and there is a need for a screening test which can be used during product development to enable different formulation options to be tested prior to selection of final candidates for pump testing. An Amsler disc machine has therefore been used as the basis for the construction of a test rig, Plate 2, with one of the discs of the original machine replaced by a phosphor-bronze or a high-tensile-brass block. The principle of the device is illustrated in Figure 1. The block is loaded onto a steel disc by a spring. Wear of the block is recorded continuously during the run by an inductive displacement transducer. The contact temperature is controlled, and can be modified, by heating the block holder. The actual applied load, disc surface roughness and contact temperature selected was determined by correlation of wear rates with wear in the piston pump test, using a range of fluids with different formulations.

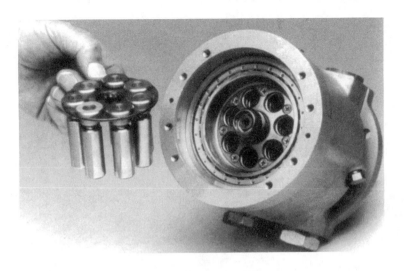

Plate 1 - <u>Piston pump test rig and disassembled piston pump</u>

Plate 2 - <u>Amsler screening test rig</u>

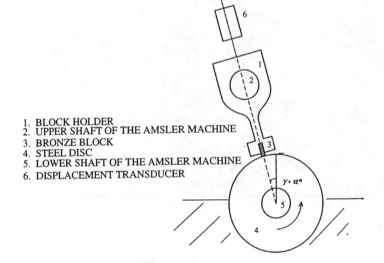

1. BLOCK HOLDER
2. UPPER SHAFT OF THE AMSLER MACHINE
3. BRONZE BLOCK
4. STEEL DISC
5. LOWER SHAFT OF THE AMSLER MACHINE
6. DISPLACEMENT TRANSDUCER

FIG. 1 - The set-up of the Amsler rig

RESULTS

<u>Steel-on-Steel Wear Tests</u>

The results of standard IP 281 vane pump wear tests, and modified tests carried out at different applied pressure and different temperature on three fluids, all meeting DIN HLP specification requirements, are compared in Table 1. These results indicate that although fluids can be of equivalent antiwear performance at the elevated pressure and moderate temperature of the IP 281 test, they can be of very different wear performance at lower pressure or higher temperature.

TABLE 1 -- <u>Results of steel-on-steel V104C antiwear tests on ISO 32/37/150 fluids meeting DIN HLP requirements</u>

Conditions: Speed 1440 rpm; Duration 250 or 1 000 hours; Temperature to give 13 mm^2s^{-1}.

	Total steel weight loss, mg		
Fluid	A	B	C
Tests at 65°C			
14 MPa	14.6	10.4	12.1
5 MPa	-	3.1	88.3
Tests at 105°C			
14 MPa	95	9	-
14 MPa	294*	33*	-

*After 1 000 hours

TABLE 2 -- <u>Comparison of V104C and 20VQ pump performance of ISO 37 fluids</u>

Conditions: Speed 1440 rpm; Duration 250 hours; Pressure 14MPa (V104C), 21 MPa (20VQ5 & 20VQ8) ; Temperature to give 13 mm^2s^{-1}.

Fluid	Total steel weight loss, mg		
	B	D	E
V104C	14.2	14.0	265
20VQ5	7.5	2.3	6.4
20VQ8	3.9	2.6	7.2

The change to higher pressure testing currently proposed by the ASTM for tests with 20VQ series pumps (from 14 to 21 MPa) will not improve the situation; for example Fluid C, tested in a 20VQ8 pump at 5 MPa also gave high wear (182 mg total steel wear after 250 hours). Even at high pressure, it can be shown, Table 2, that the 20VQ series pumps will be less discriminating than the V104C pump.

<u>Steel-on-Yellow-Metal Wear Tests</u>

The correlation between piston pump and Amsler wear tests for the optimum set of Amsler test conditions is shown in Figure 2. Of 12 oils tested, one is an outlier so

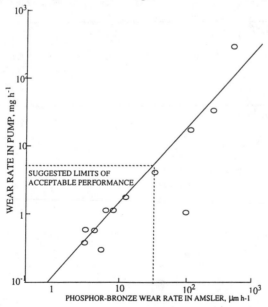

FIG. 2 - Correlation between the results of pump and Amsler tests

although the correlation is not perfect, the Amsler rig test developed is a powerful tool for screening yellow metal wear performance; only 150 cm^3 oil is required per test, and one oil can be tested per day. The piston pump test requires 150 litres of oil and is run for 500 hours.

There is no universally accepted steel-on-yellow metal wear test, yet a fluid having good steel-on-steel antiwear performance can not be assumed to have good steel-on-yellow-metal antiwear performance, as illustrated in Table 3. Fluids F and G, both of which meet DIN HLP requirements, have markedly different steel-on-yellow-metal wear performance with one fluid being judged to have unacceptably high wear in the Amsler test with high-tensile-brass. The results in Table 3 also confirm that the poor wear performance in the Amsler screening test is matched by poor wear performance in the piston pump test with unacceptably high total slipper lift increase (the pump manufacturer specifies less than 0.4 mm at end of test) caused by excessive wear of the slipper head / slipper contact.

TABLE 3 -- Comparison of steel-on-steel and steel-on-yellow-metal wear performance of two DIN HLP ISO 32 fluids

Fluid	F	G
V104C (IP 281).		
Total steel weight loss, mg	11.1	24.1
Amsler.		
Wear rate, μmh^{-1}		
Phosphor Bronze	33	40
High Tensile Brass	15	0.5
Piston Pump (High Tensile Brass).		
Piston weight loss, mg	256	177
Total Slipper Lift Increase, mm	0.6	0.1

CONCLUSION

Testing has shown that fluids having good steel-on-steel antiwear performance under the conditions of the IP 281 V104C antiwear test may not be of adequate wear performance at other temperatures and pressures. It has also been shown that it is possible to develop a steel-on-yellow metal wear test which correlates with wear obtained in an axial piston pump and that fluids with good steel-on-steel antiwear performance do not necessarily give good antiwear performance with yellow metals.

Good IP 281 or (by analogy) ASTM D2882 performance does not 'guarantee' that a hydraulic fluid will have good lubrication performance; it is necessary to carry out tests under a variety of conditions and with different materials in contact. The test work reported here has been carried out in constant temperature/pressure tests, and although more comprehensive, still does not completely simulate the conditions experienced by hydraulic pumps in practical operating environments. Field trials are therefore normally carried out on new fluids.

Pump test rigs which test the performance of hydraulic fluids under cycling conditions are currently being designed and will be installed in the author's laboratory in the near future. These rigs will help to ensure that new fluids are developed using test conditions which simulate field experience as closely as is practicable.

It is recommended that the ASTM consider the development or adoption of both vane and piston pump tests which more closely simulate conditions experienced in practical hydraulic systems.

ACKNOWLEDGEMENT

The author would like to acknowledge the extensive contribution of his colleagues in the design and development of the in-house test methods discussed in this paper.

Lev A. Bronshteyn[1] and Donald J. Smolenski[1]

ENERGY EFFICIENCY SCREENING TEST FOR HYDRAULIC FLUIDS.

REFERENCE: Bronshteyn, L. A. and Smolenski, D. J., *"Energy Efficiency Screening Test for Hydraulic Fluids,"* ASTM STP 1310, George E. Totten, Gary H. Kling, and Donald J. Smolenski, Eds., American Society for Testing and Materials, 1996.

ABSTRACT: A novel technique for measuring very small variances in friction coefficient due to a change in oil or additive type is presented. A standard pin-on-disk tribometer is used for test oil friction measurement in comparison with a reference oil. The "OCWS" (Oil Change Without Shutdown) test procedure has been used to increase test precision and eliminate the baseline variations that are unavoidable with separate test measurements.

KEYWORDS: friction modifiers, energy efficiency, screening test, hydraulic oils.

INTRODUCTION

The 1973 oil embargo stimulated studies of engine oil energy-conserving properties. Since then a large number of energy-conserving engine oil formulations, friction modifiers and test procedures have been developed and introduced into practice.

The problem of improvement of hydraulic oil energy efficiency has not been given the same attention nor studied as extensively as for engine oils. Because of the availability of the ASTM Sequence VI Equivalent Fuel Economy Improvement test [1] as a yardstick, energy-conserving engine oils have been widely available in the marketplace for many years. However, no such definitive yardstick exists for measuring the energy conserving properties of hydraulic oils. Different laboratory friction and wear tests are used to characterize oil energy efficiency [2-5], making it difficult to compare results from different investigations. In addition, plant trials for measuring a lubricant's effect on energy efficiency are exceedingly difficult to conduct. Differences among machines, cycle to cycle variation and questionable accuracy of energy consumption measurements all

---

[1] Staff Fellow and Staff Research Engineer, respectively, General Motors Research and Development Center, 30500 Mound Road, Warren, Michigan 48090-9055

make accurate, in-plant determination of energy efficiency difficult.

One recent version of a screening test was designed by Moore (2). Pin-on-disc and reciprocating test rigs have been used to evaluate friction coefficients for engine oils in boundary and hydrodynamic lubrication regimes. A significant friction reduction effect has been shown only for boundary lubrication conditions. Test "run-in" under these conditions is slow, typically taking two to three days to complete. There is no difference between friction coefficients detected for reference and friction modified engine oils in the hydrodynamic lubrication regime, for sliding speeds 100 cm/s and higher. Within the mixed lubrication regime (sliding speed 10-100 cm/s) friction is unstable, very sensitive to changes in surface topography and other factors, and is not sufficiently repeatable to make reliable comparisons.

It is difficult to maintain a stable friction reference baseline during different test measurements or even during a single run because of unavoidable variations in initial friction conditions. Therefore a reference oil test is often required to precede and follow a candidate oil test to check for baseline change. Due to different sources of random variability, friction measurement errors may be several times greater than the desired positive effect that must be reliably determined in the presence of substantial "noise". A well-designed bench test procedure must minimize this noise and allow a valid assessment of product performance at minimum cost and time.

EXPERIMENTAL

Test Rig

A commercially available pin-on-disk tribometer, ISC-200PC, from Implant Science Co. was used for a screening bench test (see Figure 1). This tribometer is equipped with a motor driven turntable, an electronic force gauge for friction measurements and a computer-controlled chart recorder.

A steel cylindrical pin specimen with a curved working surface was loaded against a rotating steel disk, creating a point sliding contact. Both pin and disk surfaces were ground and polished.

A cup-like fixture on the tribometer turntable permitted the use of liquid lubricants during a test. The test oil was supplied by means of a commercially available syringe pump. A flow rate of 0.1 ml/min was quite enough to have a fully flooded contact for modest disk rotation speeds.

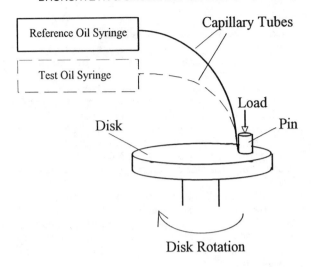

Fig. 1 Schematic of a pin-on-disk test rig.

Table 1. Tested Products.

Product	Chemical Formulation	Tested Treatment Level, mass %
Reference oil	Antiwear Hydraulic Oil of 46 cSt at 40°C with R&O and ZnDTP additives	100
PTFE	Polytetrafluoroethylene (PTFE) powder in petroleum oil	2
MoS$_2$	Molybdenum disulphide (MoS$_2$) powder in petroleum oil	0.3
Ester 1	Synthetic ester	10
Cl	Chlorinated hydrocarbons	10
Ester 2	Synthetic ester	5
MoDTC	Molybdenum dithiocarbamate	2
EP Metal Treatment	Unknown	10
Synthetic Oil	Unknown	100

Reference Oils and Tested Lubricants

A series of commercially available antiwear hydraulic oils have been selected as reference lubricants. All of these oils have the same additive package, containing antiwear, rust and oxidation (R&O) preventives, defoamant and pour point depressant additives blended in mixtures of the same hydrocarbon base stocks of different viscosity. Candidate supplements were usually blended in an ISO-46 Viscosity Grade reference oil (46 cSt at 40°C) and evaluated in comparison with this oil without the supplement. A 32 (32 cSt at 40°C) and 68 (68 cSt at 40°C) ISO Viscosity Grade oils of the same series were used to verify the test regime described below.

The treatment level used for each candidate supplement in the oil was the dosage recommended by the manufacturer of the supplement. Table 1 lists all tested lubricants, the treatment level used, and the information that was available regarding each chemical formulation. All tested products are commercially available.

Test Regime

In order to define the boundary, hydrodynamic and elastohydrodynamic lubrication regimes for the selected friction pair pin-on-disk configuration, and to design the test procedure, preliminary experiments were carried out. The results are shown in Figures 2-4.

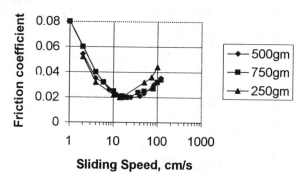

Fig. 2 Friction coefficient vs sliding speed for different loads (46 cSt oil viscosity, 50.8 mm pin radius of curvature).

As can be seen in Figure 2, the sliding velocity associated with the transition from boundary to elastohydrodynamic (EHD) lubrication regimes is located at about 10 cm/s and has very little dependence on the load. The hydrodynamic regime occurs at speeds of 30 cm/s and higher.

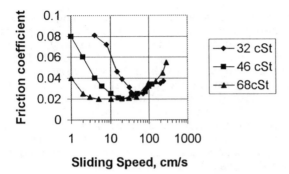

Fig. 3 Friction coefficient vs sliding speed for different
       oil viscosities (750 gm load, 50.8 mm pin radius of
       curvature).

    As shown in Figure 3, there is a significant difference
in boundary regime friction coefficients and the transition
sliding speeds among low and high viscosity oils. They vary
from 5 cm/s for a 32 cSt viscosity oil to about 40 cm/s for
a 68 cSt viscosity oil. Thus, comparable efficiency
measurements should only be made with a reference oil of the
same viscosity grade as a candidate oil. In our case of
supplements testing, we neglected their effect on viscosity
because the supplements were either liquids of approximately
the same viscosity as the reference oil (Ester 1, Ester 2,
Cl) or were tested in small concentrations (see Table 1).

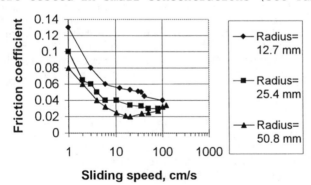

Fig. 4 Friction coefficient vs sliding speed for different
       pin radius of curvature (46 cSt oil viscosity, 750 gm
       load).

    The calculation of maximum Hertz pressure in the pin-
disk contact for a 750 gm load gives 130, 200 and 330 MPa
for a pin curvature radius of 50.8, 25.4 and 12.7 mm,
respectively. The results shown in Figure 4 indicate that
the transition point from boundary to hydrodynamic regime is

located within the tested range of sliding speeds only for a 50.8 mm pin radius of curvature. For smaller pin radiuses of curvature, the results are in the boundary lubrication area even for high tested sliding speeds of about 100 cm/s.

All measurements described below were performed with a 50.8 mm radius of curvature pin, at a speed of approximately 4 cm/s and a 750 gm load. These conditions correspond to mild boundary - EHD lubrication regime. This test regime is not severe enough to generate a high wear rate and cause a long running-in, but at the same time it is close to practical lubrication conditions corresponding to the main energy losses in real hydraulic systems.

Run-in and Baseline Repeatability

The typical run-in curves are shown in Figure 5. The "bumps" on the curves represent the smallest increment of friction variation (equal to 0.0024) that can be measured and stored digitally by the tribometer computer system. Though all three curves in Figure 5 were recorded for the same friction conditions, their final value can vary significantly for different test runs because of unavoidable variations of initial conditions, such as pin and disk surface roughness and hardness, ambient temperature and humidity, sliding velocity and other sources of random error.

Figure 6 illustrates the long-term baseline stability. Contrary to the large run-to-run random baseline variations, long-term baseline variations following one hour run-in and 3.5 hours of test are no more than one or two of the smallest measurable increments of the friction coefficient. That corresponds to friction coefficient variations of from 0.0024 to 0.0048 or about 10 percent of the baseline friction coefficient. This level of variation is acceptable as a random error for lubricant-related friction measurements.

Thus, the main source of random errors in friction measurements is run-to-run baseline variations. It is very difficult, if not impossible, to keep a reference friction baseline stable during a series of tests. To reduce this error, one might try to increase the run-in period, as has been done by Moore (2) for up to 2 or 3 days, or perform many parallel tests averaging random variations. In both cases the test will be significantly more time and labor consuming. However, there is another, much more advantageous way to increase test repeatability by using the test procedure described below.

"OCWS" Test Procedure

The "OCWS" ("Oil Change Without Shutdown") test procedure uses syringes filled with different oils during a test conducted continuously without shutdown (see Figure 1, dashed lines).

Figures 7 and 8 illustrate the typical friction curves obtained for energy efficient and non-efficient products. The test starts with the reference oil. Running-in takes no more than 0.5 to 1 hour. When the friction coefficient has stabilized, the syringe with the reference oil is replaced by the syringe with a test oil without shutting down the friction machine. It takes no more than 10 seconds to change the syringes. Since the tribometer disk is always covered with an oil film, a short oil supply interruption during the change of the syringes has no influence on the friction behavior of the test. After about 0.5 - 1 hour of operation with the test oil (the time duration depends on the test oil behavior and the experiment may take as long as necessary) the syringe with test oil is changed back to the syringe with the reference oil.

The main advantage of the "OCWS" procedure is that it does not require the same baseline for different test runs. For screening tests, the absolute value of friction coefficient is not so important as the difference between the reference and test oils. This procedure eliminates the main source of lubricant efficiency measurement errors - variations of friction baseline which occur due to different run-in friction conditions.

RESULTS AND DISCUSSION

Several commercially available supplements and a synthetic oil have been screened using the "OCWS" test (see Table 1). The test results are shown in Figure 9. The percent friction change for each supplement has been calculated as an average result of two or more tests.

The measurements indicate that PTFE containing supplements have no effect on friction, while MoS_2, Cl-containing supplements and one of the synthetic ester based products increased friction by 8, 27 and 30 percent, respectively. EP Metal Treatment showed the worst results almost doubling the baseline friction coefficient.

Two products, Ester 2 and molybdenum dithiocarbamate (MoDTC) supplements, each provided approximately a 20 percent friction reduction. These additives have also shown a carry-over effect. Such effects become evident when the friction coefficient does not return to its baseline level after the test oil is replaced by the reference oil.

The best results were provided by the synthetic oil which also showed a strong carry-over effect.

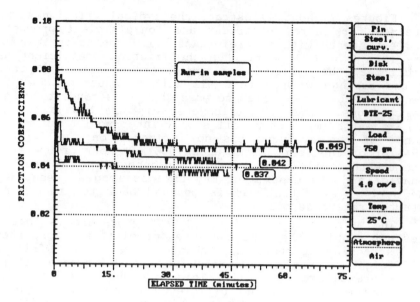

Fig. 5 Run-in curves.

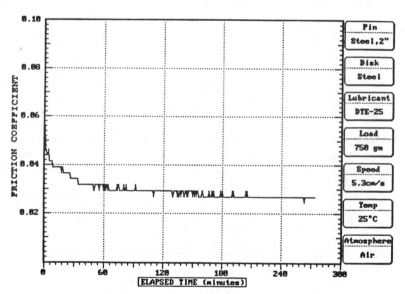

Fig. 6 Long-term baseline stability.

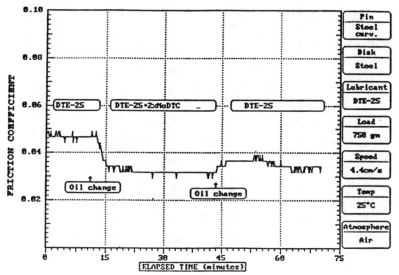

Fig. 7 Sample of energy efficient lubricant.

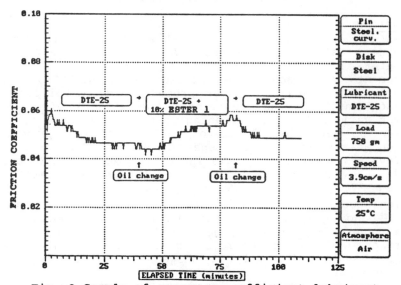

Fig. 8 Sample of energy non-efficient lubricant.

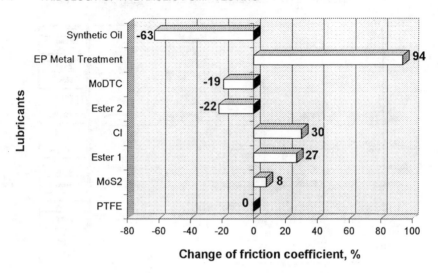

Fig. 9 Screening test results.

CONCLUSIONS

A novel, simple and reliable energy efficiency screening test is presented. A standard pin-on-disk tribometer is used for comparing test oil friction measurements with those of a reference oil. An "OCWS"("Oil Change Without Shutdown") test procedure was developed to increase sensitivity and eliminate the effects of random baseline variations. The screening test takes about 2 hours, requires no more than 20 ml of a test oil and provides friction measurements with a repeatability of 10 percent.

Several commercially available supplements and oils have been evaluated using the new screening test method. Powdered PTFE or MoS_2 based supplements did not reduce friction coefficients, while one synthetic ester based product and molybdenum dithiocarbamate supplement did provide approximately a 20 percent friction reduction in comparison with a conventional hydraulic oil. The most efficient synthetic based oil showed 63 percent friction reduction.

The screening test developed in this work can be used as a quick and reliable tool for evaluating and selecting the most energy conserving lubricating oils.

REFERENCES

[1]   ASTM   Research   Report   RR-D: 2-1204,   "Standard
        Dynamometer Test Methods for Measuring the Energy
        Conserving Quality of Engine Oils (Sequence VI)",
        ASTM, August 1985.

[2]   Moore, A.J., "Fuel Efficiency Screening Tests for
        Automotive Engine Oils," SAE Paper No. 932689, 1993.

[3]   Morecroft D.W., "The Shell Low Velocity Friction
        Machine for Evaluating Fuel Economy Motor Oils", Wear,
        v. 89, 1983, pp. 215-223.

[4]   Benchaita, M.T., Lockwood, F.E., "A Reliable Friction
        Model   of   Lubricant   Related   Friction   in   Internal
        Combustion Engines", Paper 11.4, Tribologie 2000,
        Technische Akadamie Esslingen, January 1992.

[5]   Suresh Babu, A.V., Martin, V., Mehta, A.K., "A New
        Test Technique for the Laboratory Evaluation of
        Energy-Efficient Engine Oils", Lubrication Science, v.
        5-4, July 1993, pp. 283-294.

Richard K. Tessmann[1], David J. Heer[2]

PUMP TESTS FOR HYDRAULIC FLUID WEAR QUALIFICATION

REFERENCE: Tessmann, R. K. and Heer, D. J., **"Pump Tests for Hydraulic Fluid Wear Qualification,"** *Tribology of Hydraulic Pump Testing, ASTM STP 1310,* George E. Totten, Gary H. Kling, and Donald J. Smolenski, Eds., American Society for Testing and Materials, 1996.

ABSTRACT: Modern hydraulic systems are operating at higher pressures and higher temperatures than ever before. In addition, the life and reliability of the hydraulic components and hence the hydraulic system is becoming more and more dependent upon the characteristics of the circulating fluid in the system. The hydraulic component which normally suffers most when the correct fluid is not used in the system is the hydraulic pump. Three different types of hydraulic pumps are widely used in hydraulic systems. Vane pumps have been the dominant type of pump for hydraulic systems. Gear pumps have replaced vane pumps in many applications. Piston pumps are rapidly gaining popularity. Therefore, the pump manufacturers are conducting tests to qualify fluids for use with their pumps and the fluids formulators are conducting tests to show that their formulations are compatible with hydraulic components. Unfortunately, the tests most widely used are either not industry standards or do not address problems encountered with some hydraulic pumps.

This paper presents an overview of the pump tests which are currently available or which are specific to certain hydraulic pumps. Since there are few standard pump tests, many companies have devised their own tests to be confident of the fluids which they qualify. Additional concepts in pump testing are proposed and discussed.

KEYWORDS: Fluid testing, pump testing, performance degradation, wear testing, lubrication, hydraulic fluids

[1]Vice President, FES, Inc., 5111 North Perkins Road, Stillwater, OK 74075
[2]Laboratory Manager, FES, Inc., 5111 North Perkins Road, Stillwater, OK 74075

INTRODUCTION

The life and reliability of hydraulic systems are heavily dependent upon the fluid used. In the past, the hydraulic fluids were used primarily for energy transfer and control of hydraulic components. However, as enhanced performance was demanded of the hydraulic system, the pressure increased and the fluid temperature rose. In this new environment, the hydraulic fluid assumed a new role and one which is not as easily understood as the transfer of energy. The role of lubrication was added to the traditional role played by hydraulic fluid in the operation of the system. However, there is a need to better understand fundamental concepts of tribology as applied to hydraulic components. In addition, there are few industry recognized standard test methods for assessing the ability of a hydraulic fluid to lubricate the critical surfaces in the components of a hydraulic system.

It is extremely important that the interaction between the hydraulic fluid and the pump be evaluated in some manner. Of course, the best situation is to have a precise analytical model of this interaction. Such a model could be used to develop and design hydraulic components and system as well as select the proper fluid for any given situation. In addition, a complete model would make life and reliability predictions very straightforward. Therefore, tests must be conducted to determine fluid/component interaction. Hydraulic pump performance suffers the most from the use of incorrect fluid and gains the most from the selection of the proper fluid.

Selection of a pump test on a bench testing procedure depends upon the end use of the data. The component manufacturer and the hydraulic system user need only to run the system pump to see that the fluid being considered will adequately protect the pump. The pump designer, however, needs much more information in order to improve the pump design and probably would not be satisfied with any of the bench tests. The fluid supplier has the most unique situation. He must make sure that his formulation performs adequately with every pump. The fluid manufacturer many times uses a bench test during the development of the formulation and then verifies the product by using a pump test. In most cases, however, the fluid supplier must still run a large number of pump tests to thoroughly qualify his fluid product.

The pump tests which have enjoyed the most widespread use require fairly complex and expensive test equipment. The pump manufacturer normally already has such a facility while the system user and the fluid manufacturer must either have an outside laboratory conduct the tests or they must purchase the necessary testing facilities. In any case, there is a substantial cost in the repair and maintenance of pump testing facilities and it requires significant time to run each test. Therefore, the use of pump tests to evaluate the performance of hydraulic fluids is costly. In addition, repeatability and reproducibility of pump testing is also of concern.

This paper first considers the type of wear surfaces which are present in the commonly used hydraulic pumps. This information is used as background to evaluate the pump test currently in use. Each of the pump tests presented are evaluated for their advantages and disadvantages. Finally, problems associated with pump testing are discussed and potential solutions offered.

Wear Interfaces in Hydraulic Pumps

The three types of pumps commonly found in hydraulic systems are the vane pumps, gear pumps, and piston pumps. From a design standpoint, each of these pumps offers a unique challenge in the improvement of life and reliability. In addition, there is a widespread belief that the lubrication role played by the fluid in the performance characteristics is vastly different from one pump design to the other. To assist in developing an understanding of these beliefs and establish a background from which pump testing can be evaluated, the wear interfaces which have proven to be the most critical from a wear standpoint will be presented for each pump type.

The Vane Pump

The component parts in a vane pump are pictorially shown in Fig. 1. As can be seen in this figure, the following areas are common wear surfaces:

- The contact between the vane tips and the cam ring.

- The contact between the vanes and the rotor

- The contact between the vanes, rotor and side bushings

- The bearing/shaft contact

Of these interfaces the first three are encountered more often than the bearing wear situation. Deterioration in any of these wear areas will produce a degradation in pump performance. This degradation mostly often results in a change in the volumetric efficiency of the pump but it may also affect the mechanical efficiency as well.

The Gear Pump

The component parts of a typical gear pump are illustrated in Fig. 2. The wear areas of the gear pump are listed below:

- The tips of the gear teeth and the pump body

- The sides of the gears and the wear or compensating plates

- The bearing and shaft interface

In a gear pump, like the vane pump, bearings seldom cause problems from a wear standpoint. In addition, the gear pump can not be balanced from a pressure standpoint. Therefore, the force vector created by the pressure acting upon the profile of the gear will tend to move the gears toward the suction port. That is, the pump is usually designed such that the gears are concentric with the shaft and the shaft is concentric with the bearings. There will be some clearance between the shaft and the bearings. This clearance will be moved to one side due to the pressure force. Therefore, except for the film necessary to carry the load there will be little clearance between the tips of the gear teeth and the inside surface of the housing. While wear in any of these critical areas may result in pump failure, interface between the side plates and the sides of the gears is typically the most wear-prone area.

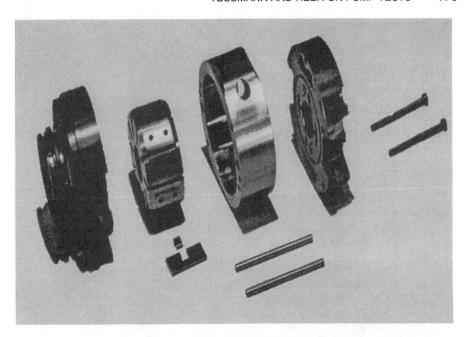

FIG. 1 – Components of a Vane Pump

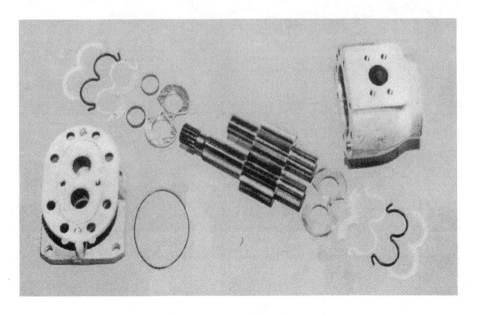

FIG. 2 – Component Parts of a Gear Pump

The Piston Pump

Normal wear surfaces of a typical piston pump are shown in Fig. 3. Critical wear interfaces of the piston pump are:

- Piston shoes and the swash plate surface.

- Outside diameter of the piston and the cylinder bore.

- Cylinder block and the valve plate.

- Trunnion bearings and bore.

- Bearings and shaft.

In the piston pump, there are two areas which tend to be involved in the wear process more than the others. One of these interfaces is the piston shoe/swash plate and the other is cylinder block/valve plate or backplate, if a valve plate is not used in the design. Wear of either of these critical interfaces will result in a loss of efficiency.

FIG. 3 – Component Parts of a Piston Pump

Pump Testing for Fluid Qualification

Wear will vary with the type of hydraulic pump. Whether or not these differences produce a large problem for the fluid formulator is a major problem in developing a single pump test for fluid evaluation and qualification. Generally, it is necessary to evaluate hydraulic fluids in the type of pump where fluid qualification is desired. Until adequate research has been performed, it is not likely that a single pump test will be universally accepted for hydraulic fluid. Most component manufacturers will not approve a hydraulic fluid until they have operational experience with a particular fluid. Therefore, fluid suppliers must produce data with their particular fluid with each and every pump.

There have been several different pump tests used to produce wear data which can be used to qualify various fluids [1]. Some of these tests used vane pumps while others utilized gear pumps and still others used piston pumps. There is no published document comparing the results of one fluid at one set of carefully controlled operating parameters as subjected to the various types of pumps. A typical schematic for the test conducted using the Vickers V-104 vane pump is shown in Fig. 4. This approximates the system normally recommended in ASTM D -2882. This figure shows that the test is a constant pressure test at a fixed temperature. The Vickers V-104 is a very old pump design and attempts have been made to change the pump and the operating conditions to those more representative of current designs.

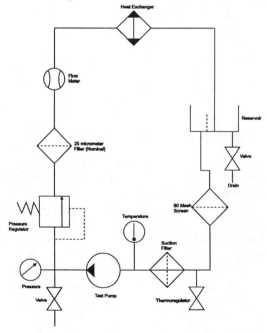

FIG. 4 -- Typical Schematic of Vickers V-104 Vane Pump Test

Gear pumps have been run using a schematic closely approximating that shown in Fig. 4. The typical schematic for testing a piston pump is somewhat different than that of the vane or gear pump to accommodate the case drain and the more sensitive inlet conditions. A typical schematic for a fixed displacement piston pump test is shown in Fig. 5.

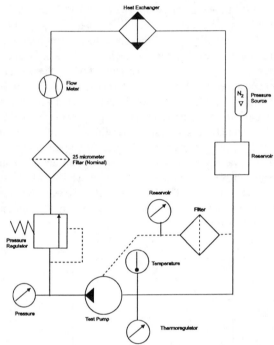

FIG. 5 -- Typical Schematic for Constant Pressure Piston Test

Various pump tests which have been reported in the literature have evaluated hydraulic fluids over a broad range of operating conditions. For example, Perez and Brenner[2] ran different vane pumps at different pressures and temperatures while Racine Fluid Power uses a cyclic pressure test on vane pumps. Toogood[3] conducted tests using gear pumps with both constant and cyclic loading conditions. Many pumps of all three types were tested at the Fluid Power Research Center of Oklahoma State University with controlled contaminant environments [4, 5, 6, & 7]. A large portion of the work reported to date utilized weight change of various parts to measure the amount of wear which occurred. When the literature is fully explored, one will find that there are many reported where one or more of the three types of pumps have been used to evaluate the performance characteristics of hydraulic fluids. Besides the type of pump, operating pressure must be considered along with the manner in which pressure is applied (cyclic or steady). Operating temperature is another parameter of great importance in qualifying hydraulic fluids as well as speed of the pump. If the fluid will protect the pump, it will protect the other components. However, there has been no industry-wide agreement on what type of pump to use and what operating conditions should be applied during the test. Reproducibility and repeatability of the data is also important.

Most of the pump tests used in the evaluation of the anti-wear properties of hydraulic fluids are pass/fail. When the test has been run, parts are analyzed and a determination is made as to whether the amount of wear is acceptable. If there is too much wear the fluid is said to fail the test. Most of these tests do not employ continuous monitoring of the performance characteristics of the pumps. Fluid qualification pump tests should incorporate continuous measurements modeling wear in hydraulic pumps.

Recommendations to Improve Pump Testing for Fluid Qualification

The use of weight loss and dimensional changes to characterize the wear of any hydraulic component does not lend itself readily to the modeling of wear. The thing which is of primary importance is component performance. In the case of the fixed displacement pump the most important performance parameter is efficiency. Most of the work reported which employed performance degradation as a measurement of wear used the flow change at a fixed pressure [7]. Since the pump is primarily a flow generator, this is insufficient. However, if input speed and torque were measured along with output flow and pressure, overall efficiency could be used as a criterion of pump wear. Secondly, to have any chance of accurate measurement when evaluating weight loss, the pump must be disassembled, critical parts must be thoroughly cleaned and initial weight must the measured. Then, after the test is run, the same procedure must be followed. Since pumps must be disassembled, measuring intermediate wear points is not very desirable because disassembling and re-assembling of the pump will cause the position of the various parts to change and therefore potentially alter the wear surfaces relative to each other. Therefore, it is very important that pump tests which are used to measure the performance characteristics of a hydraulic fluid employ continuous performance monitoring. This does not mean that dimensional changes or weight loss can not be used in addition to performance parameters, however, attempts to measure weight loss or dimensional changes at intermediate points during the test should be discouraged.

A second consideration concerns the contamination level of the fluid being tested. There have been reliable estimates made that about 70 % of the field failures of hydraulic systems can be traced to the presence of particulate contamination [7]. However, pump tests used to establish the performance of the hydraulic fluid do not report the contamination level of the fluid nor is there any qualification of the system to assure that a selected contamination level is maintained. This observation leads to two possible considerations. One is that fluids should be clean enough that particulate contamination will not influence test results. However, if particulate contamination is not part of fluid qualification pump tests, then contamination levels of circulating fluids must be measured. A second approach is to conduct the test with a known contamination level. This is very close to the contaminant sensitivity test developed at the Fluid Power Research Center (NFPA T3.9.18). In addition to accounting for the contamination, the contaminant sensitivity test accelerates the test such that useful results can be obtained in a much shorter period of time than in most of the tests currently being used for fluid qualification.

Should hydraulic pump tests be performed using constant or cycled pressure? In the field, the pumps are subjected to a cyclic pressure load. However, how high should the peak pressures be and what cycle rate should be used? Pump endurance tests normally involve cyclic pressure, but there is little agreement concerning the shape and frequency of the pressure curve. In addition, the level of pressure can be used as a means to accelerate the test. Great care should be exercised when using a pressure higher than that recommended by the pump manufacturer. The same thing applies to temperature and rotational speed.

Based upon all of the information at hand, it very doubtful that one pump test can be developed for the qualification of hydraulic fluid. Each pump manufacturer will maintain that the particular pump manufactured must be tested with the candidate fluid before qualification. Pump tests can be made more useful in the modeling of the wear phenomena by using performance parameters for continuously monitoring test situation. Data may be presented in terms of a degradation curve instead of a pass/fail conclusion. The use of cyclic pressure introduces a whole new situation. The equipment necessary to conduct cyclic pressure tests is more complex and will require more expertise and maintenance than constant pressure tests. There is little evidence that cyclic pressure will provide more definitive results. Testing times can be reduced by introducing a known and controlled contamination level. However, contaminant sensitivity is not employed, the presence of contamination should be controlled to the extent that it is measured and reported as part of the test parameters.

References

[1]   Totten, G. E. and Bishop, R. J. Jr. "Hydraulic Pump Testing Procedures to Evaluate Lubrication Performance of Hydraulic Fluids," Paper 952092, SAE Technical Paper Series, International Off-Highway & Powerplant Congress and Exposition, Milwaukee, WI, Sept. 1995.

[2]   Perez, R. J. and Brenner, M. S., "Development of a New Constant Volume Vane Pump Test for Measuring Wear Characteristics of Fluids, *Lubrication Engineering*, May, 1992.

[3]   Toogood, G. J., "The Testing of Hydraulic Pumps and Motors", *Proc. Natl. Conf. Fluid Power*, 37th ,35,(1981).

[4]   Bensch, L. E. and Tessmann, R. K., "Verification of the Pump Contaminant Wear Theory", Paper No. P76-5, *Basic Fluid Power Research Program, Annual Report No. 10,* Oklahoma State University, October 1976.

[5]   Tessmann, R. K., "Assessing Pump Contaminant Wear Through Ferrographic Analysis," *The Basic Fluid Power Research Journal*. 1978, 11:1-7.

[6]   Fitch, E. C. and Tessmann, R. K., "Controlling Contaminant Wear Through Filtration," Third International Tribology Conference, Tribology for the Eighties, Aberdeen, Scotland, September 22-25, 1975.

[7]   Fitch, E. C., "Fluid Contamination Control", FES, Inc., Stillwater, OK, 1988.

J. Matthew Jackson[1] and Steven D. Marty[2]

STANDARDIZED HYDRAULIC FLUID TESTING — AN OVERVIEW AND HISTORY

REFERENCE: Jackson, J. M. and Marty, S. D., "Standardized Hydraulic Fluid Testing—An Overview and History," *Tribology of Hydraulic Pump Testing, ASTM STP 1310*, George E. Totten, Gary H. Kling, and Donald J. Smolenski, Eds., American Society for Testing and Materials, 1996.

ABSTRACT: In order for a hydraulic fluid to be approved by the various pump manufacturers for use in their products, it must qualify for acceptance according to certain test methods. Chief among these tests are the Denison HF-0, the Vickers 35VQ25, and the John Deere Sundstrand piston pump test.

The Denison HF-0 test evaluates hydraulic fluid performance based on flow, deposits, and wear in both a Denison P46 axial piston pump and a Denison T5D vane pump. Both pumps are run for 100 hours and are periodically disassembled for inspection. Obtaining satisfactory pump results is the largest step in the process for a fluid to qualify under the HF-0 Standard. In the past, the P46 and T5D pumps could be tested separately or "bootstrapped" with a 46 series motor to reduce horsepower requirements.

The Vickers 35VQ25 test measures the anti-wear characteristics of a hydraulic oil according to cam ring and vane weight loss in a 35VQ25 vane pump. A minimum of three cartridges are tested for 50 hours each. The first three, or four of five cartridges, must perform satisfactorily in order for a fluid to pass this test.

The John Deere piston pump test screens oils that cause corrosion of copper containing metals. A Sundstrand 22 series axial piston pump is operated for 25 hours.

---

[1]Engineer, Speciality Fluids Section, Automotive Products and Emissions Research Division, Southwest Research Institute, 6220 Culebra Road, P.O. Drawer 28510, San Antonio, Texas 78228-0510.

[2]Research Engineer, Specialty Fluids Section, Automotive Products and Emissions Research Division, Southwest Research Institute, 6220 Culebra Road, P.O. Drawer 28510, San Antonio, Texas 78228-0510.

Water is then added to the test oil. The test is then run for 200 more hours. Over this time multiple parameters are monitored to determine the effect of the added water.

KEYWORDS: piston pump, vane pump, Denison P46, Denison T5D, Vickers 35VQ25, Sundstrand, JDQ-84, testing

---

HISTORICAL PERSPECTIVE–DENISON HF-0

The Denison HF-0 specification came into being as a result of fluid-related field problems in Denison's vane and piston pumps. In the early 1980's, a major aircraft manufacturer began using a hydraulic fluid containing zinc-based antiwear additives in one of its wing stress test fixtures. The test rig employed both Denison vane and piston pumps in its hydraulic system. The antiwear fluid performed well in the vane pumps, but caused severe failures in the piston pumps. In response to these field phenomena, Denison developed the hydraulic fluid test that led to the HF-0 specification.

The original SwRI test stand was driven by an 8V92 Detroit Diesel engine. The rig was bootstrapped to lower its input power requirements using a four-pad gearbox coupled to a Denison P46 piston pump, a Denison T5D vane pump, a Denison M46 piston motor, and a Denison T1C vane pump (used as a charge pump). The stand configuration was designed to be similar to Denison's test stand. Bootstrapping involves the conversion of some of the hydraulic fluid energy to mechanical energy. This is done by running some of the high pressure flow through a hydraulic motor. The motor, which is mounted ot the same gearbox as the pump, then returns mechanical energy to the system.

Early hardware and stand configuration problems plagued SwRI's first Denison test rig. Pistons and piston shoes in the P46 pump occasionally separated. Denison resolved the problem by modifying the crimping process used to secure the pistons and shoes. There were multiple configuration problems. The P46 pump and M46 motor both loaded the same side of the gearbox center bearing, causing it to fail prematurely. This was corrected by changing the pump/motor positons on the gearbox. Maintaining a positive pressure T5D inlet condition free of entrained air was difficult. In an attempt to address this problem, the reservoir was pressurized with nitrogen. This solved the inlet pressure problem, but may have introduced nitrogen into the hydraulic fluid. This method was used for a short time, but was rejected in favor of a hydraulically pressurized reservoir.

As testing progressed, a second HF-0 test stand was purchased by SwRI from FMC. This second stand eventually replaced the original test stand. The second stand was later destroyed by fire in August of 1994. It has since been replaced by separate Denison P46 piston pump and T5D vane pump test rigs. The test pumps were separated under Denison's diretion, since it was determined at the March 1994 hydraulics industry meeting (at the Denison plant in Marysville, Ohio) that SwRI was the only lab running the test pumps simultaneously, as Denison had originally done.

TEST CONFIGURATION–HF-O

The SwRI Denison T5D vane pump test circuit is shown in Figure 1. The main components in the circuit are a test pump, a high pressure pilot-operated relief valve, dual canister 3-micron filters, a heat exchanger, an 80-gallon main reservoir, a 5-gallon secondary reservoir, a 3-gpm gear pump and a 3-15 psig manual relief valve. The 80-gallon reservoir is pressurized via the gear pump and low pressure relief valve so the test pump inlet is maintained at a small positive pressure. The 5-gallon reservoir is at atmospheric pressure and allows for thermal expansion of the test fluid. A benefit of this configuration is that the small gear pump can push fluid throughout the system, purging any

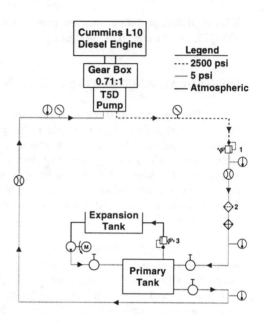

Fig. 1--Denison T5D vane pump test circuit

air in the lines or in the test pump before the main pump is engaged. Also, this configuration does not require that the reservoir be located above the test pump. A limitation of this configuration is that the relatively small volume of fluid in the small reservoir is not circulated as rapidly as the fluid in the 80-gallon reservoir. The test fluid is also subjected to the low flow gear pump. Lastly, this design does not allow for flushing with small quantities of fluid. The reservoir must be filled to the same level for flushing as for testing. It is felt that this configuration is better than either applying a nitrogen blanket in the reservoir or simply elevating the reservoir. Unlike the nitrogen blanket, the gear pump does not push any gas into solution, which would likely result in increased cavitation within the pump. Simply elevating the reservoir does not provide a way to purge all of the air from the circuit before engaging the test pump. The initial wear that might occur in the test pump caused by air entrained in the hydraulic fluid could ruin a test from the beginning.

The SwRI Denison P-46 test stand layout is shown in Figure 2. This circuit configuration is mandated by Denison. The main components in the circuit include the test piston pump, a vane charge pump, a piston motor, a high pressure pilot-operated relief valve, a low pressure manually operated relief valve, a heat exchanger, dual canister 3 micron filters, and a surge suppressor. The test stand also uses an 80-gallon reservoir with the same configuration previously described for the T5D vane pump. The 80-gallon reservoir is pressurized in order to maintain a small positive pressure at the charge pump inlet. The charge pump (a Vickers 35VQ25 vane pump) then pressurizes the inlet to the test pump to 125 psig via the low-pressure manual relief valve. The flow to the test pump inlet is also passed through the pressure attenuation device, which reduces anomalous pressure spikes. The test pump pressurizes the fluid to 5000 psi, and the high-pressure flow is routed in two directions. Approximately 15 percent of the flow is directed across the relief valve (this provides the necessary heating for the circuit) and the other 85 percent of the flow is directed to the piston motor. The piston motor is coupled to the same gearbox as the test pump and charge pump. The piston motor returns power to the gearbox, thus reducing the total power needed to drive the test stand.

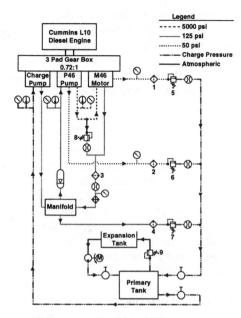

Fig. 2--Denison P46 piston pump test circuit

COMMON MODES OF FAILURE

The Denison T5D and P46 tests have come to be known as "no harm" tests. Wear or distress other than the normal polishing of sliding parts is of great concern.

Denison T5D Vane Pump:

In the T5D vane pump test, the original grinding marks are wavy in nature (See Figure 3). The vanes in the pump have both a leading and trailing edge that contact the cam ring surface. The test hardware is shown in Figure 4. During one rotor revolution, the vane/cam ring contact surface changes from leading vane edge to trailing vane edge four times. The location where these change-overs occur are called step-over points. The unpolished cam surface is visible at these points. Each step-over point

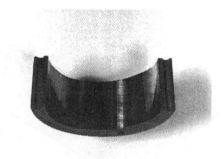

Fig. 3--Denison T5D cam ring grinding marks

appears as a mark across the cam ring, but is actually untouched cam surface. These change-over areas are diametrically opposed on the cam ring surface. These markings occur in normal pump operation and cause little concern.

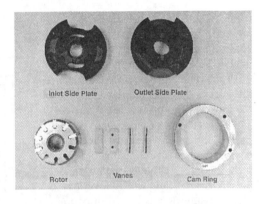

Fig. 4--Denison T5D test hardware

One common area of distress on the cam surface is found at each pressure ramp. The pressure ramps are the areas of the cam surface nearest the rotor. At these locations the leading edge of the vane is in contact with the cam surface. As the cam surface approaches the rotor, the fluid is compressed. The common failure modes at the pressure ramps are pitting and "chatter" marks.

In addition to pressure ramp distress, "streaking" or "smearing" can occur anywhere on the cam ring surface. This streaking can be described as a smearing of the cam ring metal, presumably due to poor antiwear properties of the hydraulic fluid.

Vane tip wear is a general indicator of overall pump wear. The leading and trailing edges of the vanes in the Denison T5D vane pump are chamfered as sketched in Figure 5. When inspecting a new vane, the length of the chamfered area is difficult to determine because the chamfer flows smoothly toward the top of the vane. After the pump has been tested, the cam/vane lip interface may develop its own chamfer. This is shown in Figure 5. In general, the larger this wear chamfer, the more the cam surface will be affected.

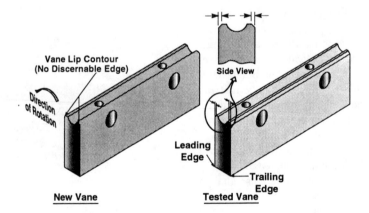

Fig. 5--Denison T5D vane lip contours

The rotor, rotor slots, and vane pins show little evidence of wear even with the most marginal of fluids. With "passing" fluids, the inlet and outlet side plates generally show slight circular scratching, due to random contact with the sides of the vanes. Occasionally the press fit of the splined section of the rotor will be slightly misaligned. If this occurs or the splines on the driving shaft are damaged, the rotor will be misaligned in the pump cartridge. As the vanes rotate, they will scratch or groove the side plates. This is a hardware or configuration problem, not a fluid-related phenomenon.

<u>Denison P46 Piston Pump</u>:

The most common mode of fluid failure is yellow metal transfer from the piston shoes to the creep/swash plate. The longstanding generality has been that zinc-based antiwear additives that work well at protecting the cam/vane interface in the vane pump lead to yellow metal transfer onto the creep plate in the piston pump due to their reaction with the copper containing metals. Historically, rust- and oxidation-inhibited hydraulic oils (R&O oils) have produced acceptable test results. The same R&O oil should have unsatisfactory results in the T5D Vane Pump test, due to its lack of antiwear additives.

The main components of the rotating group of the P-46 piston pump consist of the swash plate, seven pistons and shoes, piston shoe retainer, face plate, port plate and a barrel that holds the seven pistons. These parts are shown in Figure 6. During normal pump operation, the piston shoes slide against the creep plate which is held at an angle. As the assembly rotates, the angle of the creep plate causes the pistons to stroke in the barrel. During operation, the creep plate should slowly rotate (due to the frictional forces from the sliding piston shoes on it) on the cam that provides its angular orientation. Hence the name "creep" plate. If the creep plate does not rotate normally, the plate may show uneven wear/distress (see Figure 7). The inward motion

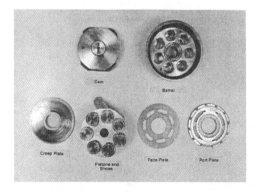

Fig. 6--Denison P46 test hardware

Fig. 7--P46 creep plate with uneven wear

of the pistons is the pressure stroke. This puts the highest normal forces on the creep plate. If the creep plate does not rotate, the high forces on the piston shoes only act on one half of the plate and may show as uneven wear.

HISTORICAL PERSPECTIVE - VICKERS 35VQ25

In the late 1970's Southwest Research Institute installed its first Vickers 35VQ25 vane pump test stand at the request of a hydraulic fluid additive supplier. The additive company and Vickers were both involved in vane pump testing at the time and desired an unbiased third party to conduct similar work. At one point, the Institute operated two 35VQ25 stands simultaneously in order to meet testing demand. As the level of testing decreased, one of the stands was disassembled. The remaining test rig was destroyed by fire in August of 1994, but has since been replaced by a new 35VQ25 test stand. The test itself is detailed in the *Pump Test Procedure for Evaluation of Antiwear Fluids for Mobile Systems,* published by Vickers.

TEST CONFIGURATION - 35VQ25

The test configuration is shown in Figure 8. The test hardware is shown in Figure 9. The pamphlet available from Vickers states that there are two ways to fail the test; weight loss and appearance. A minimum of three test cartridges must be run for 50 hours each. After the 50 hours, the cartridge is cleaned and weighed. The weight loss of the cam ring must be less than 75 mg, and the combined weight loss of the vanes must be less than 15 mg. If for any reason one of the first three cartridges does not meet the weight loss or visual criteria as outlined by Vickers, another two cartridges should be run. If they perform acceptably, four of the five tested cartridges must meet

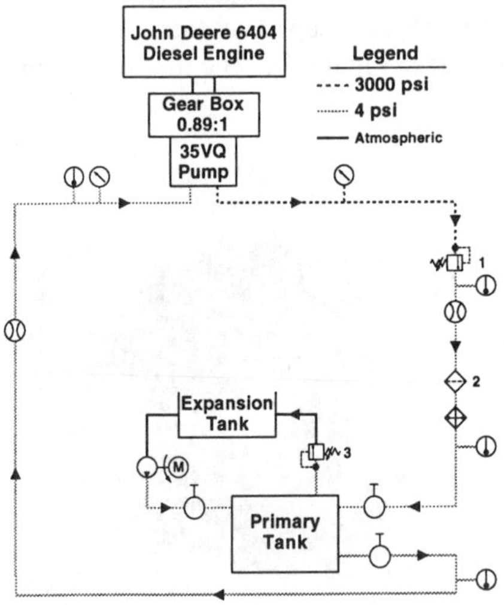

Fig. 8--Vickers 35VQ25 vane pump test circuit

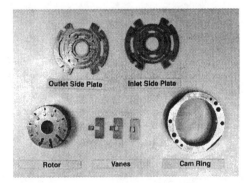

Fig. 9--Vickers 35VQ25 test hardware

accept/reject guidelines. It is noted in the pamphlet that "This procedure is offered only as a fluid screening method. Successful completion of this test does not constitute endorsement or approval of fluids by Vickers."

COMMON MODES OF FAILURE

The most common failure mode observed for this test is for the ring and vanes not to meet the weight loss criteria. It has been SwRI's experience that the levels of visual distress depicted in the pamphlet have only occurred in conjunction with severe weight loss and wear.

HISTORICAL PERSPECTIVE-SUNDSTRAND

Standardized hydraulic fluid testing began at Southwest Research with the installation of the Institute's first Sundstrand test stand in 1977. Based on a test stand in existence at another testing facility at the time, the hydrostatic configuration employed a 350-cid Chevrolet V8 engine as its prime mover. (The Chevrolet engine was later replaced by a Ford V8, which in turn was replaced by the current John Deere 6.9L six-cylinder turbocharged diesel engine.) The 22-Series Sundstrand pump used in that stand is the same series currently in use today, with some notable hardware differences. In the late 1970's, the bearing plates used in the Sundstrand pumps were made of bronze, as opposed to the bimetallic bearing plates in use today. The switch was made to bimetallic plates due to the fact that they provide longer service life and eliminate the galling problems associated with the bronze plates. A makeup flow loop was later added to the test circuit to allow monitoring of flow degradation.

The purpose of the JDQ-84 Sundstrand Dynamic Corrosion Test is to assess a fluid's performance after it has been contaminated with a small amount of water. The hydrostatic transmissions and hydraulic pumps found on agricultural and industrial off-highway equipment are often faced with the problem of water contamination from rainfall or other sources. The reaction between the water and the additives in the hydraulic fluid tends to cause the corrosion of the softer yellow metals that make up some of the vital internal components, resulting in flow degradation and loss in volumetric efficiency.

TEST CONFIGURATION - SUNDSTRAND

The Sundstrand test stand currently in use at SwRI employs a John Deere diesel engine to drive a speed-increasing gearbox which provides the 3100 rpm input speed to the pump at 1850 rpm. (The test circuit is shown in Figure 10.) The Sundstrand test pump is rebuilt prior to each test and all components are inspected prior to use. Those not found in good condition are replaced, along with the cylinder block kit, valve plate, and seals, which are replaced for each test regardless of condition. The

test hardware is shown in Figure 11.
The test circuit is drained and flushed
prior to each test with the fluid to be
tested in order to remove any traces of
previously tested fluids. Following the
25-hour wear-in period, 1% water (by
volume) is blended with the test fluid in
order to simulate water contamination
experienced in field applications. The
test is then continued for an additional
200 hours at a load pressure of 5000 psi,
main loop temperature of 180°F, and a
flow rate of 24 gpm in accordance with
the John Deere J20C and J20D
specifications. Flow readings are taken
throughout the duration of the test to
assess flow degradation.

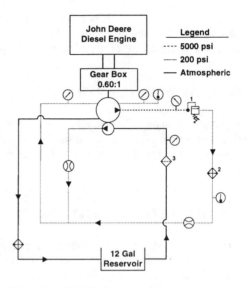

Fig. 10--JDQ-84 Sundstrand piston pump
test circuit

COMMON MODES OF FAILURE

Component condition, post-test fluid
composition, and flow degradation
are all factors used to determine fluid
performance in the Sundstrand test.
Pump components in good condition
at the conclusion of the test will
appear darkened from a protective
oxide layer that forms on the
component surface. Those in poor
condition will appear etched or
polished and usually exhibit
measurable amounts of material
wear. When the switch was made to
bimetallic bearing plates, the
dynamic corrosion problem moved to
the brass-lined piston bores. A

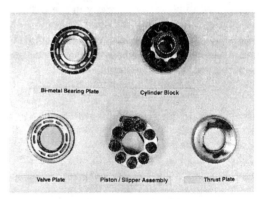

Fig. 11--Sundstrand pump test hardware

poorly performing fluid may result in a bore increase of 0.001 to 0.002 inch. This
typically results in discernable leakage from the hydrostatic loop to the case drain.
This leakage is expressed in terms of percent flow degradation, which can be used to
approximate the loss in volumetric efficiency of the pump during the course of the
test. Flow degradation of 10-15 percent (or more) is generally considered a failure.
Pass/fail decisions concerning the qualification of a fluid evaluated using the
Sundstrand test may be made only by John Deere. Chemical analyses will show
increased levels of copper in a poorly performing fluid, confirming the dynamic
corrosion of the copper-containing pump components.

Sundstrand test may be made only by John Deere. Chemical analyses will show increased levels of copper in a poorly performing fluid, confirming the dynamic corrosion of the copper-containing pump components.

SUMMARY

This paper discussed the procedures involved in conducting the aforementioned tests, as well as the operating conditions that must exist and specifications that must be met by each test in order for hydraulic fluids to qualify for industry acceptance. Typical failure modes for each test were presented, along with some general historical background concerning test development and operation. It is hoped that those seeking an increased knowledge of these hydraulic pump test methods found this paper of significant interest.

BIBLIOGRAPHY

1.   Brownlow, Amond D., Manager (retired), Fleet, Gear Oil and Hydraulics Fluids Section of SwRI, Floresville, TX, Personal Interview, September 29, 1995.

2.   "Denison Pump Test Procedure for the HF-O Standard (Petroleum Base Oil with Anti-Wear Additive for Severe Duty)," Marysville, Ohio, Revised February 2, 1980. Pages 1-7.

3.   Garcia, John A., Manager, Lubrizol San Antonio Liason Office, San Antonio, TX, Personal Interview, October 10, 1995.

4.   Lochte, Michael D., Senior Research Engineer, Specialty Fluids Section of SwRI, San Antonio, TX, Personal Interview, November 2, 1995.

5.   McLeod, Kenneth L., Senior Research Engineer, Oronite Additives - San Antonio Test Group, San Antonio, TX, Personal Interview, October 19, 1995.

6.   Moreland, Gary G., Assistant Manager, Fleet and Specialty Fluids Sections of SwRI, San Antonio, TX, Personal Interview, September 29, 1995.

7.   "Pump Test Procedure for Evaluation of Anti-wear Fluids for Mobil Systems, Vickers Incorporated, Troy, MI, Revised August 1988.

8.   **Sauer-Sundstrand Series 20 Axial Piston Pumps and Motors Repair Manual.** Sauer-Sundstrand, Ames, IA, February 1990.

9.   **Vickers Industrial Hydraulics Manual.** Copyright 1992. Rochester Hills, MI, pages 17-1 thru 17-28.

APPENDIX - HYDRAULIC SYMBOLS

1. Filter (3 Micron)
2. Filter (3 Micron)
3. Filter (3 Micron)
4. Filter (3 Micron)
5. Pressure Relief Valve (50 psi)
6. Pressure Relief Valve (50 psi)
7. Pressure Relief Valve (125 psi)
8. Pressure Relief Valve (5000 psi)
9. Pressure Relief Valve (Tank Pressure)

Gate Valve

Flow Meter

Thermocouple

Pressure Gauge

Filter

Heat Exchanger

(M) Electric Motor

Gear Pump

Gas Charged
Shock Supressor

Pressure Relief Valve

Gear Pump

Reservoir, Vented

Reservoir, Pressurized

Denison P46 Hydraulic Pump
        Inlet Pressure: 125 ± 10 psi
        Outlet Pressure: 5000 ± 100 psi
Case Drains: 50 ± 5 psi

1. Pressure Relief Valve (3000 psi)
2. Filter (3 Micron)
3. Pressure Relief Valve (4 psi)

 Gate Valve

 Flow Meter

Thermocouple

Pressure Gauge

 Filter

 Heat Exchanger

(M) Electric Motor

Gear Pump

Pressure Relief Valve

Vickers 35VQ25 Hydraulic Pump
Inlet Pressure: 4 ± 1 psi
Outlet Pressure: 3000 ± 25 psi

1. Pressure Relief Valve (2500 psi)
2. Filter (3 Micron)
3. Pressure Relief Valve (5 psi)

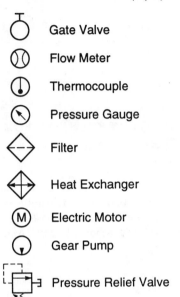

Gate Valve

Flow Meter

Thermocouple

Pressure Gauge

Filter

Heat Exchanger

Electric Motor

Gear Pump

Pressure Relief Valve

Denison T5D Hydraulic Pump
        Inlet Pressure: 5 ± 1 psi
        Outlet Pressure: 2500 ± 25 psi

Hans M. Melief [1]

PROPOSED HYDRAULIC PUMP TESTING FOR HYDRAULIC FLUID QUALIFICATION

REFERENCE: Melief, H. M., **"Proposed Hydraulic Pump Testing for Hydraulic Fluid Qualification,"** *Tribology of Hydraulic Pump Testing, ASTM STP 1310,* George E. Totten, Gary H. Kling, and Donald J. Smolenski, Eds., American Society for Testing and Materials, 1996.

ABSTRACT: The current ASTM D-2882 hydraulic vane pump test does not provide the necessary correlation required for the prediction of the lubricating properties of a hydraulic fluid in various piston pump operations. All too often, a fluid will exhibit excellent wear properties in the Vickers V-104 vane pump used in the ASTM D-2882 test, yet produce catastrophic failure at various wear interfaces in a piston pump which may consist of different material pairs, contact loading, configuration or speed. In this paper, a new piston pump test, which is conducted under cycled pressure testing conditions, is proposed. The new test will provide an excellent assessment of the lubricating properties of a hydraulic fluid under a wide variety of wear conditions.

KEYWORDS: hydraulic fluid, piston pump test, lubrication

---

1) The Rexroth Corporation, Industrial Hydraulics Division, P.O. Box 25407 Leheigh Valley, PA 18002-5407.

INTRODUCTION

Mineral oil based hydraulic fluids are excellent lubricants and have served the fluid power industry well for many years. However, there are currently various changes underway in the marketplace which will dictate the use of alternative hydraulic fluid media. For example, the new testing protocol being introduced by Factory Mutual Research Corporation will require reconsideration of the fire-resistant hydraulic fluid being used if the user is to continue to receive optimal insurance rates [1]. In addition, there is also pressure within the fluid power industry to convert to hydraulic fluids that are more biodegradable than mineral oils, such as vegetable oil based hydraulic fluids [2,3]. By the year 2000, the use of mineral oil based hydraulic fluids will be drastically reduced due primarily to safety, disposal and environmental considerations.

These factors have created much market turbulence since current hydraulic components are often designed to be used with mineral oils and may not operate at their full rating with alternative fluids such as vegetable oil, or water-glycol based hydraulic fluids. This problem is exacerbated by user requests to use these, and other alternative fluids, in equipment such as axial piston pumps since performance standards for these fluids typically do not exist. The unavailability of performance standards has led to:

- Indiscriminate use of alternative fluids without hydraulic component manufacturer input or approval,

- Inability of a component manufacturer to provide satisfactory guidance on the use of alternative fluids due to either, the manufacturers non-familiarity with the fluid, or the unavailability of adequate performance standards, or both.

- Incorrect assumption that all fluids of a certain type, for example HFC, perform equivalently which has led to performance problems when this assumption was found to be untrue in practice.

Currently the component manufacturer has two available choices for granting hydraulic fluid use approvals. One is to grant an approval based on available field experience data. The second is to conduct an application-specific test appropriate for the component in question. However, these practices are no longer acceptable because of: product liability risk, application safety risks, resulting production downtime, and the cost of conducting appropriate tests. The unacceptability of these risks is further validated by increasing warranty costs, particularly those associated with the use of non-mineral oil hydraulic fluids.

The objective of this paper is to describe a standard piston pump testing procedure that will:

1. Provide a "standard fluid-suitability test", that will be appropriate for use by component manufacturers, to grant hydraulic fluid use approvals for axial piston equipment,

2. Reduce the burden of defining, specifying and requiring the test, that currently is residing solely with the axial piston equipment manufacturer,

3. To provide a definition of a standarized test stand and test conditions for fluid evaluations,

4. Make available a clear definition of pass or fail criteria and clasification,

5. Be recognized and accepted by ASTM, VDMA, ISO, SAE and DIN. This would create a worldwide standard.

DISCUSSION

A. Why A Piston Pump Test?

Currently most of the recognized pump tests are vane pump tests that are operated in the medium pressure range. The sliding vane-on-ring wear contact typified by the vane pump test is significantly different from the rolling contacts characterizing some of the wear in a piston pump. In addition, the medium pressure tests provide inconclusive results relative to the performance of a fluid at high pressures. Therefore, a piston pump test best models the overall wear performance of a hydraulic fluid in a piston pump.

B. What Hydraulic Fluids Should Be Included?

In principle, the piston pump test that is being proposed is suitable to evaluate the performance of all classes of hydraulic fluids. However, for mineral oil based hydraulic fluids, a piston pump test may not be prioritized since there is considerable data available that shows many easier to conduct, less expensive bench tests will provide an acceptable correlation with expected field experience. Therefore, the initial emphasis of test development should be on alternative, non-mineral oil based hydraulic fluids where such correlation do not exist and, in fact, actual pump test or application data is unavailable. These alternative fluids include:

- Environmental Evaluated Fluids (EE-Fluids),
- Special Application Fluids such as Food Grade Fluids, and
- Fire Resistant (HF) Fluids.

C. Standard Fluid Suitability Test

The proposed standard fluid suitability test is based on a Rexroth A4VSO 125 DR tandem pump arrangement which has a maximum displacement each of 125 cc/rev (7.63 in^3/rev) using the schematic as in Fig. 1. An illustration of the A4VSO pump is provided in Fig.2. This tandem pump arrangement is driven by a 45 kW (60 hp) prime mover at 1500/1800 r/min. The regenerative drive power is 130 kW (173 hp). This allows load pressures up to 345 bar (5000 psi). The minimum auxiliary hardware includes a load valve, a test fluid reservoir of 150 L (40 gallon), a cooler, and a filter with a $\beta_{10} \geq 100$ rating. The resulting filtration level is be a NAS class 9/ISO 18/15. The fluid quantity required per test is one standard size drum, 150 - 200 L (40-52 gallon). Monitoring equipment can be added for continuous operation.

D. Test Procedure
1. Fluid Characterization, Pre - Pump Test.

The first part of the test sequence is to establish a fluid qualification protocol, and verify the physical properties of the hydraulic fluid being tested. In addition, the fluid formulation specifications will be obtained and ability of the fluid being tested to meet these specifications will be verified. Fluid seal compatibility and corrosion resistance information will be obtained. The required formulation and compatibility information is:

1. Seal compatibility is to be clarified by the fluid and seal manufacturer.

2. Fluid compatibility with respect to: lead, zinc, tin, copper, brass, and special DU-bushing material shall be obtained. For non-aqueous fluids, test shall be completed with small quantities (<2%) of water present.

3. Compatibility with bonding materials, paints and plastic bearing cages shall be established.

4. The filterability characteristics shall be determined.

5. Air-release and corrosion test data shall be obtained.

6. Miscibility tolerance with HLP, HLP-D and preservation oils shall be determined.

7. Bearing life test with roller bearings shall be conducted.

8. Aging characteristics, without and with up to 2% water shall be determined.

9. Where possible, the fluid formulation shall be determined by infra-red spectroscopy.

10. .The pressure and temperature capabilities are to be defined by the fluid manufacturer.

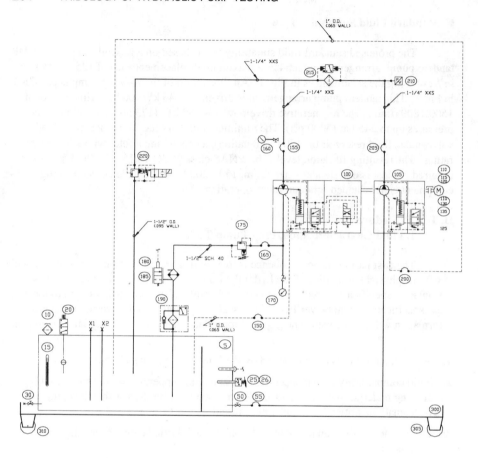

Fig. 1 -- Proposed "Standard Fluid-Suitability Test" Axial Piston Pump Test Stand

Item 100, 105 Tandem pump assembly   Item 220 Load valve
Item 190, 215 Fluid filters   Item 175 Pre load valve
Item 180, 185 Cooler assembly   Item 5 Test fluid reservoir
Item 110 Prime mover, E-motor assy.

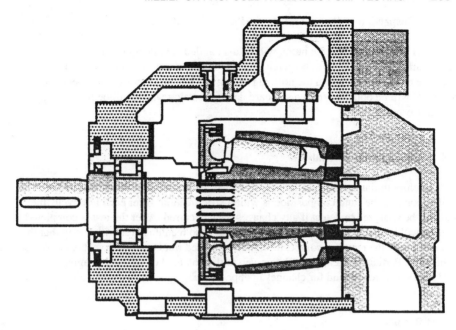

Fig. 2 -- Section View of a Rexroth A4VSO Axial Piston Pump

2. Pump Testing Sequences

The continuous load portion of the test will be conducted at maximum operating pressure and temperature and at the minimum viscosity specified for the fluid type being tested. The test duration will be 250 hours at which time the pump will be dismantled and visually inspected.

A second portion of the test sequence will include a pulsed pressure operation of the pump at maximum displacement. The load pressure will vary from 0 to the maximum load at maximum temperature and minimum viscosity as defined by the fluid specification. The total test duration will be 1 million cycles at 2-3 Hz (100-150 hours) at which time the pump will be dismantled and inspected.

The pump will also be operated in a variable displacement mode with the displacement varying from V_{min} to V_{max} at maximum pressure., maximum temperature and minimum viscosity permissible by the fluid specification. The total test duration will be 1 million cycles, 1 cycle/s, taking 280 hours, at which point the pump will be dismantled and visually inspected.

The cycle times are to be selected to obtain a total running time of approximately 3 weeks.

E. Visual Inspection of Pump Test Parts (Pass/Fail Criteria)

As indicated above, the pump will be dismantled and inspected after each test cycle. Particular attention will be given to the contact and running areas where no wear should be observed other than the normal running marks. Indications of the degree of "pass/fail" will be obtained from a standard damage catalog which will be provided by Rexroth Corporation.

F. New and Used Fluid Analyses

Before initiating the test, the new fluid was analyzed by the fluid manufacturer for certification that the fluid composition is within specification. After each test cycle, the fluid will be analyzed to detect if any changes have occurred. After the test is completed, the fluid will be analyzed again at which point the fluid manufacturer will certify that the fluid still meets the required composition limits for used fluids and is, in fact, suitable for further use. Particular attention will be given to the potential of significant aging (oxidation) occurrences and for changes in appearance.

G. Test Results

At the conclusion of the "Standard Fluid Suitability Test", the fluid will be rated a "Pass" if : 1. the pump is still fully functional, 2. the wear assessment meets the pre-established wear limits as described in the Standard Damage Assessment Catalog and 3. the used fluid at the conclusion of the test meets the required use specifications.

CONCLUSIONS

In this paper, a Standard Fluid Suitability Test for axial piston equipment has been proposed. In addition to an overview of the test, new and used fluid characterization and the establishment of a pass/fail criteria using a Standard Damage Assessment Catalog were described. The application of this standard test protocol will be of great value in addressing the ever-increasing problems related to the granting of alternative hydraulic fluid approvals for use in standard axial piston equipment. The objective of this test protocol is for the development of an international standard axial piston pump fluid performance classification and approval test.

REFERENCES

[1] M.M. Khan and A.V. Brandao, "Method for Testing the Spray Flammability of Hydraulic Fluids", SAE Technical paper Series, Paper No. 921737, 1992.

[2] R.A. Padavich and L. Honary, "A Market Research and Analysis Report on Vegetable-Based Industrial Lubricants", SAE Technical Paper No. 952077, 1995.

[3] T. Mang, "Environmentally Friendly Biodegradable Lube Base Oils - Technical and Environmental Trends in the European Market", Adv. Prod. Appl. Lube Base Stocks, Proc. Int. Symp., 1994, pp 66-80.

Karl-Heinz Witte, [1] and David K. Wills, [2]

Tribological Experiences of an Axial Piston Pump and Motor Manufacturer with Todays Available Biodegradable Fluids

REFERENCE: Witte, K.-H. and Wills, D. K., "Tribological Experiences of an Axial Piston Pump and Motor Manufacturer with Todays Available Biodegradable Fluids," *Tribology of Hydraulic Pump Testing, ASTM STP 1310,* George E. Totten, Gary H. Kling, and Donald J. Smolenski, Eds., American Society for Testing and Materials, 1996.

ABSTRACT: Since the late 1970's, biodegradable fluids have been used in hydrostatics with limited success. Performance has not matched expectations during extensive field testing. For example, fluids reported as having excellent lubricating properties, by bench and field testing, produced excessive wear. Design modifications successfully resolved some of these problems, but wear problems with sealing and bearing surfaces still exist, suggesting that fluid properties may need to be modified. New test procedures measureing wear e.g. may need to be developed as well because of the correlation problem between reported fluid properties and axial piston pump performance like good lubricity resulted in excessive wear.

KEYWORDS: biodegradable fluids, needle bearing wear, radial lip seal shaft wear, friction coefficient, rape seed fluid, synthetic ester, spline wear, DU-Bearing wear

Biodegradable fluids have been used in hydrostatics for several years now. Fluid performance has been satisfactory in many applications, but has been lacking in many others. Numerous field and laboratory tests have provided valuable experience, regarding excessive wear of transmission components.

[1] Manager, Technical Sales Literature, Fluid Specialist, Sauer-Sundstrand GmbH & Co, D-24531 Neumünster, Germany

[2] Chief Metallurgist, Test Lab and Metallurgy Testing, Sauer-Sundstrand Company, 2800 East 13th Street, Ames, IOWA 50010, USA

Experience

<u>Bearing plate</u>

A rapeseed fluid was applied in a snowgroming machine used to groom ski trails on steep mountain slopes (picture 1).

Picture 1: Snow Groomer

The propel drive of this vehicle consists of a diesel engine and a splitter box which drive two series 20 variable displacement axial piston pumps (Pictures 2 + 3). Each pump powers a series 20 fixed displacement axial piston motor which, through planetary gearboxes, drives the tracks. The maximum pump speed and pressures are 2800 rpm and 420 bar (6000 psi) with average pressures of 250 to 300 bar (3600 - 4300 psi).

Poor performance was reported after less than 100 hours of operation. Subsequent examination of the pumps and motors determined the bearing plates to have heavy wear in the slot used to locate and time the bearing plate. Picture 4 shows a cross section of the pump with the bearing plate being a part of the "cylinder block assembly". Picture 5 shows the details of the "cylinder block assembly" in which a ring pilots the bearing plate while a pin times it. Picture 6 shows excessive wear of the slot which led to incorrect timing, resulting in poor performance. For comparison purposes, a bearing plate slot without excessive wear is shown in Picture 7.

Picture 2: Plumbing Installation (Variable Displacement Pump - Fixed Displacement Motor)

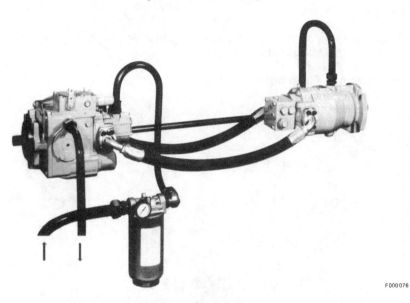

F000076

Picture 3: System Circuit Description

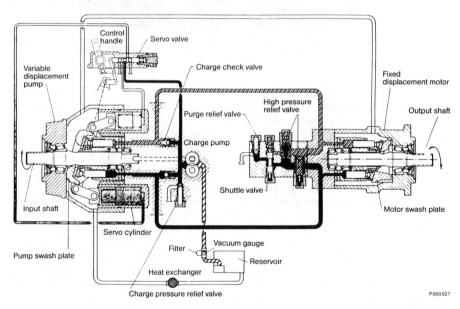

P000027

Picture 4: Sectional view - Variable Displacement Pump

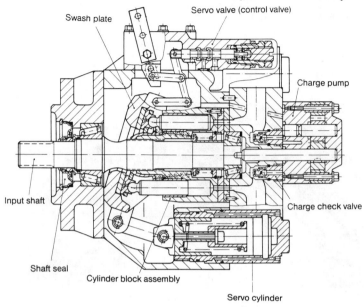

Picture 5: Rotating group assembly

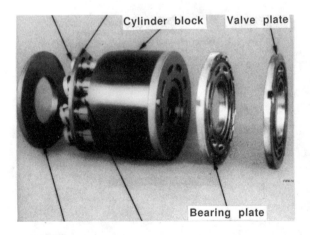

Picture 6: Bearing plate - rape seed fluid

Picture 7: Bearing plate without wear

An investigation into the cause of the wear from the relative motion between the bearing plate and the pin ensued. The rotating cylinder block accelerates and decelerates each time the piston changes from high pressure to low pressure which occurs multiple times per revolution. The bearing plate pin is designed only to locate and time the bearing plate and not to transfer torque. The frictional forces at the interface between the cylinder block and bearing plate have apparently been compatible with the friction properties of mineral based fluids allowing the bearing plate to follow the acceleration and deceleration of the cylinder block. The apparent unique lubricating properties -low coefficient of friction - of rape seed fluid allows the bearing plate to slip on the cylinder block and impact the pin producing excessive wear. For the pump, which rotates in one direction, wear occurred on one side of the groove. For the motor which changes direction when the machine is operated forward and backward, wear occurred on both sides of the groove.

Alleviation of the excessive slot wear by increasing surface hardness was considered to be inadequate because of the magnitude of the impact forces. The interface between the cylinder block and bearing plate were therefore modified by eliminating relative slip between the bearing plate and the cylinder block. Since applying this modification to the snow groomer, the problem has been eliminated.

The same problem existed in a combine application (picture 8) but with a different rapeseed base fluid. The modification sucessfully eliminated this problem as well.

Picture 8: Combine

Needle bearing

Cyclic flywheel tests with a series 51 bent axis motor (picture 9) and a series 90 axial piston pump were run at the Technical University Hamburg Harburg (TUHH) [1] in Germany. The test conditions were: forward and reverse, speeds up to 3700 rpm, system pressure relief valve setting of 450 bar (6400 psi), reservoir temperature 90°C (70°C for rapeseed based fluid), and 300 hours total test time.

After completion of one test with a rapeseed base fluid, the centering shaft of the cylinder block and the needle bearing experienced significant pitting and spalling. The cause and significance of this failure was not apparent until a similar failure occurred during development testing of a new product in the USA using rapeseed base fluid. This product had been sucessfully tested previously using a mineral base fluid. For this new product, the solution to the problem was to change to a ball bearing.

During subsequent testing at the TUHH, the tests were interrupted every 50 hours for examination. After 100 hours the needle bearing and the shaft were replaced because the shaft showed excessive pitting and spalling.

After another 100 hours the problem repeated itself. A solution to this wear problem was not as easy as redesigning the bearings and remains under investigation.

Picture 9: Sectional view - Variable Displacement Motor

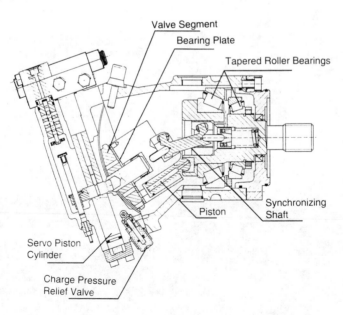

DU-bearing wear

A test set up in England [2] to test the compatibility of biodegradable fluids with gear pumps consisted of a gear pump (picture 10) running against a high pressure relief valve (HPRV) set at 230 bar (3300 psi) for a period of 500 hours. The gear pump is driven at 3000 rpm and the reservoir temperature is controlled to approximately 80°C. The gear pump shaft uses a DU-bearing, whose location is shown in the picture.

During testing of several synthetic ester fluids, the DU-bearing, as with mineral based fluids, did not show any appreciable wear. But one synthetic ester fluid did show after less than 100 hours, severe wear of the Teflon (PTFE) coating. The PTFE-coating was nearly completely removed in the load area.

It was not immediately apparent wether this was mechanical wear or wear accelerated by corrosion. In any case, it fits in the "wear" box.

This accelerated DU-bearing wear with synthetic esters has also seen in testing at TUHH. Under the same test conditions, a mineral based oil did not show any wear.

Picture 10: Sectional view - Gear Pump

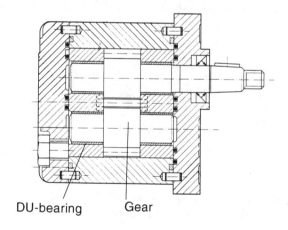

DU-bearing   Gear

Radial lip seal and shaft spline

A second snow groomer application using a series 90 axial piston pump design shown in picture 11 experienced wear at the pump shaft underneath the lip seal. The shaft had run 1600 hours at 2800 rpm maximum speed with synthetic ester. Picture 12 illustrates the measured roughness of the shaft surface. The lip seal has cut a 20 μm (.0008 inches) deep groove into the shaft. On the left and right hand side of the groove the shaft material has actually been deformed. A similar test run for 1800 hours at 1500 rpm with mineral base fluid had no indication of wear (see picture 13).

Picture 11: Sectional view - Variable Displacement Pump

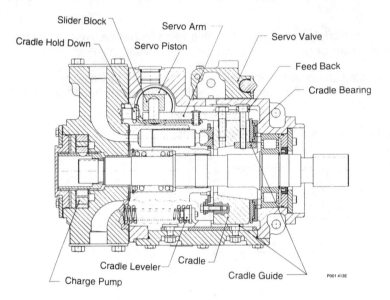

Tests run at the TUHH have shown leakage in the area of the lip seal after less than 300 hours, suggesting exessive shaft wear.

Picture 14 shows a foto of a shaft shaft which ran with a synthetic Ester.

A third snow groomer application using a series 90 motor with a rapeseed base fluid (see picture 15) experienced wear, approximately 38 μm (.0015 inches) deep at the motor shaft underneath the lip seal after 1500 hours. This motor shaft was a quenched and tempered 4140 (42CrMo4V) material with a hardness of approximately 32 - 37 Rc in the lip seal area. Another customer snow groomer test vehicle using the same rapeseed base fluid, but with a series 90 motor with a shot peened ETD 150 shaft, experienced excessive wear in the motor shaft spline after 1000 hours. The spline area is shot peened and is 32 - 37 Rc. It appears that either the shot peened spline or material difference is causing a fretting type wear in the spline.

It is worth noting that pump shafts of 8620 (21MnCr5) carburized material did not experience any wear in this application.

Picture 12: Pump shaft surface - synthetic Ester

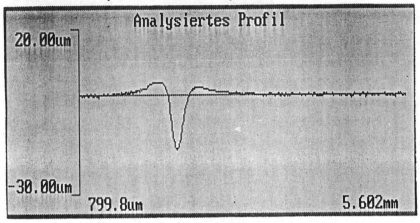

Rt = 20.3 μm (.0008 inch)

Picture 13: Pump shaft surface - mineral based oil

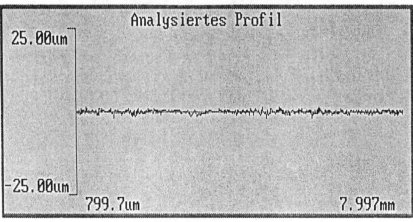

Rt = 3.1 μm (.00012 inch)

Picture No. 14: Pump shaft - synthetic Ester

Picture No. 15: Sectional view - Fixed Displacement Motor

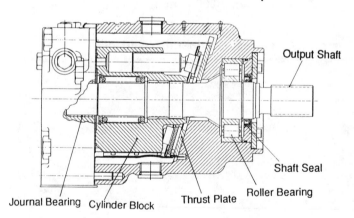

CONCLUSION

Both field experience and bench testing of hydrostatic transmissions with biodegradable fluids the last several years have shown that they are not equivalent to mineral base fluids in all aspects of their performance. Wear problems have surfaced with both rapeseed base and synthetic ester base fluids as either the direct result of a particular fluid or susceptibility of a specific design. Solutions to all wear problems have not been found. Where there has been success in eliminating wear, it has been by selecting those fluids which perform satisfactorily, or redesigning components to eliminate wear. We believe that there is a need to create test procedures that will help identify acceptable fluids and fluid characteristics that are important to wear.

REFERENCES

[1]   Feldmann, D.G. and Hinrichs, J., Biodegradable fluids in high loaded hydrostatic transmissions, 9th International Colloquium Tribology, Technical Academy Esslingen, Germany, 1994

[2]   Mercer, Keith D., SAUER-SUNDSTRAND internal test report, Swindon, England, 1994

[3]   SAUER-SUNDSTRAND, Applications Technical Information ATI 9101, Neumünster, Germany, 1995

Dierk G. Feldmann[1], Jan Hinrichs[2]

EVALUATION OF THE LUBRICATION PROPERTIES OF BIODEGRADABLE FLUIDS AND THEIR POTENTIAL TO REPLACE MINERAL OIL IN HEAVILY LOADED HYDROSTATIC TRANSMISSIONS

REFERENCE: Feldmann, D. G. and Hinrichs, J., **"Evaluation of the Lubrication Properties of Biodegradable Fluids and Their Potential to Replace Mineral Oil in Heavily Loaded Hydrostatic Transmissions,"** *Tribology of Hydraulic Pump Testing, ASTM STP 1310,* George E. Totten, Gary H. Kling, and Donald J. Smolenski, Eds., American Society for Testing and Materials, 1996.

ABSTRACT: Increasing public interest in the environmental impact of technical machinery has led to the development of new hydraulic fluids. In case of leakage these fluids pose less of an environmental threat than mineral oil, because they degrade faster and are less toxic or non-toxic.
The following paper describes method and results of laboratory tests with these new, so called biodegradable fluids, in a hydrostatic transmission on a flywheel testrig under high load conditions.

KEYWORDS: biodegradable fluid, hydrostatic transmission, component wear, fluid aging

INTRODUCTION, STATE OF ENGINEERING

With an increasing public sensibility to the environmental impact of technical machinery and the force on users of hydrostatic equipment to minimize the negative effects of leakage, environmentally more acceptable or better " biodegradable and non-toxic" hydraulic fluids have established their position in the market and are used more and more. Good reasons for the use of such fluids are trends in hydrostatic applications, which promote the potential of leakage, these are
- the use of higher pressures (450 bar and more),
- higher system temperatures (up to 120 °C),
- more duty hours and harsh treatment of the hydrostatic driven device,
- less maintenance, especially in the case of complex systems, and
- more hoses and hose couplings within the systems.

There is special interest to use those new fluids in agricultural machinery, earth moving machinery, tunnel machines, snow crawlers, deck machinery on ships, waste trucks and road cleaners, where the loss of fluid is not absolutely controllable and on the other hand unacceptable. State authority, but also private investors, increasingly connect an order with the contribution to biodegradable and non-toxic fluids in all equipment used in the project.

Commercially available, common biodegradable fluids for hydrostatic applications are based on three types of basic fluids: native esteroils (in most of the cases rapeseed oil), synthetic esteroils and polyglycols (Fig. 1). The three fluids have different chemical structures and by this, different features of technical relevance. Esteroils in general are not soluble in water, but soluble in mineral oil; this does not mean, that such a solution is an acceptable hydraulic fluid. Polyglycols on the other hand are soluble in water and not in mineral oil; a polyglycol with a high water content again is not a hydraulic fluid, which leads to good system performance. Water in an esteroil causes hydrolytic cracking and destroys the fluid. All of the biodegradable fluids have a smaller change of viscosity with temperature

[1]Professor, Arbeitsbereich Konstruktionstechnik I, Technical University Hamburg-Harburg, 21073 Hamburg, Germany, e-mail address feldmann@tu-harburg.d400.de

[2]Dr.-Ing., research assistant, Arbeitsbereich Konstruktionstechnik I, Technical University Hamburg-Harburg, 21073 Hamburg, Germany

Biodegradable Fluids		
basic fluids	natural esteroil (triglycerid)	synthetical esteroil	polyalcylenglycol (polyglycol)
formulation	rapeseed oil additivpackage	diester polyolester complexester additivpackage	polyglycol additivpackage
type	HETG	HEES	HEPG
H = hydraulic E = environmental TG = triglycerid ES = esteroil synthetic PG = polyalcylenglycol		

FIG. 1--Types of biodegradable fluids

compared to a mineral oil without VI-improver, and they have a higher polarity, which means that they have a higher adhesive power to surfaces of machine parts. For the application it is important to know that the fluids have a different aging stability; rapeseed oil has the lowest. Having this in mind one should not expose rapeseed oil to as high temperatures as can be accepted for synthetic esteroils, polyglycols and mineral oil.

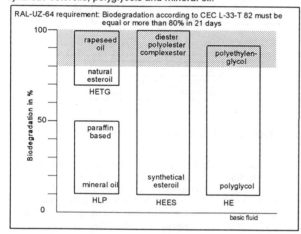

FIG. 2--Biodegradability of basic fluids according to CEC L-33-T82 test

The significant feature of the new fluids is their biodegradability, which can differ very much (Fig. 2); only native esteroils generally have a high biodegradability. To fulfill the requirements of RAL-UZ-64, a fluid must degrade equal to or more than 80% within 21 days under the test conditions set in CEC L-33-T82 [1].

A complex item is the toxic behaviour of the fluids, therefore the authors don't want to go into details. A fluid with the label "biodegradable" has, according to a German standard, to fulfill a number of conditions, especially the requirement of "Wassergefährdungsklasse" (WGK), i.e. to be within the limits of WGK 0 or 1. The WGK [2] depends on the "Wassergefährdungszahl" (WGZ), which is the mean value of the results of three laboratory tests describing the toxic impact of the fluid on mammals, fish and bacteria. Using a mean value suppresses the information about the toxic impact on the specific test object, this should be kept in mind.

Hydrostatic transmissions, which are subject of this investigation, are typically used to transmit mechanic power from the prime mover to wheels or tracks of working equipment and change transmission ratio; sometimes they do also other jobs as drive the drum of a concrete mixer or a ship winch. A hydrostatic transmission exists of a pump, one or more motors (i.e. cylinders or motors with rotating shaft) and a control system, pump and motor are volumetric machines. A typical solution for offroad vehicles is a transmission working in a closed circuit (see Fig. 3), where a variable displacement pump delivers its flow to a variable displacement motor. As both ports of pump and motor are direct connected, pump flow can be delivered at either port and by this, direction of rotation of the motor can be controlled. This type of circuit allows reverse power flow: in a load braking condition the hydraulic motor acts as a pump and the pump as motor, and the prime mover within its braking capability slows down the hydraulic motor. The fluid in the system has two main objectives: at first it transmits power between the units, pump and motor, at second it has to lubricate surfaces, which move relative to another; if this happens under high loads, the lubrication capability has a significant impact on general function, friction and wear.

Fig. 4 shows an axial piston swashplate type pump with variable displacement; all functional groups, where lubrication capability of the hydraulic fluid may have an impact, are named in the figure. The hydraulic motor (Fig. 5) used for the tests is an axial piston bent axis type motor.

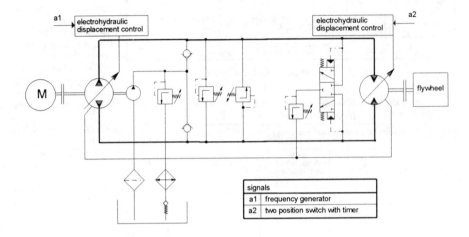

FIG 3.--Closed circuit hydrostatic transmission of the flywheel-testrig

With the increasing interest in the off-highway industry in the use of fluid alternatives to mineral oils it is necessary to determine the lubrication properties of these fluids in hydrostatic transmissions. The best method of making this assessment is to evaluate these fluids in a hydrostatic transmission test. The objective of this paper is to discuss this test and to illustrate the effect of fluid lubrication under the conditions that this test provides on parts of system components (pump and motor) as well as the change of fluid properties and the effect of both on the overall system efficiency.

As the subject of this paper is of high relevance to the present especially in Germany and northern Europe, there have been some presentations in conferences about Fluid Power [3], [4] and Tribology [5]. Results of a project, which was carried out from about 1990 to 1995, are presented in the PhD-thesis of Hinrichs [6].

EXPERIMENTAL METHOD

Laboratory tests are necessary to confirm field experience and figure out the potential of the fluids under extreme, accurately controlled conditions. Tests are carried out at several places, our institute works on the subject for about 10 years and has collected a huge amount of test results and an acceptable amount of experience within this time. Our main testrig is a flywheel-testrig, the circuit diagram is shown in Fig. 3; in addition we squeeze the fluids on an aging-testrig, where a gear pump delivers its flow to a pressure-relief valve. At third we use a cylinder testrig, where we test 4 cylinders at the same time.

As it is typical for a mobile application, pump and motor see cyclic pressure on the flywheel-testrig. The pump is driven with constant speed, the motor runs from zero rpm to maximum, back to zero, changes direction of rotation, accelerates again to maximum speed and slows down again to zero within one cycle. Motion of the motor is controlled by pump displacement; 90% of the time the motor

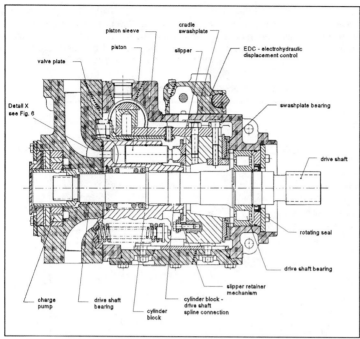

FIG. 4--Axial piston swashplate type pump (SAUER-SUNDSTRAND)

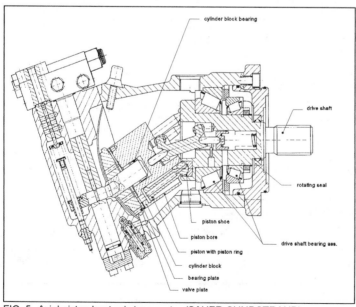

FIG. 5--Axial piston bent axis type motor (SAUER-SUNDSTRAND)

is at full displacement, 10% at reduced swashplate angle, which means higher speeds of the motor-shaft. The duty cycle as described has the consequence, that power flow in the transmission changes direction two times within the cycle, pump becomes motor and motor becomes pump; this also is a typical situation for mobile applications, where vehicles are accelerated and decelerated via the transmission. Data of the test equipment and the mode of operation are given in Table 1.

TABLE 1--Data of the flywheel-testrig

swashplate pump	
displacement control, a1:	electrohydraulic with pressure limiter
displacement:	75 cm³
displacement, charge pump:	20 cm³
speed:	3000 rpm
bent axis motor	
displacement control, a2:	electrohydraulic 2 position control
maximum displacement:	110 cm³
configuration of the hydr. system:	closed circuit
oil volume in the hydraulic system:	70 liters
system pressure:	450 bar (absolut)
case pressure:	3 bar, ± 0,1 bar
loadprofile 1:	25 minutes
control frequenz pump:	0,070 Hz (cycletime 14,3 sec.)
displacement motor:	110 cm³
theor. motor speed:	± 2045 rpm
loadprofile 2:	5 minutes
control frequenz pump:	0,025 Hz (cycletime 40 sec.)
displacement motor:	60 cm³
motor speed:	± 3700 rpm
fluid temperature in the reservoir:	HETG 70°C, HEES, HEPG, HLP 90°C
max. oil temperatur drain line:	reservoir temperature plus 15°C
test time:	300 hours
number of tests:	2 tests per fluid

TABLE 2--Comparison of the load on a pump on the flywheel-testrig and in a field application (combine harvester)

Flywheel - Testrig	example 1: Combine machine 092 04336	example 2: Combine machine 093 01783
P_{CH} = 160 kW n = 3000 rpm t = 300 hours p_m = 300 bar	P_{CH} = 120 kW n = 2000 rpm t = 390 hours p_m = 130 bar	P_{CH} = 150 kW n = 2000 rpm t = 924 hours p_m = 136 bar
The mechanic load on the fly-wheel-testrig equals a duty time of the combine of:	5200 hours	4800 hours
*)	The field report doesn't mention the pump speed, so 2000 rpm constant have been used for the calculation	

Subject of investigation are changes of the performance of pump and motor and wear on their parts, aging of the fluid and the behavior of system components as filters, reservoir etc. Test results are compared with results of tests with a high quality mineral oil under the same load conditions. The size of pump and motor have been chosen as typical for mobile vehicles, equipped with prime movers in the range of 60 to 100 kW. Pressures, speeds, accelerations and oil temperatures have been set high, and tank volume low, to produce heavy load for the system and shorten test time. As test time seems low with 300 hrs we tried to find a measure to compare testrig runs with field operation. Assuming that rolling contacts have a remarkable impact on service life of a pump or motor, we defined a load factor using the life calculation theory for ball bearings, which is the product of speed, running time and pressure raised to the power of three. Taking now the test data and data of two field applications, i.e. combine transmissions, one can see in Table 2, that the 300 hrs testrig are compatible with a reasonable service time of a combine.

The investigation on parts is directed to elements of function groups, which have an impact on transmission overall function, component life and transmission efficiency. These are mainly parts of the rotating groups as cylinderblock and valveplate, piston and bore, slipper and swashplate, ball-, roller- and needle-bearings, journal bearings and rotating (shaft) seals.

TEST RESULTS

The test results presented demonstrate the interaction of specific products, i.e. components and fluids of certain manufacturers. Therefore it is not correct to tranfer conclusions out of the presented test esults to other product combinations, even of same basic type, without a close look to design details and - for a mechanical engineer very difficult - to the internals of the fluid. Nevertheless it is allowed to make some general remarks at the end of this presentation and to extract hints for individual tests to be carried out. In the tests which are subject of this article we tested swashplate-type pumps and bent-axis-type motors from SAUER-SUNDSTRAND, filters from EPPENSTEINER, and hydraulic fluids from RAISIO, Finnland, MOBIL OIL, Hamburg, FRAGOL, Mülheim and FUCHS/ WENZEL &WEIDMANN, Eschweiler. Each fluid was tested in two identical test runs to exclude accidental results; both test run normally showed same results. A new rotating group was used for each test and we tested two products of different suppliers in each group of fluids. Due to the limited space in this presentation, in Table 3 and 4 the test results of only three fluids are presented.

Table 3 shows how the technically relevant features of the tested fluids compare to the reference mineral oil (1. row = fresh oil) and what magnitude and direction of change can be detected after the test run (2. row = change). One can see, that for a fluid with a similar viscosity (at operating conditions) density, air release potential and filterability are close to the values of mineral oil, and that the test run itself has no significant effect on the majority of the fluid properties. One remarkable exception shows rapeseed oil where one can find an improvement of the air release potential after the test, but at the same time a degradation in regards to foam production.
A look to the total efficiency of the transmission shows, that after a running time of 24 hrs the values are slightly lower as found with HLP-oil, and after 300 hrs, depending on the fluid, there is a small increase (smaller as in case of HLP) or a moderate decrease. It seems that the changes in any case are small and have no significant impact on the transmission performance.

In Table 4 we try to give a summary of the wear investigations; due to space reasons again three fluids were selected, the two rows of symbols mark two test runs with the same product. It can be seen that the impact of the fluids on function groups differs from fluid to fluid. Attention should be spent to the straight downward arrows, they indicate the potential of trouble. The HETG-and HEES-products do not have those arrows, the HEPG-product would have to be improved. The moderate downward arrows indicate more wear as in the case of HLP mineral oil, but, having the extreme test conditions in mind, it is not said, that this would lead to an early breakdown of the machines.

TABLE 3--Deviation of technically relevant properties of biodegradable fluids compared with the values of HLP and change of properties relativ to the values before the test run on the flywheel testrig

Meaning of symbols:

← significant better
↗ moderate better
○ no relevant difference
↘ moderate worse
→ significant worse

Fluid property	HETG B	HEES B	HEPG A
ISO VG	32	46	46
reservoir temperature	70°C	90°C	90°C
viscosity, fresh oil *	↗	↗	→
viscosity, change **	↗	○	→
density fresh oil *	○	○	○
density change **	○	↗	→
neutralization number fresh oil *	○	○	—
neutralization number change **	↗	○	—
water content fresh oil *	↗	○	○
water content change **	○	○	○
air release fresh oil *	←	←	○
air release change **	→	→	○
foaming fresh oil *	○	←	○
foaming change **	←	→	○
flow resistance of filters fresh oil *	↗	○	↗
flow resistance of filters change **	↗	↗	→
Transmission Total Efficiency			
after running in motor torque: 600 Nm	↗	○	↗
change relativ to HLP	↗	→	○
after running in motor torque: 300 Nm	↗	↗	↗
change relativ to HLP	○	↗	○

* deviation of properties of fresh oil compared with HLP
** change of properties relativ to the values before test

TABLE 4--Deviation of selected parts of swashplate pump and bent axis motor by test runs with biodegradable fluids compared with HLP

Fluid	HETG B	HEES B	HEPG A
Function Groups Swashplate-type Pump			
valveplate /	←	←	←
cylinderblock	○	○	○
swashplate /	↗	○	↗
slipper	↗	○	↗
piston /	○	○	○
bore (sleeve)	○	←	→
lip seal /	↗	↗	→
shaft	↗	↗	→
journal bearing	○	○	○
shaft bearing	○	○	○
Function Groups Bent-axis-type Motor			
valveplate /	○	↗	↗
bearingplate	○	↗	↗
piston ring /	○	○	→
bore	○	○	○
lip seal /	○	○	↗
shaft	○	○	↗
cylin. block needle bearing	↗	○	○
shaft bearing	○	○	○

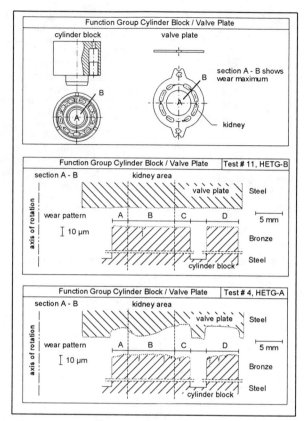

FIG. 6--Wear patterns in the cylinder block / valveplate ass.

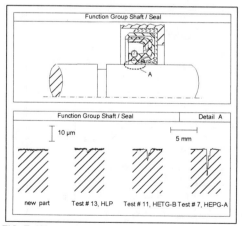

FIG. 7--Wear patterns on the pumpshaft / lipseal ass.

Wear in function groups is an indicator of the quality of the lubricant, the next figures show wear patterns and according measurements on selected parts. Fig. 6 shows surface measurements on the faces of the cylinder block and the valveplate of the pump. Looking to the picture, one should realize the different scales in either direction, µm in axial and mm in radial direction. Both surfaces show wear, but the amount is very different, although both tests were carried out under the same conditions and the different fluids belonged to the same group. This underlines the difficulty mentioned before to transfer test results from one configuration to another.

Fig.7 shows how different the fluids act on the pumpshaft / lip seal assembly. All tested biodegradable fluids led to lea-

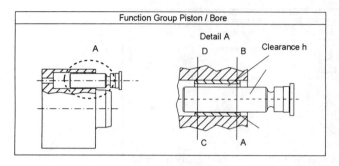

FIG. 8--function group piston / bore

kage, although FPM, the recommended seal material, was used.

Wear in the piston / bore function group causes an opening of the clearance between piston and bore (or sleeve) and increasing leakage flow. Fig. 8 shows the assembly, Table 5 the results of measurements in the crossections AB and CD. As one can see, there is a significant change in some cases and none or nearly none in others. At least two of the fluids are as good as mineral oil in this part of the test.

TABLE 5--Clearance change caused by wear in the piston / bore assembly

Test #	Fluid	section	mean value [%]	change of clearance between piston and bore [%]
1	HLP	AB	3	
		CD	3	
13	HLP	AB	11	
		CD	1	
2	HETG-A	AB	17	
		CD	18	
4	HETG-A	AB	21	
		CD	15	
10	HETG-B	AB	11	
		CD	4	
11	HETG-B	AB	6	
		CD	7	
5	HEES-A	AB	35	
		CD	36	
6	HEES-A	AB	17	
		CD	23	
12	HEES-B	AB	0	
		CD	0	
7	HEPG-A	AB	37	
		CD	21	
8	HEPG-A	AB	26	
		CD	15	

(scale markers: -7 0 7 10 20 30 40 50)

▨ spread of measures   ▮ arithmetic mean value
inaccuracy of measures: ± 2μm ≈ ± 7 %

CONCLUSION

In this paper, a test procedure to determine the lubrication properties, their effect on wear of parts, and the effect of both on overall transmission efficiency and fluid aging have been presented.
Generally spoken testing on a flywheel testrig with "real size" units provides an excellent insight into the interaction between fluid and mechanical components. The repeatability of the test results is satisfying regarding the big number of parameters, which influence the test and its results.

The test runs with biodegradable fluids in hydrostatic transmissions under high loads show that high quality products based on rapeseed oil, synthetic ester and, with some reservations, polyglycol work satisfactorily in the system and do not cause major function- or life-problems. The fluids do not show significant change in their technically relevant properties, and wear on parts is small. Nevertheless there are two areas of concern: bearings, which fail or show significant wear patterns, although the water content in the test fluids was very small, and lip seals.

It seems to be acceptable to state, that products of all three types, delivered in high quality and used within their specified limits, should have the potential to serve system needs in the same manner as is known for mineral oil. Specified limits points especially to content of water, maximum operating temperature (case drain) and cleanliness, i.e. no mixup with other fluids.

Biodegradable fluids are now on the market and have been in applications for a number of years, experiences with the fluids are generally positive, as long as the application keeps within the limits mentioned above.

REFERENCES

[1]   RAL-UZ-79, "Biologisch schnell abbaubare Hydraulikflüssigkeiten". Deutsches Institut für Gütesicherung und Kennzeichnung e.V., Bonn, Germany, 1994

[2]   Katalog wassergefährdender Stoffe, Bekanntmachung des Bundesministeriums des Inneren vom 1. März 1985 (GMBI 1985, S. 175) -U III 6523074/3, und Änderung vom 8. Mai 1985 (GMBI 1985, S. 369)

[3]   Feldmann, D. G., "Biologisch schnellabbaubare Hydraulikflüssigkeiten für Fahrantriebe von Landmaschinen - Ergebnisse von Prüfläufen im Labor zur Beurteilung ihres Leistungsver-mögens", VDI Berichte Nr. 1211, Düsseldorf, Germany, 1995

[4]   Feldmann, D. G. and Hinrichs, J., "About the Potential of Biodegradable Fluids to replace Mineral Oil in heavily loaded Hydrostatic Transmissions", Proc. of the Fourth Scandinavian Conference on Fluid Power, Tampere, Finland, 1995

[5]   Feldmann, D. G and Hinrichs, J., "About the Potential of Biodegradable Fluids to replace Mineral Oil in heavily loaded Hydrostatic Transmissions", Proc. of the 10th International Colloquium Tribology - Solving Friction and Waer Problems, Esslingen, Germany, 1996

[6]   Hinrichs, J., "Gebrauchseigenschaften von Druckflüssigkeiten für hydrostatische Verdränger-maschinen auf Basis von Rapsöl, synthetischem Esteröl und Polyalkylenglykol". Doctoral thesis, Technical University Hamburg-Harburg, 1995

Andreas Remmelmann[1]

TESTING METHOD FOR BIODEGRADABLE HYDRAULIC PRESSURE MEDIA BASED ON NATURAL AND SYNTHETIC ESTERS

REFERENCE: Remmelmann, A., "Testing Method for Biodegradable Pressure Media Based on Natural and Synthetic Esters," *Tribology of Hydraulic Pump Testing, ASTM STP 1310,* George E. Totten, Gary H. Kling, and Donald J. Smolenski, Eds., American Society for Testing and Materials, 1996.

ABSTRACT: This article describes a special test procedure for biodegradable fluids based on natural and synthetic esters and gives results of their examination. The test allows the evaluation of the special properties of these fluids, e. g. hydrolytic stability. By using a long endurance test, the fluid's behaviour during its use can be examined. The test rig works with serial hydraulic components. The loads are higher than in practice applications in order to reduce testing time. Therefore the results are transformable to practice and it is possible to give a precise description of the fluid's behaviour. The change of the fluid's properties and the ageing behaviour can be determined by analysing samples. With this knowledge the technical performance of the fluids can be improved, by changing additive packages or base fluid types.

KEYWORDS: Biodegradable hydraulic pressure media, oxidation stability, ageing mechanism, hydrolytic stability, deep temperature performance, ageing test procedure, tribology

---

INTRODUCTION

Disregarding the use and ageing properties of hydraulic pressure media can cause premature breakdowns of the hydraulic system. Choosing the right fluid requires a precise knowledge of the fluid's characteristics and their changes during the use of the fluid. Especially the properties of the new biodegradable hydraulic fluids are mainly unknown. So there is the need of closer research and investigation. Testing these fluids requires a testing procedure, that takes the fluids special characteristics, which are different than mineral oil, into consideration. Unfortunately most tests are designed for mineral oil based fluids and the results of testing other fluid types can only be transformed to practice with limitations. Also the ageing conditions in hydraulic systems are usually not taken into consideration, since they have a huge influence on the changes of properties during the use of the fluid. A suitable test contributes to the improvement of the fluids and the knowledge in which hydraulic applications the fluids are usable. [1],[2]

---

[1] Dipl.-Ing., Academic employee at the Institute of Fluidpower Transmission and Control, University of Technology Aachen, RWTH.

MAIN FUNCTIONS OF HYDRAULIC PRESSURE MEDIA

Pressure conveyance media are confronted with manifold demands. These can be divided into primary and secondary functions. The main tasks of a pressure conveyance medium are, above all, to transmit pressure forces, to ensure the flow connection between pump and consumer, and to transport kinetic energy [3]. These days, all pressure conveyance media accomplish these tasks. These primary demands are complemented by a number of secondary tasks, which, however, become increasingly important. Falling under these are especially the wear protection and the environmental compatability [4] of the fluid as well as the compatability with the components, Figure 1.

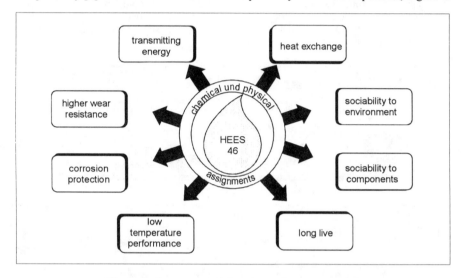

Fig. 1: Assignments and functions of pressure media

The requirement for biodegradable fluid types, however, can be complied with only by new, better decomposable basic fluid types, which are no longer based on mineral-oil but are e. g. obtained from animal or vegetable fat or respectively from chemical modifications of these kinds of fat. The new base fluid types possess new chemical and physical characteristics which differ considerably from those of the conventional mineral-oil based fluid. These new fluid properties require the conduction of extensive tests by which the applicability for practical use can be determined. The test procedures have not only to consider by the demands which are generally made on the hydraulic fluid but also to consider the special characteristics of the fluids. [5]

In order to accomplish these demands, the investigations have to include laboratory as well as test rig examinations before the fluids can be employed in practical operation.

With this article the investigation method, especially regarding the examination of biologically rapidly decomposable pressure fluids at the Institute of Fluidpower, Transmission and Control, (IFAS) shall be presented. One major problem of the new fluids consists in their unsatisfactory ageing stability, which has to be improved by appropriate provisions. The ageing stability is, however, also directly connected to the rapid biological decomposability, so that a compromise between stability and biodegradability has to be made. In order to find an optimal solution to this conflict, an exact analysis of the different ageing mechanisms and their impact on the properties for use of the fluids is required. [6]

AGEING MECHANISM OF PRESSURE MEDIA

To minimize the ageing of the fluids during the operation time, first, all the causes of ageing of the fluid have to be analysed. With respect to pressure fluids these are particularly the four reasons presented in Figure 2. These give rise to the four main ageing mechanisms, oxidation, polymerisation, cracking and hydrolysis. The hydrolysis, however, assumes a special position in this survey, because it only occurs when the pressure medium is contaminated with water.

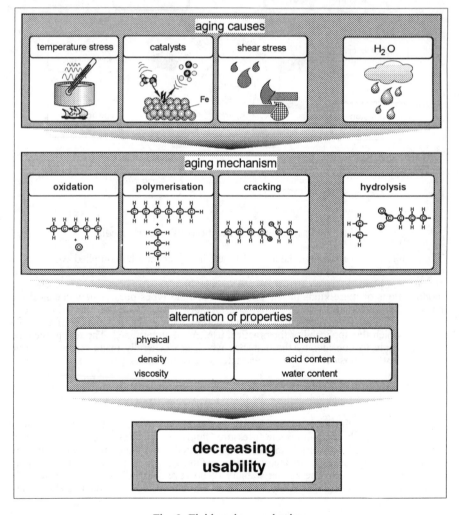

Fig. 2: Fluid ageing mechanism

Inside the pressure medium, these ageing mechanisms effect a change of the fluid characteristics. These changes can be so severe that the applicability for practical use is seriously restricted. This can be expressed in an intensified attack on non-ferrous metals or sealing materials, so that the functioning of the whole hydraulic plant is questioned. In order to confront these problems by means of an efficient modification of the base fluid or the additive system, it is thus imperative to subject the fluids to a defined ageing process. However, the ageing has to take place as differenciatedly as possible in order to be able to analyse the single ageing mechanisms systematically and to take optimizing measures subsequently.

Thus, two courses for the investigation of the suitability of pressure media are followed at the IFAS, Figure 3.

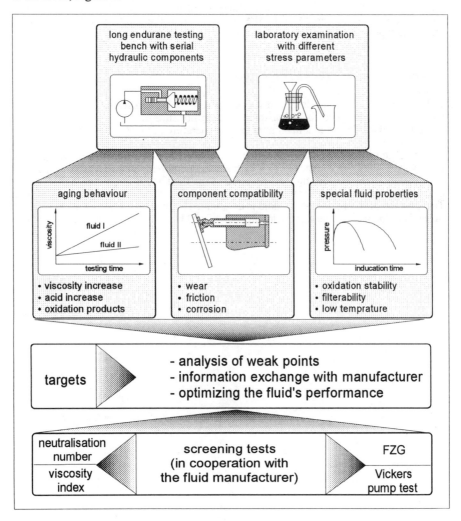

Fig. 3: Testing procedure

In long term examinations in loaded test rigs, which consist of production hydraulic components and represent a simple hydraulic circuit, the fluid ageing is conducted under conditions close to actual use.

By means of the examinations on these test stands, the fluids are aged under conditions close to reality, so that conclusions on the ageing behaviour in practical operation can be drawn from this. Moreover, the use of production hydraulic components allows the determination of comparability of the component parts with the fluid. In parallel to these test rig examinations, laboratory investigations by which particular properties of the fluids (e. g. the low temperature performance or the oxidation stability can be determined), are conducted. Combined with the screening tests of the fluid producers, this test method makes an efficient and systematic analysis of pressure fluids possible. By the help of this analysis potential weaknesses of the fluids can be exposed in a very short time and can be eliminated by the optimization of the additive packages or of the base fluid types.

TESTING BENCH CONFIGURATION

In Figure 4 the schematic design of an ageing test rig is represented in detail. The test stand has been designed for a maximum system pressure of 300 bar with an effective pump flow of 20 l/min and a maximum tank temperature of 90 °C.

By throttling of the complete hydraulic power to a serial plot composed of two pressure relief valves, the fluid is exposed to a high shearing and temperature load. Compared to a hydrostatic transmission, this test rig stresses the fluid about five times higher, so that a test duration of 1000 h corresponds to a practical application time of about 5000 h. The advance of the ageing of the fluid is determined by means of fluid samples which are taken from the tank of the test rig in intervals of 100 h.

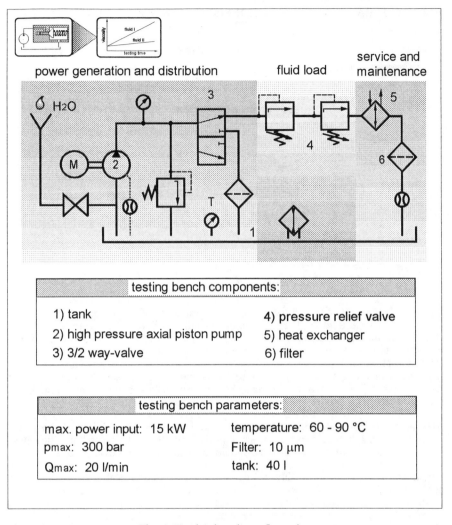

Fig. 4: Testing bench configuration

Furthermore, the test stand offers the possibility to inject different fluids, particularly water, into the test fluid, to mix them carefully, and thus be able to investigate the impact of these foreign fluids on the ageing behaviour. By the help of extended control- and safety components a continuous operation of this test stand over a period of 1000 h is also ensured.

In the following, the results from the investigation of biologically quickly degradabile pressure fluids on the basis of synthetic esters are presented.

FLUID CHARACTERISTICS AFTER LONG ENDURANCE TEST

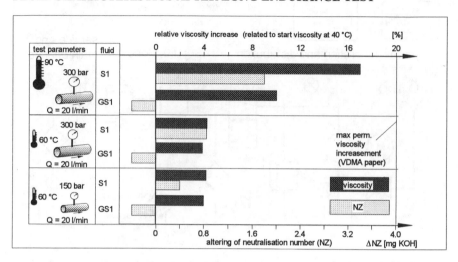

Fig. 5: Altering of fluid properties

Figure 5 demonstrates the variation of the parameters viscosity and neutralisation number for a base fluid GS1 and an additized fluid S1 formulated from it under the condition of different load parameters. Both fluids are based on synthetic esters. These two characteristics were chosen for the description of the ageing, because they reflect most accurative the processes inside the fluid. The different loads, and after the termination of the test runs also the different ageing situations, are clearly reflected in the changes of these two parameters. Particularly remarkable are the insignificant changes of these parameters with regard to the basic fluid.

If regarded in isolation this result would suggest that the basic fluid had the better properties. The subsequent examinations, especially on the oxidation stability, however, show, that a stabilization of the fluids by additives is absolutely necessary. [6]

For the purpose of being able to evaluate the oxidation stability of the fluids differentiatedly, the modified Rotary-Bomb-Test (ASTM NR 2272-85) is applied at the IFAS. The modifications consist in the fact that, on the one hand, an electronic pressure transformer is used to determine the pressure drop, and, on the other hand, these tests dispense with the addition of water into the bomb, because the water would immediately bring about a hydrolytic splitting of the fluid. Thus a differentiated evaluation of the results would no longer be possible. The criterion for the oxidation stability in the context of this test procedure is the period of time in which the oil sample consumes a certain quantity of oxygen.

Figure 6 shows the differences regarding the oxidation stability between the additized and the base fluid.

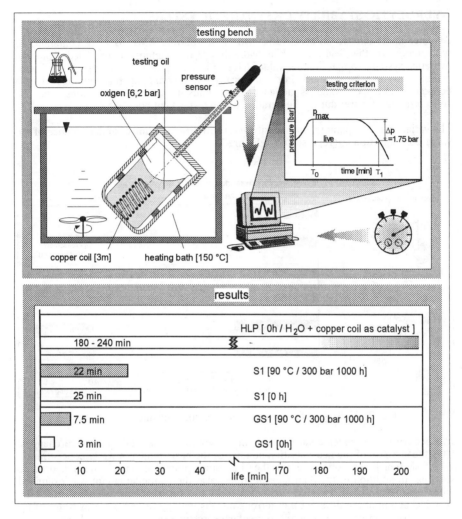

Fig. 6: Oxidation resistance

The positive effect the additives have on the oxidation stability can be gathered from the considerably longer service life of the additized fluid compared with the basic fluid. The comparison with a HLP oil demonstrates, however, the significantly lower oxidation stability of the fluids which are readily biodegradable. With the help of additives or base oil modifications it can, no doubt, be improved, but here a compromise between biodegradability and oxidation stability has to be found, because these two parameters are in conflict with eachother.

HYDROLYTIC STABILITY OF SYNTHETIC ESTERS

All ester fluids tend to an accelerated hydrolytic splitting when they are contaminated with water. In the field of mobile hydraulics water cannot be excluded from the hydraulic system, so that limiting values have to be set up to which an operation free of malfunctions can be ensured. For the purpose of determining this limiting value, a defined quantity of water is added to the fluids in another test series, and the concentration of water during the duration of the test is kept constant.

The variation of the parameters viscosity and neutralisation number in dependence of the water concentration are represented in Figure 7.

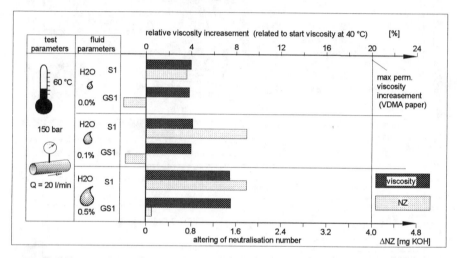

Fig. 7: Altering of fluid properties after contamination with water

Irrespective of the water concentration, they show no extraordinarily severe changes. Only the increase of the neutralisation number of the basic fluid at a water concentration of 0,5%, which was observed for the first time, points to an advanced hydrolytic splitting. In addition to the fluid investigations, also the hydraulic circuit was examined with regard to variations. Here no changes in the form of corrosion or attack by chemical action at the different non-ferrous metals with a concentration of water of up to 0,1% were observed. Furthermore, neither were the sealing materials damaged, nor were any sediments found within the hydraulic circuit.

The water content of 0,5%, however, lead to evident reactions in both fluids. In both cases soapy sediments were found on the bottom of the tank which caused a premature blocking of the filter elements. The quantity of the sediments in case of the fluid which had not been additized, however, was considerably greater. The failure of the filter plant in the circuit brought about an intensified abrasive wear, particularly at the tribosystems of the pumps due to the higher concentration of the solid particles in the fluid.

TRIBOLOGICAL CHARACTERISTICS OF SYNTHETIC ESTERS

Apart from these fluid investigations, also its tribological properties are of extraordinary importance for the application of a fluid as hydraulic medium. The favourable wear resistance characteristics of the esters are manifested, e.g. in the insignificant efficiency variations of the pump after a total operating time of 2000 h compared with the condition as new.

These results are affirmed by the surface measurements at the tribosystems sensitive to wear. These tribosystems do not show significant wear directly connected to the fluids. As an example the tribosystem cylinder block / valve plate is shown in Figure 8. In the illustration both tribopartners are represented as new and after a total service life of 2000 h. The pump even has operated 1000 h with a water concentration of 0,1% in the pressure medium. [7]

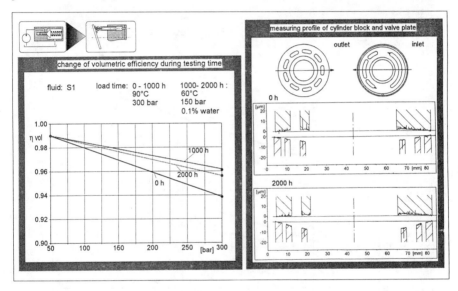

Fig. 8: Wear behaviour of tribologic system cylinder block / valve plate

In addition to these test rig investigations in the laboratory, the viscosity pressure behaviour of the fluids is determined, because it can decisively influence the service life of tribosystems with distinct Hertzian stresses. Figure 9 demonstrates this dependence for the cases of a synthetic ester and a mineral oil. Clearly recognizable is the significantly lower viscosity increase of the synthetic ester (HEES) compared with the mineral oil (HLP). This lesser-pronounced viscosity pressure dependence thus can, under unfavourable conditions, lead to a premature failure of tribosystems with Hertzian stresses.

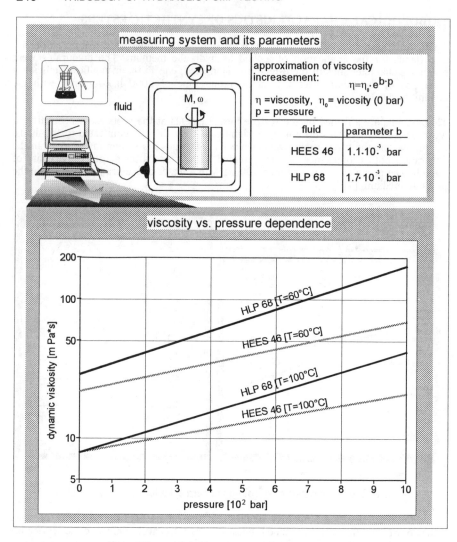

Fig. 9: Viscosity pressure dependence

For trouble-free operation of hydraulic plants, sealing plays a central role. The swelling behaviour of the different sealing materials are especially important for secure functioning. However, the sealing materials have to be provided with a sufficient swelling stability not only in fresh media, but also in aged ones. Figure 10 shows that the materials NBR and FKM are appropriate for the employment in synthetic esters of different types, because they shrink or swell respectively in the fresh as well as in the aged fluids only to a small degree.

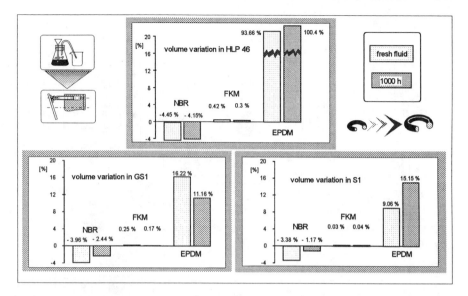

Fig. 10: O-ring swelling

For the purpose of comparison, a completely unsuitable material (EPDM), which after the storage time was heavily swelled and could no longer accomplish its task, was included in this schedule.

LOW TEMPERATURE PERFORMANCE

As the employment of readily biodegradable fluids is particularly efficient in mobile working machines, these fluids have to have of a viscosity as low as possible even at low temperatures, and even after a longer down time of the machine at low temperatures, a flocculation from the fluid must not occur. These would block the filters of the plant in no time, so that the solid particles contained in the fluid could cause intensified abrasive wear at the components of the hydraulic system.

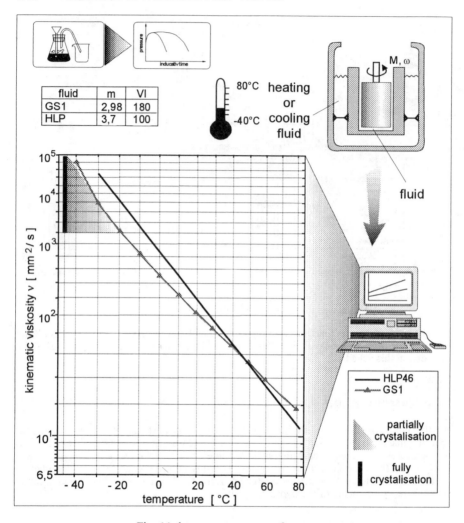

Fig. 11: low temperature performance

In Figure 11 the viscosity temperature dependence of the synthetic ester is contrasted with that of the mineral oil. The considerably higher viscosity index and thus the considerably lower viscosity are clearly evident, e.g. at a temperature of -20°C, so that the starting behaviour of a hydraulic plant filled with these fluids is much better than that of plants filled with ordinary HLP-mineral oil at these low temperatures. As a matter of fact, at these temperatures a tendency to crystallization could be observed, which increased at even lower temperatures and symbolysied by the grey shade in Figure 11.

For the analysis of the filterability, a diaphragm filter with a screen size of 5 μm, which was pressurized with a differential pressure of 6 bar, was chosen. The test temperature had to be raised to -10°C, because at lower temperatures the oil could no longer be filtered. Here, too, the favourable low temperature properties of the synthetic ester become evident, because these can still be filtered without problems even at these temperatures whereas the filter paper is blocked in no time with mineral oil, Figure 12.

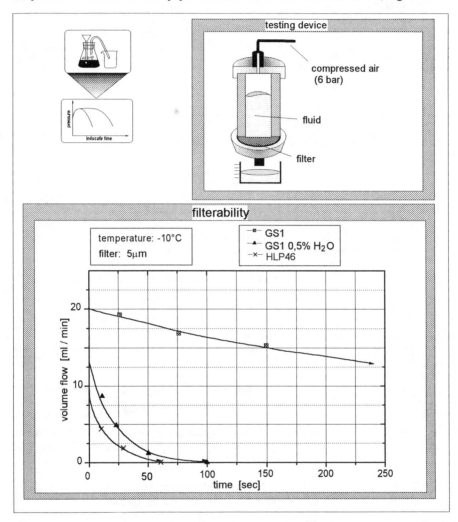

Fig. 12: Low temperature performance - filterability

In order to examine the impact of water on the pressure medium at lower temperatures, the filter test was conducted with different concentrations of water in the base fluid. Up to a water content of 0,1% this lead only to a very slight impairment of the filterability. A higher water concentration of 0,5%, however, causes an accelerated blocking of the filter elements, so that also for this reason a water concentration in the pressure should be maintained as low as possible.

By means of the methods presented in this article an extended analysis of hydraulic fluids under reproducible conditions is possible. The combination of test rig- and laboratory investigations allows a systematic analysis of the fluids. The test procedure considers not only the behaviour of the fresh but also of the aged fluids, so that it is possible to make statements about the behaviour of the fluids in actual operation.

The efficiency of the method has been pointed out distinctly by means of the examination of readily biodegradable hydraulic fluids on the basis of synthetic esters. By the investigations it could be demonstrated that these fluids are absolutely able to substitute mineral oil based in many cases.

LITERATURE

[1]   Busch, C., "HETG-Flüssigkeiten vor dem Hintergrund der neuen VDMA-Richtlinien," o + p Nr. 7 1994

[2]   NN, "DIN 51 524," Beuth Verlag Berlin

[3]   Backé, W., "Grundlagen der Ölhydraulik," Vorlesung RWTH Aachen

[4]   NN, "Grundlage für die Umweltzeichenvergabe für biologisch schnell abbaubare Hydraulikflüssigkeiten," UBA, Berlin

[5]   Eichenberger, H. F, "Biodegradable lubricant, an overview of current developments in Central Europe," SAE Technical Paper Series, 42nd Earthmoving Industry Conference, Peoria, Illionois, 9-10.04.1991

[6]   Feßenbecker, A.,Korff, J., "Additives for biodegradable lubricants," 59th Anual Meeting, The National lubricating grease Institute, 25-28.10.1992, Hilton Head Island, USA

[7]   Höhn, B. R., Michaelis, K., "Reibungsverhalten biologisch leicht abbaubarer Schmierstoffe," Tribologie und Schmierungstechnik 42, 01.03.1995

[8]   Feldmann, D. G. Hinrichs, J., "Biologisch schnell abbaubare Hydrauliflüssigkeiten - ein neuartiges Konstruktionselement für hydrostatische Getriebe," Konstruktion 47, 1995

[9]   Feßenbecker, A., Korff, J., "Additive für ökologisch unbedenklichere Schmierstoffe," Tribologie und Schmierungstechnik 42/1/1994

[10]   Backé, W., "The present Future of Fluid Power," Proceeding of the Institution of Mech. Engineers 12/93

[11]   Randels, S. J., Wright, M., "Enviormentally cosiderate Ester for the Automotive and Engineering Industries," Synthetic Lubrication, Vol 9 No. 2, 01.07.1992

Fluid Cleanliness

Robert H Frith,[1] Will Scott,[2]

RELATING SOLID CONTAMINANT PARTICLE SIZE DISTRIBUTION TO FLOW DEGRADATION IN HYDRAULIC PUMPS

REFERENCE: Frith, R.H. and Scott, W., **"Relating Solid Contaminant Particle Size Distribution to Flow Degradation in Hydraulic Pumps,"** *Tribology of Hydraulic Pump Testing, ASTM STP 1310,* George E. Totten, Gary H. Kling, and Donald J. Smolenski, Eds., American Society for Testing and Materials, 1996.

ABSTRACT: Power hydraulic pump wear is difficult to quantify. Measuring the actual material lost due to wear is impossible in a practical sense. An available simulated procedure is the contaminated sensitivity performance test which measures flow degradation with increasing amounts of contamination in the fluid. It is shown that wear can be related to flow degradation and knowing the mass of contamination causing wear, models for pump wear can be explored. An approach to calculating damaging contaminant mass from particle size distribution is given.

KEYWORDS: Hydraulic Pumps, Contamination, Wear

NOMENCLATURE

c^* constant due to pressure and gap geometry

c_l $c_h/\Delta h$.

c_h geometrical constant

$F(x)$ cumulative size distribution and function of x

h leakage gap height

---

[1] Chief Engineer, Pinnacle Engineering Pty Ltd, Formation St (cnr Boundary Rd) Carole Park, Qld, Australia 4300

[2] Professor of Tribology, Tribology Research Concentration, School of Mechanical, Manufacturing and Medical Engineering, Queensland University of Technology, GPO Box 2434, Brisbane, Qld, Australia 4001

h_o gap height of the unworn pump

Δh increase in gap height due to wear

L^* dimension related to the gap length

M_{1-x_c} total contaminant mass

N_x number of particles greater than a diameter, x

q_i leakage flow at any time during the ith test period

q leakage flow

q_o leakage flow of the unworn pump

Q_i flow at any time during the ith test period

Q_o,   unworn pump flow

Q_{oi},   pump flow at the start of the ith test period

Qv theoretical volumetric pump flow

S_w total wear mass

$(Sw)_i$ wear mass produced during the ith test period

u integration variable (= In (x))

v integration variable $\left(= \sqrt{2\chi} \cdot \left(u - \dfrac{3}{2\chi} \right) \right)$

Vol_{1-x_c} total volume of particles between $1 < x < x_c$ $((\mu m)^3)$

w^* dimension related to the gap width

x particle size (μm)

x_c 'cut off' particle size (μm)

x_{crit} critical particle size for wear to occur (μm)

Greek letters

η_v volumetric efficiency (= Q_o/Q_v)

θ_i flow degradation ratio (= Q_i/Q_o)

ξ particle shape correction factor

ρ_i i^{th} particle type density

ρ_w wear debris density

χ constant

INTRODUCTION

Practical measurement of flow degradation in hydraulic pumps due to solid contamination has been undertaken for many years [1]. Despite an unresolved dispute over how to interpret test results, the basic test has gone unchallenged and is now nearing international standardisation [2]. Very little fundamental understanding of the pump wear process has been advanced, with the result that no model exists to describe it. Reported here is a method to relate flow degradation test results to the generation of wear debris. This will form the basis of an attempt to compare pump flow degradation results to a model for the wear process in pumps.

PUMP WEAR MECHANISM

With the pursuit for high efficiency, hydraulic pumps must operate with minimal clearances. Any increase in clearance is seen as a loss in flow due to an increase in leakage. It is this reduction in flow or 'flow degradation' that is measured during a pump test. Although knowledge of how the pump flow degrades due to solid contamination is important in selecting wear tolerant pumps, it provides no direct measurement of wear.

Pumps are typically configured in a system to operate at constant pressure with the excess flow passing over a relief valve or, in the case of pumps having a variable displacement, flow is provided only on demand. In either case, as the pump wears, leakage increases to a point where the required flow can no longer be met. The pump has then degraded.

It has been demonstrated [3] that pump wear occurs in regions of close contact. Contaminant borne with the leakage fluid abrades the close coupled surfaces as it passes through, resulting in the generation of wear debris. With the loss of material both the gap between the surfaces and the leakage flow through the gap are increased. Clearly, the increase in pump leakage or alternatively the pump flow degradation is related to the mass of wear debris generated. Moreover, the wear debris mass is related to the mass of damaging contaminant passing through the gap. It is argued that not all contaminant causes wear, only that greater than a certain critical size, x_{crit} [3] which in turn will be related to the gap size.

If a relationship can be established between pump flow degradation and wear debris mass, and the damaging contaminant mass causing the wear can be determined, results from pump flow degradation may become useful in exploring pump wear models.

This paper relates the particle size distribution of solid contaminants to the total contaminant mass before deriving an expression for the damaging contaminant mass in relation to a critical particle size. On the assumption that wear debris is directly related to flow degradation, a relationship is developed between the two in terms of a simple geometry pump. The expressions so developed may be validated against test data and this is done in [8].

RELATING PARTICLE SIZE TO TOTAL CONTAMINANT MASS

Contaminant distributions are conveniently expressed cumulatively with numbers of contaminant particles greater than a size being recorded against that size. Considerable work has been done in measuring contaminant distributions and have led to the development of automatic particle counters and supporting international standards [1]. Total particulate mass is difficult to measure directly. The gravimetric method [4] will in principle measure the mass directly, but can be subject to considerable uncertainty, For low concentrations of contaminants as occurs in hydraulic oils, automatic particle counting is proving to be both accurate and reproducible and is becoming the recommended method of measuring cleanliness level in hydraulic oil [5].

Because of its use as a standardised test dust for various purposes including pump flow degradation testing, the relationship of particle count to total particulate mass is well known for Air Cleaner Fine Test Dust (ACFTD) [6]. For general contamination, establishing a relationship is complicated by its unknown composition. Particles will not only be of different size but have a differing chemical composition.

Nonetheless, assume the particles to be spherical and to have a cumulative size distribution given by,

$$N_x = F(x) \qquad (1)$$

where N_x = number of particles greater than a diameter, x. The number of particles of a given size x will be given by,

$$\text{number of particles} = \lim_{\delta x \to O} \left[N_x - N_{x+\delta x} \right]$$

It follows that the total solids volume of particles less than x_c will be given by,

$$Vol_{0-x_c} = \lim_{\delta x \to 0} \sum_{x_0}^{x_c} \left(N_x - N_{x+\delta x}\right) \cdot \frac{\pi x^3}{6} \cdot \delta x$$

$$= \int_{x_0}^{x_c} - \frac{d}{dx} F(x) \frac{\pi x^3}{6} dx$$

(2)

where x_c = 'cut-off' or sieve size (maximum value of x)
and x_0 = minimum size of particle.

It would appear that many distributions, where minimum value x_o is 1 μm, can be represented by,

$$F(x) = N_1 . e^{-\chi(\ln x)^2}$$

(3)

or,

$$ln(F(x)) = -\chi.(ln\ x)^2 + ln(N_1)$$

(3a)

where, N_1 = number of particles greater than or equal to 1 micron and χ = constant.

Expanding (2) and for $x_0 = 1$,

$$Vol_{1-x_c} = \left[-F(x) \cdot \frac{\pi x^3}{6} \right]_1^{x_c} + \int_{x=1}^{x_c} \frac{\pi x^2}{2} \cdot F(x).dx$$

(4)

or, $$Vol_{1-x_c} = \int_{x=1}^{x_c} \frac{\pi x^2}{2} \cdot F(x).dx - F(x_c) \cdot \left(\frac{\pi x_c^3}{6} \right) + \left[N_1 \frac{\pi}{6} \right]$$

(4a)

ACFTD is well represented by (4) and oil samples taken from typical underground mining machinery also demonstrate a good fit. Seen in Figure 1 is the distribution for ACFTD and two typical oil samples taken from underground mining machinery plotted on log-(log)^2 scale. The 'goodness of fit' to (4) or (4a) is apparent. No doubt there will be distributions that deviate from (4). If the deviation is substantial, an alternative expression to (4) would be necessary or (3a) solved using other methods.

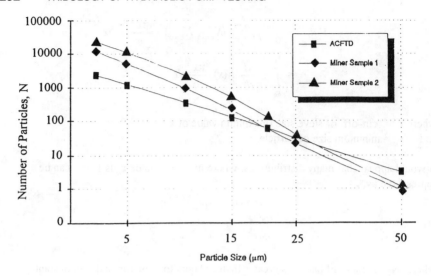

Figure 1 Distributions of ACFTD and several typical oil samples taken from continuous miners. Note scale is log-(log)^2

Substituting (3) into (4a),

$$Vol_{1-x_c} = \frac{\pi}{2}N_1\left[\int_{x=1}^{x_c} x^2 \cdot e^{-\chi(\ln x)^2} \cdot dx - \frac{x_c^3}{3} \cdot e^{-\chi(\ln x_c)^2} + 1/3\right] \quad (5)$$

By letting u = ln x it follows that dx = x.du and (5) becomes,

$$Vol_{1-x_c} = \frac{\pi}{2} \cdot N_1\left[\int_{x=1}^{x_c} e^{2u} \cdot e^{-\chi u^2} \cdot e^{u} \cdot du - \frac{x_c^3}{3}e^{-\chi(\ln x_c)^2} + 1/3\right]$$

$$= \frac{\pi}{2}.N_1\left[\int_{x=1}^{x_c} e^{-(\chi u^2 - 3u)} \cdot du - \frac{x_c^3}{3}.e^{-\chi(\ln x_c)^2} + 1/3\right]$$

$$= \frac{\pi}{2}.N_1\left[e^{\frac{9}{4\chi}}\int_{x=1}^{x_c} e^{-\left(\sqrt{\chi}\,u - \frac{3}{2\sqrt{\chi}}\right)^2} \cdot du - \frac{x_c^3}{3}.e^{-\chi(\ln x_c)^2} + 1/3\right] \quad (6)$$

Further, letting, $v = \sqrt{2\chi} \cdot \left(u - \frac{3}{2\chi}\right)$, it follows that $dv = \sqrt{2\chi} \cdot du$ and (6) becomes,

$$Vol_{1-x_c} = \frac{\pi}{2} \cdot N_1 \left[e^{\frac{9}{4\chi}} \cdot \frac{1}{\sqrt{2\chi}} \int_{x=1}^{x_c} e^{-\frac{v^2}{2}} \cdot dv - \frac{x_c^3}{3} e^{-\chi\left(\ln x_c^2\right)} + 1/3 \right]$$

$$= \frac{\pi}{2} \cdot N_1 \left[\sqrt{\frac{\pi}{\chi}} \cdot e^{\frac{9}{4\chi}} \cdot \left(\frac{1}{\sqrt{2\pi}} \cdot \int_{x=1}^{x_c} e^{-\frac{v^2}{2}} \cdot dv \right) - \frac{x_c^3}{3} e^{-\chi\left(\ln x_c^2\right)} + 1/3 \right] \tag{7}$$

It will be noted that the normal probability distribution appears bracketed in (7). This integral cannot be found in terms of known functions and so is tabulated. Tables of the normal probabiity function can be found in books on statistics and probability, and in collections of mathematical tables. Equation (7) may be written as

$$Vol_{1-x_c} = \frac{\pi}{2} \cdot N_1 \left[\sqrt{\frac{\pi}{\chi}} e^{\frac{9}{4\chi}} P\left(\sqrt{2\chi} \left(\ln x_c - \frac{3}{2\chi} \right) \right) - \frac{x_c^3}{3} e^{-\chi(\ln x_c)^2} + 1/3 \right] \tag{7a}$$

where P( ) is the normal probability function for $V = \sqrt{2\chi} \left(\ln x_c - \frac{3}{2\chi} \right)$ having a standard deviation of 1 and zero mean.

Strictly one should use (7a) for each particle type in determining the total mass. Moreover, by introducing a shape correction factor, ξ to account for the deviation of the actual particle shape from spherical, the total contaminant mass will be,

$$M_{1-x_c} = \rho_1.\xi_1.Vol_{1-x_c}|_1 + \rho_2.\xi_2.Vol_{1-x_c}|_2 + .. + \rho_n.\xi_n.Vol_{1-x_c}|_n \tag{8}$$

where M_{1-x_c} = total contaminant mass, ρ = particle density and subscripts (1,n) identify each particle type.

A problem is that (8) requires extensive knowledge of the distribution. With present technology, it is difficult enough to determine the different particle types and composition (i.e. ρ_i). Establishing $N_1|_i$, x_i and ξ_i for each particle type is a major undertaking. A simplifying assumption needs to be made.

Data from particle counting can be interpreted as providing an averaged N_1 and χ for the total distribution. If the averaged density, ρ and shape factor, ξ can also be established or assumed, the contaminant total mass, M_{1-x_c} can be calculated simply as,

$$M_{1-x_c} = \rho\xi Vol_{1-x_c} \tag{9}$$

Clearly the accuracy of (9) is dependent on ρ and ξ. While a reasonable judgement for density can be made based on spectrometric analysis of the oil sample and even then, certain elements are not observed and reading accuracy is suspect at the relatively low concentrations experienced in typical hydraulic oil systems, the shape factor is completely unknown. Extensive analysis using a visual technique such as electron microscopy may provide an estimate, but even this is suspect since it is only two dimensional. Moreover the effort is daunting and does not lend itself to routine analysis. More work needs to be done to establish typical values for ξ possibly as a function of ρ or the hydraulic system.

Here, interest is concentrated on relating ACFTD as the contaminant mass to wear. ACFTD is well known and referring to data given in [6] and fitting a straight line on a log-$(\log)^2$ scale, the parameters N_1, χ and ξ can readily be evaluated as,

$$
\begin{array}{rcl}
N_1 & = & 1751.95 \ ml^{-1} \\
\chi & = & 0.47 \\
\xi & = & 0.422
\end{array}
$$

ESTIMATION OF DAMAGING CONTAMINANT MASS

Illustrated in Figure 2 is the standardised particle size distribution for ACFTD used in pump flow degradation testing. To obtain the various 'cuts', the raw ACFTD is treated by sieving or centrifuging. Ideally the cut should exhibit a sharp cut off at the size of

interest as illustrated in Figure 2. As would be expected in practice, the ideal 'step' is smoothed at the extremities. Since the degree of smoothing is unknown and to facilitate ease of computations, the assumption will be made that the cut off is sharp.

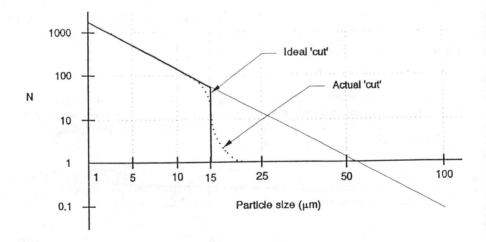

Figure 2   Particle size distribution of standardised Air Cleaner Fine Test Dust (ACFTD). Also shown is a theoretical 'cut'.

Moreover each cut has the same mass/fluid volume of 300 mg/litre and it follows that N_1 must differ. That is, using (9) and for $M_{1-x_c} = M_{1-x_-} = 300$ mg/litre, it follows that,

$$N_1 = \bar{N}_1 \cdot \left(\frac{F(\chi, \infty)}{F(\chi, x_c)} \right) \tag{10}$$

where,

$$F(\chi, x_c) = \left[\sqrt{\frac{\pi}{\chi}} \cdot e^{\frac{9}{4\chi}} \cdot P \left(\sqrt{2\chi} \left(\ln x_c - \frac{3}{2\chi} \right) \right) - \frac{x_c^3}{3} e^{-\chi (\ln x_c)^2} + 1/3 \right]$$

= factor relating N_1 and x_c to total particle volume

$\bar{N}_1$ = number of 1 micron particles and greater/fluid volume corresponding to 300 mg/litre of the raw ACFTD.

= 300 x 1751.95 particles/litre

= 525585. particles/litre

Finally, D is defined as the portion of M_{1-x_c} which is greater than x_{crit}. Hence

$$D = M_{1-x_c} - M_{1-x_{crit}} \quad \text{for } x_{crit} < x_c$$
$$D = 0 \quad \text{for } x_{crit} > x_c \tag{11}$$

Alternatively, (11) can be written for $x_{crit} < x_c$, as ,

$$D = \rho_1 \xi_1 \frac{\pi}{2} N_1 \left[F(x_c) - F(x_{crit}) \right]$$

$$= \rho_1 \xi_1 \frac{\pi}{2} \bar{N}_1 F(\infty) \left[\frac{F(x_c)}{F(x_c)} - \frac{F(x_{crit})}{F(x_c)} \right] \tag{12}$$

$$= M_{1-\infty} \left[1 - \frac{F(x_{crit})}{F(x_c)} \right]$$

$$= 300 \left[1 - \frac{F(x_{crit})}{F(x_c)} \right] \text{ mg/litre}$$

Values for D are seen in Table 1.

TABLE 1 -- Values for the damaging contaminant portion, D in mg/litre. Values shown in brackets are D as a percentage of the total contaminant mass of 300 mg/litre

x_c (µm)	5	10	20	30	40	50	∞
$F(\chi, x_c)$	11.1	51.28	144.48	219.72	275.5	316.78	479.78
x_{crit} (µm) 5	0.	235.06 (78)	276.95 (92)	284.84 (95)	287.91 (96)	289.5 (97)	293.06 (98)
10	0.	0.	193.53 (65)	230.0 (77)	244.16 (81)	251.44 (84)	267.94 (89)
15	0.	0.	94.30 (31)	164.74 (55)	192.12 (64)	206.18 (67)	238.05 (79)
20	0.	0.	0.	102.73 (34)	142.67 (48)	163.17 (54)	209.66 (70)
25	0.	0.	0.	47.69 (16)	98.77 (33)	125.0 (42)	184.45 (62)
30	0.	0.	0.	0.	60.74 (20)	91.91 (31)	162.6 (54)

RELATING WEAR MASS TO FLOW DEGRADATION

Assuming wear mass S_w results only from an increase in the leakage gap height h, it follows that,

$$\Delta h \; \alpha \; S_w \tag{12}$$

or,

$$\Delta h = c_h.S_w \tag{12a}$$

where c_h = geometrical constant.

The constant, c_h, has physical significance. It is the inverse of the wetted area that is being worn. For example, if the gap was of a simple rectangular configuration having a length L and width w, the constant c_h would be given by,

$$c_h = \frac{1}{\rho_w} \cdot \left(\frac{1}{L.w} \right) \tag{13}$$

where,

ρ_w = wear debris density.

Gap geometry in a pump will be more complex and wear will not be uniformly distributed over the leakage flow wetted area. Nonetheless, the general form could be similar. Wear debris density will be variant, depending on the surface material being worn and an average density would be used. The generalised form could be,

$$c_h = \frac{1}{\rho_w} \cdot \text{function} \left(\frac{1}{L^*} \, , \, \frac{1}{w^*} \right) \tag{13a}$$

where,

L^*	=	dimension related to the gap length,
w^*	=	dimension related to the gap width.

Leakage flow is a complex relationship depending on the geometrical arrangement of each gap in the pump, the pressure difference across the pump as well as the pump rotational speed. It is demonstrated [3] that only for conditions of very close clearance will the pump speed (Couette flow) have a significant influence on leakage flow. Accordingly the simplifying assumption will be made that leakage is due to the pressure difference only (Poiseuille flow). Referring to [7], for a simple gap with a pressure difference between entrance and exit, flow is related to the cube of the gap height . That is,

$$q = c^* . h^3 \tag{14}$$

where, q = leakage flow, c^* = constant due to pressure and gap geometry and h = leakage gap height. In the case of a pump, gaps are of varying height and the leakage flow expression is more complex [3]. Here, (14) will be accepted as being representative of that occurring in a pump.

Pump degradation testing involves the addition of contaminant to the clean fluid. After a 30 minute test period, the pump is unloaded and the contaminant and subsequent wear debris are removed from the fluid. The pump flow Q_{0i}, is recorded at the commencement of the ith 30 minute test period and periodically throughout the test (Q_i). Several consecutive test periods will be run until the pump has degraded substantially below the unworn pump flow, Q_0, usually 70% of the unworn flow.

Letting,

$$h = h_o \left(1 + \left(\frac{\Delta h}{h_o} \right) \right) \tag{15}$$

$$= h_o \left(1 + c_1 S_w \right)$$

where, h_o = gap height of the unworn pump and $c_1 = c_h / h_o$.

Substituting (15) into (14),

$$q_i = c.h_o^3 (1 + c_1.S_w)^3 \tag{16}$$

$$= q_o(1 + c_1.S_w)^3$$

where q_o = leakage flow of the unworn pump.

Knowing that,

$$q_i = Q_v - Q_i$$

and,

$$q_o = Q_v - Q_o$$

where, Q_v = theoretical volumetric pump flow, (16) becomes,

$$c_1.S_w = \sqrt[3]{\frac{1 - Q_i / Q_v}{1 - Q_o / Q_v}} - 1 \tag{17}$$

or,

$$c_1.S_w = \sqrt[3]{\frac{1 - \dfrac{Q_i}{Q_o} \cdot \dfrac{Q_o}{Q_v}}{1 - \dfrac{Q_o}{Q_v}}} - 1 \tag{17a}$$

Considering c_1, as the pump size increases, all parameters, L^*, w^* and h_o will increase. It is probable there will be a relationship between c_1 and pump size or displacement. It may be reasonable to assume an inverse relationship to pump displacement. That is, as the pump displacement increases, c_1 decreases proportionately.

It should be recognised that S_w as given in (17) or (17a) is the cumulative wear mass. Following each test period, wear debris is removed. Hence, referring to (17a), the debris produced in the ith test period will be given by,

$$c_1 \cdot (S_w)_i = \sqrt[3]{\frac{1 - \theta_i . \eta_v}{1 - \eta_v}} - \sqrt[3]{\frac{1 - \theta_{i-1} . \eta_v}{1 - \eta_v}} \tag{18}$$

where, θ_i = flow degradation ratio ($= Q_i/Q_o$), η_v = volumetric efficiency (= Q_o/Q_v) and $(S_w)_i$ = wear mass produced during the ith test period.

CONCLUSION

The reasonable assumption has been made that there is a relationship between contaminant mass and wear mass although neither parameter is directly measured in a pump flow degradation test. Contaminant mass can be deduced from particle count, provided knowledge of the particle composition and its deviation from a spheroid is known. Fortunately ACFTD is used in pump degradation testing as the contaminant and knowledge of its properties is well established.

Wear mass is less certain. A number of simplifying assumptions were made to arrive at (17) or (18) as an expression for wear mass given flow degradation results. What is now required is to demonstrate the validity of the expressions against test data. This is done in [8].

REFERENCES

[1]   Frith, R.H. and Scott, W., "Control of Solids Contamination in Hydraulic Systems - an overview", Wear, 165, (1993), 69-74.

[2]   ISO/DIS, 9632, "Hydraulic Fluid Power - Fixed Displacement Pumps - Flow Degradation due to Classified AC fine test dust Contaminant - Test Method (draft standard)", International Organisation for Standardisation, (1990).

[3]   Frith, R.H. and Scott, W., "Wear in External Gear Pumps - A simplified model", Wear, 172 (1994) pp 121-126

[4]   Hunt, T.M., "A Review of Condition Monitoring Techniques Applicable to Fluid Power Systems", Proc 7th. Inter Fluid Power Symposium, (16-18 Sept 1986), 285-294.

[5]   Day, M.J and Tumbrink, M., "Options for Contaminant Monitoring of Hydraulic Systems", Condition Monitoring '91, Pineridge Press, (1991), 246-268.

[6]   BS 5540: part 2 , "Evaluating Particulate Contamination of Hydraulic Fluids-method of calibrating liquid automatic particle-count instruments', British Standards Institution, (1 978).

[7]   Thoma, J.U., <u>Modern Oil hydraulic Engineering</u>, Trade and Technical Press Ltd., Morden, (1970).

[8]   Frith, R.H. and Scott, W., "Comparison of an External Gear Pump Wear Model with Test Data", (Submitted to <u>Wear</u>, 1995)

Stephan Lehner, [1] and Georg Jacobs [2]

CONTAMINATION SENSITIVITY OF HYDRAULIC PUMPS AND VALVES

REFERENCE: Lehner, S., and Jacobs, G., "Contamination Sensitivity of Hydraulic Pumps and Valves," *Tribology of Hydraulic Pump Testing, ASTM STP 1310,* George E. Totten, Gary H. Kling, and Donald J. Smolenski, Eds., American Society for Testing and Materials, 1996.

ABSTRACT: The article describes the methods and results of the examinations of the wear behavior of pressure controlled pumps and proportional valves. For the purpose of examination, pressure controlled pumps are exposed to a reproducible wear load in highly contaminated oil by means of a special developed test procedure. An analysis of the damage process that occurs in several tribological systems shows that the wear behavior of pumps and valves is highly dependent on the constructive design and materials of single tribological systems. On the basis of examples, provisions for wear design to minimize wear are pointed out.

KEYWORDS: Wear, wear mechanism, solid contamination, wear test procedure, tribology, pressure controlled pumps, proportional controlled valves,

INTRDUCTION

The contamination sensitivity of hydraulic components is a general problem, and oil filtration is not an adequate provision in order to avoid the damage due to wear by particle contaminated oil. This unsatisfactory situation requires further investigation. The purpose of the examinations is to achieve fundamental knowledge about the influence the solid particles exert in the characteristic tribological systems of hydraulic pumps and valves. The wear behavior of the pumps and valves can be deduced from the results of a suitable test procedure. It is then possible to identify and to analyze the critical tribological systems. The wear test results illustrate the impact of the design and the material properties. Knowledge of the wear interactions is a basic precondition for optimizing wear resistance of hydraulic components.

---

[1] Dipl.-Ing., Academic employee at the Institute of Fluidpower Transmission and Control, University of Technology Aachen, RWTH.

[2] Dr. Ing., Cheif Engineer at the Institute of Fluidpower Transmission and Control, University of Technology Aachen, RWTH.

WEAR OF HYDRAULIC COMPONENTS

What happens if solid particles intrude into a hydraulic circuit? - What are the main interactions? - What are the influencing parameters?

The first figure shows the typical tribological systems of hydraulic components, the characteristic wear interactions in case of solid contamination, and the most important influence parameters.

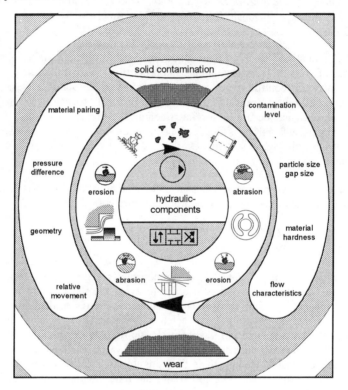

Fig. 1: Wear of hydraulic components - influence parameters

According to the tribological system under consideration, i.e. the system structure and the dominating kinematics, respectively the tribological stress, different kinds of wear occur in hydraulic systems.

Erosion and 3-body abrasion are the main kinds of wear with regard to the solid contamination in the field of hydraulics. The 3-body abrasion is distinguished by the system structure solid to solid and particle. The tribological stress is produced by the relative movement between basic- and counter-body.

Erosion provides a different kind of wear. Its characteristic system structure consists of solid to fluid with particles. In this process, the tribological stress is caused by particles which are dragged along by the flow and carry off material when they collide with the surface of the solid.

WEAR EXAMINATION

There are different possibilities to examine the wear behavior of hydraulic components. The results gained by means of component examinations under actual operating conditions are, no doubt, easily transferable, but at the same time they are very expensive and costly. A precise analysis of the active kinds of wear and their influence parameters is not possible. Therefore, special test procedures for the different hydraulic components were developed. One of the fundamental demands. of these test procedures is a defined and reproducible wear load.

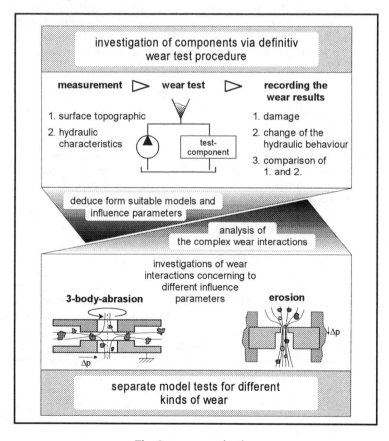

Fig. 2: wear examination

In parallel, test models for the kinds of wear which are typical of hydraulics were developed. These are the 3-body-abrasion and the erosion. Thus, it is possible to investigate the different kinds of wear separately. Influence parameters like geometry, material pairings, and viscosity of the pressure medium can easily be varied.

The results of these examinations provide the basis for the analysis of the wear mechanisms and for the development and application of appropriate measures for the reduction of wear.

The two wear test procedures for pressure-controlled pumps and proportional controlled valves are described in the following.

Wear Behavior of Control Pumps

For the purpose of examination, pressure-compensated pumps are exposed to a reproducible wear load in highly contaminated oil by means of a newly developed test procedure. The damage the test pump sustains due to wear and the impact it exerts on the behavior during operation is analyzed. Moreover, a comparative evaluation of the contamination sensitivity of various test pumps becomes possible[1].

Test-Procedure

The test conditions are set up in compliance with the agreements regarding the fixed displacement pump wear test (ISO/DIN 9632). The same provisions are valid for the run-in program which the test-pumps pass before the actual wear test. The single steps of the run-in program are evident in Figure 3, which represents the entire test-procedure for the examination of the wear behavior of pressure-compensated variable displacement pumps. After the run-in program or, respectively, the pretests are terminated, the rheostat is adjusted in such a manner, that the test-pump regulates at 50% of the maximum stroke at the upper test pressure of 210 bar. Then, the test circuit is contaminated with the first dust injection, and the load program is started. The load program is supposed to evoke wear at the control valve and the adjusting piston by certain movements. For this purpose, two different alternating nominal pressure values are given at the pressure-compensation valve of the pump, which is working against the set resistance. An external oil pressure is responsible for the setting of the nominal pressure values at the pressure compensation of the test-pumps. The load pressure p is switched between an upper value (p_o = 210 bar) and a lower value (p_U = 100 bar). Meanwhile, the test-pump changes its displacement quantity as described before. The cycle takes 10 sec., which is equal to a frequency of f = 0,1 Hz. The switching is conducted according to the displayed ramps which were selected in a way that they are as steep as possible in terms of load and that they can also be achieved by dynamically slower test-pumps in terms of comparability. As a consequence of high displace-velocities, undesirable pressure-failures can occur in the suction port of the test-pumps. Therefore, the displacement must not be increased too rapidly.

When the load program has run for 30 min. the next dust injection is given. When all fractions have been injected at 30 min intervals or when the upper test pressure cannot be adjusted any longer due to high leakage losses, the test is terminated.

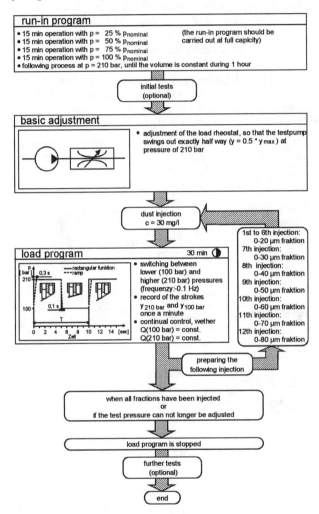

Fig. 3: Test procedure for the examination of the wear behavior of pressure-compensated variable displacement pumps

Wear Behavior of Proportional Control Valves

Apart from the wear test for pressure-compensated pumps, the IFAS (Institut of Fluidpower Transmission and Control, RWTH Aachen) developed a wear test for proportional valves. The aim of this research project consists in the examination of damages occurring due to wear and their effects on the operation- and transference-behavior of the valves. In practice, basically two symptoms of wear proportional directional control valves are observed. These are firstly erosion, which leads to a washing-out and rounding off in the control edge area, and secondly abrasion, which causes a widening of the clearence between valve spool and housing. Both wear symptoms occur as consequences of the valve operation in particle-contaminated oil. The presence of particles in the oil does not only affect damages due to wear in proportional directional control valves but can also lead to direct operation failures. A phenomenon especially feared is the clamping of the valve spool caused by silting. Despite the fact that the silting of a valve does not directly relate to its wear behavior, attempts were made to integrate respective silting measuments into the wear test. In order to evoke wear symptoms and silting effects in the wear test, the test valves are exposed to extremely critical operation situations in contaminated oil.

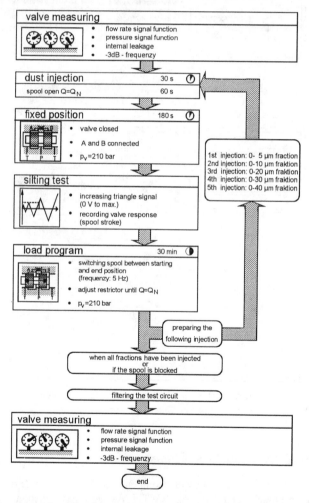

Fig. 4: Test-procedure for the examination of the wear behavior of proportional directional control valves

Test-Procedure

A highly effective method of bringing about both wear symptoms, erosion and abrasion, on the test valves in such a way that they can be distinguished, consists of a combined load program. The spool continuously oscillates between the starting position and the final position. During this movement, two control-edges of the spool are lapped with the housing. These valve areas are primarily loaded by abrasion and only secondarily loaded by erosion. The starting position of the spool is defined in such a way that both control-edges are not cover up with the housing. As a consequence, the corresponding areas of the spool and the housing can only be damaged by erosion. By means of pretests, this combined load program proved to be successful. The differently loaded control-edges showed clearly distinguishable damages due to wear. The load program was carried out with ACFTD-dust fractions of 0,5 µm, 0-10 µm, 0-20 µm, 0-30 µm and 0-40 µm sized particles. The amount of dust of the cumulatively added fractions was adjusted in such a way that high - in course of the test, continuously increasing - damage due to wear of the test-valves resultes from the contamination. An increase of the system contamination by 30 mg/l per injection turned out to be favourable (cf. pump wear-test). Figure 4 represents the entire test-procedure.

Measurements of silting at such high dust concentrations were found difficult because spools often tend to total blocking. Moreover, it is not possible to determine the dependency of the silting behavior and the particle size due to the cumulative dust addition. The basic examination of the silting behavior requires an independent test procedure[4] and oil contaminant levels specifically in the 5-µm or less particle size range[6].

RESULTS FROM THE WEAR INVESTIGATIONS

By means of the pump wear test, pressure controlled variable displacement pumps of different construction type and design were examinated. In these tests, serious impairments of the hydraulic behavior in the form of leakage increase and control malfunctions were observed at the test-objects damaged by wear. The leakage increase is above all brought about at the displacement assemblies and at the control valves. The increase of the valve leakage depends on the respective working points. Especially with regard to governed pumps, it is often higher than the displacement leakage increase. The increase of setting system leakages can throughout be described as insignificant. About 50% of all examined pumps showed faults in the control behavior which are caused by wear damage at the control valves. Extended examinations of the surface topography of the tribosystems damaged by wear indicate that the wear behavior of the tribosystems is extremely dependent on the design!

The wear test for proportional directional valves allows a separate evaluation of the valve damage caused on the one hand by erosion, on the other hand by 3-body-abrasion. Thus, it appeared that the wear damage brought about by 3-body-abrasion clearly prevail. With regard to the conveyance behavior of the valves, particularly the pressure signal function is impaired by the occurring wear processes. The abrasive damage of the valves is also responsible for the high increase of the internal leakage.

The flow signal function merely shows slight changes. These wear tests support the designer in determining the interdependencies between wear behavior and design and in deducing from this instructions for a design which takes the wear processes into consideration.

INSTRUCTIONS FOR THE WEAR CONTROLLED DESIGN OF HYDRAULIC COMPONENTS

The results presented in the following were derived partly from model tests, partly from component tests. They are supposed to make the designer realize the relevance of the wear problem, as the wear sensitivity of the components is decided or rather determined in and by the design. Thus, the geometry of the tribosystems exerts a significant impact on the wear result.

Support and Sealing Ridges

The first example in figure 5 considers the design of hydrostatically relieved bearing points in hydraulic components. Wear and changes of the hydraulic properties (e.g. volumetric losses) are considerably influenced by the use of so-called support ridges. The model test demonstrates that support ridges to which no pressure difference is adjacent (which distinguishes them from the sealing ridges) simply do not wear. In this they differ from the sealing ridges.

Due to the lack of a pressure difference, none or only a very small number of particles intrude the gap. Thus, there can be no or only minimal wear. The sealing ridge, on the contrary, widens more and more. The consequence of this consists of an obvious increase of the internal leakage. The volumetrical efficiency decreases.

In this, the support ridges impede the resetting process which would otherwise compensate the clearance widening and reduce the internal leakage.

This is illustrated by means of the two examples shown in the figure

- piston shoe/swashplate and
- cylinder/valveplate

However, also the material choice of the tribopartners (basic- and counter-body) has an important impact on the wear behavior.

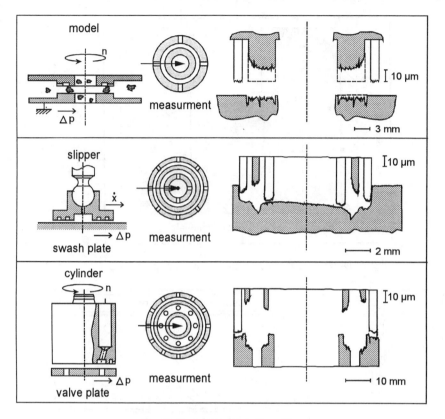

Fig. 5: Support and sealing ridge

Material Choice of the Tribopartners

The model test of figure 6 shows that especially the soft materials, e.g. bronze, are exposed to wear. This high wear rate can be reduced by hardening the counter-body. Even better, however, is the substitution of soft bearing materials by modern hard material coats.

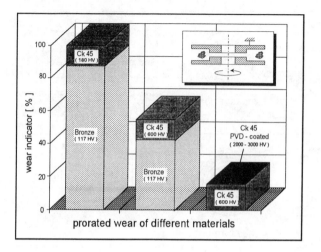

Fig. 6: Different materials have different wear properties

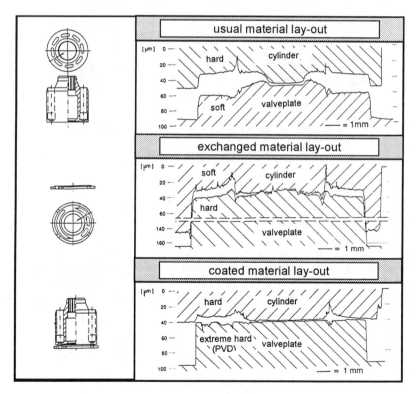

Fig. 7: Wear reduction by new materials in the tribo-system: cylinder/valveplate[3]

The results of the component tests illustrate the different possibilities of wear reduction by the adequate choice of materials here by the example of the tribo-system cylinder/valve-plate (Figure 7).

The figure shows the condition of the surface topography after the wear test for the standard material pairing, for the interchanged material pairing, and for the version with PVD-coated valve-plate. If the wear-dependent clearance widening still exceeds 30 μm in the first case - (usual material layout) -, this can be significantly reduced by interchanging the materials (second case). Nevertheless, the wear causes an amount of materials which should not be neglected to be carried off (= particle generation). This process can only be stopped by means of the substitution of the bronze. In the third case, on the basis of the standard case, the soft valve-plate (bronze), was substituted by a hard, PVD-coated valve-plate.

Thus, clearance widening and material abrasion can be reduced to a warrantable level. The same is valid for the tribo-system valve-spool/housing, as the following example shows.

After the wear test, a standard valve with cast housing and a spool made of hardened steel shows an evident widening of the clearance in the area of the sealing ridge.

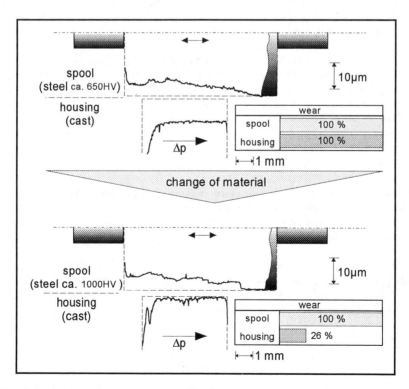

Fig. 8: Reduction of wear by means of hardened material for the tribo-system: valve

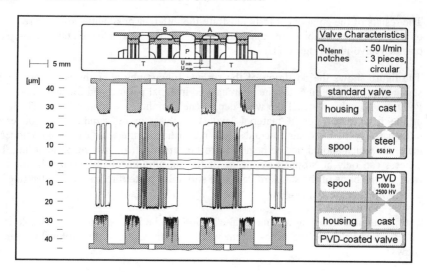

Fig. 9: comparison PVD/standard valve

This can be reduced by means of further hardening the spool. Particularly the wear at the housing web can be decreased. This is due to the fact already shown by means of the model test (Fig. 6) that a hardening of one of the tribopartners prevents the storing of the particles and thus reduces the wear of the other unchanged tribopartner.

PVD-coats can also improve the wear behavior[2]. Figure 9 presents the surface of a proportional directional valve after the wear test. The upper spool is a standard one the lower spool is PVD coated. Parallel to the reduction of the component damage, also the hydraulic performance improves. Apart from the choice of the materials, the geometry of the tribo-systems is decisive for the wear behavior.

Geometry

Particularly the length of the clearance is directly connected to the wear result. It determines the contact surface in the clearance, which interacts with the wear mechanisms. Furthermore, it influences the flow form or velocity, respectively, and thus the portions of erosion and abrasion.

Adequate model tests with different clearance lengths (Figure 10) allow us to conclude that, above all, the contact surface determines the wear. A reduction of the clearance length leads to the diminution of the contact-, and, thus, efficient wear-surface. This results in an evident increase of wear or a clearance widening.

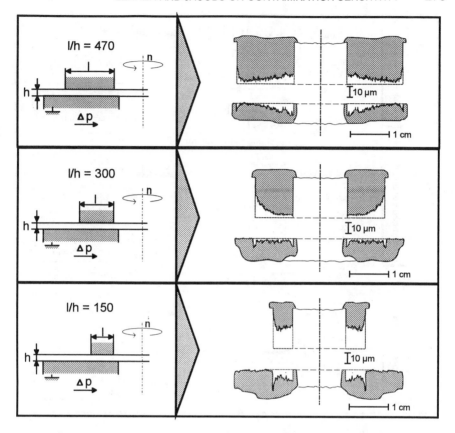

Fig. 10: Increase of wear due to the deminuition of the wear area

Corresponding tests at a valve confirm these results. Figure 11 shows the geometry variation of a valve in the area of the sealing ridge between high pressure- and low pressure-side.

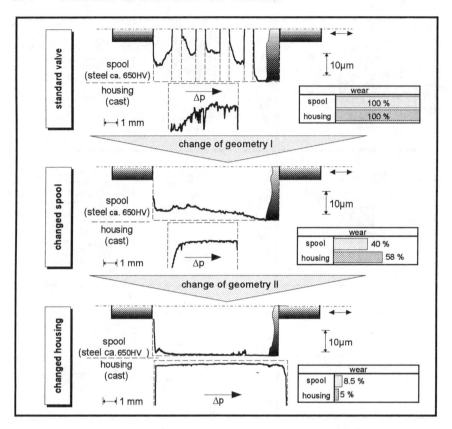

Fig. 11: Eenlargement of the wear area leads to reduction of clearence widening

On the basis of the standard valve, the contact- and thus the wear- surface between spool and housing is continuously enlarged. To this purpose, firstly the relief grooves are omitted ("changed spool") and subsequently, in another version, the housing web is elongated ("changed housing"). What the surface measurment presented in the example already indicates is illustrated by the comparison of the clearance widening middled above the extent. With regard to the standard valve for spool and housing, it is set to 100%. By means of variation of the piston/spool, the wear can be reduced by 50% and even by 90% if the housing web is additionally lengthened. It makes no difference whether the contact surface is enlarged by means of variation of the geometry or indirectly by the form of movement (stroke), as the following example shows.

By the variation of the control notches (number and size), a different stroke is necessary with equal nominal flow. Parallel to the stroke, the contact surface and thus the wear result changes proportionally. The clearance widening, which is directly connected with the hydraulic behavior, is presented on the basis of the smallest valve (100%). Generally it can be stated that an enlargement of the control notches leads to a worsening of the wear behavior.

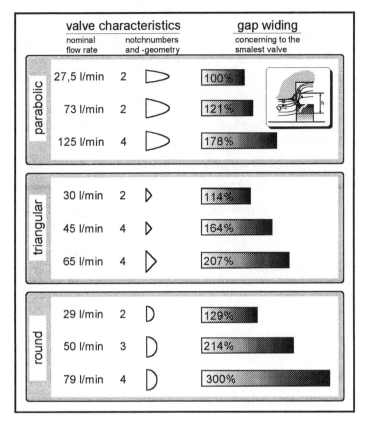

Fig. 12: Comparison of the geometry of the control notches with regard to the wear
behavior of a proportional valve

CONCLUSION

This contribution demonstrates that the design determines the wear sensitivity of
hydraulic conponents. Special test methods can point out constructive weaknesses and
can thus support the designer in rendering the systems more resistent to wear. As the
wear, however, depends on a great variety of impact parameters the applied test
procedure has to accomplish particular demands. The wear load ought to be
reproducible and the different kinds of wear should be investigated separately if possible.
The exact measurment of the components before and after the wear test constitutes a
precondition for the correlation of component damage and changed hydraulic behavior.
A detailed analysis of the wear interactions or the tribological correlations respectively
can be further supported by systematic model tests.

REFERENCES

[1]   Backé, W., "Verschleißempfindlichkeit von hydraulischen Verdrängereinheiten durch Feststoffverschmutzung", <u>Ölhydraulik und Pneumatik</u> Nr. 6, 1989

[2]   Berger, M., "PVD - Schichten in der Fluidtechnik, Verschleißprobleme und Schichtauswahl", <u>Metalloberfläche</u> Nr. 10, 1991

[3]   Jacobs, G., "Verschleißverhalten hydraulischer Pumpen und Ventile beim Betrieb mit feststoffverschmutztem Öl", <u>Dissertation RWTH Aachen</u>, 1993

[4]   Inoue, R., "Contaminant effects: Look what happends to proportional valves", <u>Hydraulics and Pneumatics</u> Nr. 11, 1984

[5]   Lehner, S., "Kleine Teilchen - Große Wirkung", <u>Ölhydraulik und Pneumatik</u> Nr. 3, 1994

[6]   Nowicki, H., "The problem of silting in sliding spool hydraulic directional valves", <u>Hydraulics and Pneumatics</u>, Nr. 11, 1993

Akira Sasaki[1]

A REVIEW OF CONTAMINATION RELATED HYDRAULIC PUMP PROBLEMS IN JAPANESE
INJECTION MOLDING, EXTRUSION AND RUBBER MOLDING INDUSTRIES

REFERENCE: Sasaki, A., **"A Review of Contamination Related Hydraulic Pump Problems in Japanese Injection Molding, Extrusion and Rubber Molding Industries,"** *Tribology of Hydraulic Pump Testing, ASTM STP 1310,* George E. Totten, Gary H. Kling, and Donald J. Smolenski, Eds., American Society for Testing and Materials, 1996.

ABSTRACT: It is known that contamination of hydraulic oil is one of the major factors causing hydraulic pump problems. Many test reports on contaminant sensibility of hydraulic pumps have been published with new oil and standard dusts but the results of these tests could not guarantee to predict in-service performance. This report describes three cases investigated.

The first investigation was done on hydraulic pumps used for injection molding machines application. The causes of pump problems were examined by analysis of maintenance records. The second investigation was performed to determine overhaul frequency of hydraulic pumps used for aluminum extruders. By introducing a new method of hydraulic oil management which reduces oil oxidation products, pump life was extended from 3000 to 15000 hours. The third investigation was done to determine the relationship between pump problems and contamination levels of hydraulic oils of 411 rubber molding machines for 20 months. The results showed that pump problems appeared at half the recommended oil lifetimes for these fluids.

These studies showed that the cause of pump problems was clogging of suction strainers leading to pump cavitation. The clogged strainers were washed with several different solvents to identify the causes of suction strainer clogging. Clogging of suction strainers was attributable to sticky oxidation products of hydraulic oils. Electrostatic oil cleaners removed not only micron range solid particles but also submicron size particles. Hydraulic pump problems have been substantially reduced by introducing this new method of contamination control.

KEYWORDS: Hydraulic fluid, cleanliness, sludge, pumps, electrostatic

---

Due to their importance, many studies on hydraulic pump problems have been performed. The results of these studies are reflected in the design, development, manufacturing and testing of hydraulic pumps, hydraulic oils and additives. However many hydraulic pump problems remain. To prevent pump problems, it is imperative that the causes be identified. In this papers, the results of long term studies of the root

---

[1]Managing Director, KLEENTEK Industrial Co., Ltd., 2-7-7, Higashi-Ohi, Shinagawa-ku, Tokyo, Japan.

causes of hydraulic pump failures which were conducted with engineers and maintenance teams of three companies in Japan will be reported.

The first study was conducted with hydraulic pump problems associated with injection molding machines. The maintenance team had kept daily maintenance records of over 50 injection molding machines for 6 years. The root causes of these pump problems have been reported previously [1].

The second study was conducted over 9 years to increase the maintenance interval of hydraulic pumps used with aluminum extruders. By introducing a new method of contamination control of hydraulic oils, the maintenance interval was extended from 3000 hours to 15000 hours [2].

The third study was performed to determine the relation between pump problems and the contamination level of hydraulic oils used in 411 rubber molding machines for 20 months. The investigation revealed that pump problems appeared at half the standard criteria of oil change that machine makers and oil makers suggested [3].

The results of these investigations were compared with the recent study by Japan Lubricating Oil Society [4]. It was confirmed that the causes of the hydraulic pump problems were attributable to contamination of hydraulic oils and inadvertent negligence of contamination control with respect to both maintenance and hydraulic circuit design. Electrostatic liquid cleaners (ELC), which could remove not only large particulate contaminants but also oil insoluble polymerized oxidation products, were introduced for contamination control of hydraulic oils. Almost all pump problems have been solved with the use of ELC.

INVESTIGATIONS

First Investigation

In order to establish the number and the type of hydraulic pump problems with Zinc (Zn)- type anti-wear hydraulic oil, a one-year primary investigation was performed from June 1978 through May 1979 on the 51 injection molding machines with conventional in-line filters, suction strainers, periodic oil changes and occasional mechanical filtration. Four ELCs were used for side-loop cleaning of the 51 injection molding machines in June of 1978. Based on the favorable results obtained, ELC units are now mounted on each machine. The results of the investigation are shown in the Table 1.

During the primary investigation period, 26 occurrences of pump problems were recorded. Twelve (12) cases were pressure failure due to excessive wear of pumps and nine (9) cases were pump cavitation due to clogging of suction strainers. The pumps which lost pressure had cavitation when the maintenance records were retroactively examined. This suggests that clogging of suction strainers was the possible cause of cavitation to pumps, which then lead to pump wear and finally to pressure failure. These failures were attributable to hydraulic circuit designs, which incorporated suction strainers on the inlet side of the pump, to provide fail-safe and reliable hydraulic systems. Pump cavitation was corrected by changing or removing the clogged suction strainers. This showed that the cause of pump cavitation was suction strainer clogging.

TABLE 1--Pump Problems.

Primary Investigation Period				The Period when Contamination Control by ELC has been applied			
Pump Problems	Period	June 1978 - May 1979		Jan. 1981 - Dec. 1981		Jan. 1983 - Dec. 1983	
Downtime			Down-time		Down-time		Down-time
Phenomena	Cause	Case	hour	Case	hour	Case	hour
Pressure Failure	Wear	12	259	1	12.4	0	0
Breakdown	Oil Starvation	1	122	0	0	0	0
Vibration	Filter Clogging	4	11.5	2	41	1	0.5
Cavitation	Strainer Clogging	9	69.5	3	20	0	0
Total		26	462	6	36.4	1	0.5
Machines		51		54		57	

Clogging of suction strainers and pressure line filters reduces oil flow and causes irregular cycle time and cycle time loss. Irregular cycle time produces defective products. Oil starvation due to pump cavitation causes high noise and poor lubrication. High noise makes the working environment unbearable and repair work produces industrial waste.

Introduction of advanced contamination control reduces pump cavitation due to suction strainer clogging. Oil samples were analyzed by a gravimetric method (SAE ARP 785). Contaminants in the oils, before and after the use of ELC, were extracted with toluene, and toluene-soluble and toluene-insoluble fractions were compared quantitatively. The results of this separation are summarized in Table 2.

Table 2 indicates that the absolute quantities of inorganic materials, such as wear metal contaminants, are low when contamination control is used with hydraulic oils and that the majority of these contaminants obtained when proper control methods are not used are toluene-soluble resinous materials. This study showed that contamination control which could remove resinous matter was very important for preventing pump problems.

TABLE 2--Comparison of Contaminants in the Oil
Before and After Contamination Control.

| Kinds of Contaminants | Oil Samples | |
	Before contamination control by ELC	After contamination control by ELC
Insoluble in Petroleum ether	13.6 mg/100 ml (100%)	0.30 mg/100 ml (100%)
Soluble in toluene	9.14 mg (67.2%)	0.02 mg (5.0%)*
Organometallic compounds	3.96 mg (29.1%)	0.12 mg (39.3%)
Inorganic compounds	0.50 mg (3.6%)	0.16 mg (55.7%)

*Obtained by taking the difference between the petroleum ether insoluble
fraction and the toluene insoluble fraction.

Second Investigation

Hydraulic pumps of aluminum extruders were examined. The number of
the extruders were eight (8) and each extruder had four (4) high
pressure (210 kg/cm^2) plunger pumps. Each pump had suction strainers to
protect pump. These pumps had been overhauled every 3000 hours to check
pistons and bearings in accordance with recommendation of pump makers
and the strainers had been cleaned at the time of overhaul of hydraulic
pumps.

A new method of oil management was introduced in 1977 to enlarge
the capacity of the oil cooler to maintain the oil temperature at 405
and to use electrostatic oil cleaners to remove not only large and solid
contaminants but also polymerized oil oxidation products. Being
confirmed that the conditions of the pump pistons and the pump bearings
were substantially improved by implementing a new contamination control,
the overhaul interval of pumps was changed from 3000 hours to 4000 hours
in 1982. Thereafter, the overhaul interval of pumps was annually
extended by 1000 hours. In 1985, the maintenance standard was revised
and the pump overhaul interval was set every 15000 hours. The number of
the overhauled pumps were listed in the Table 3. The success was
attributable to reduction oil oxidation by lowering oil temperature and
removing oil oxidation products by electrostatic oil cleaners. Pump
cavitation was substantially reduced by reduction and removal of oil
oxidation products.

Third Investigation

The pump problems of 411 rubber molding machines were investigated
in relation to contamination levels of the hydraulic oils for twenty
(20) months. All of the molding machines had suction strainers. Criteria
of oil change recommended by oil makers and machine makers had been 10
mg/100 ml. But pump cavitation had already been reported at
contamination levels of 5 mg/100 ml and lower. Pump cavitation was
corrected by removing or cleaning the suction strainers. It suggested
that pump cavitation was caused by clogging of the suction strainers.
One of the suction strainers was cut for investigation, Figures 1 and 2.

Figure 3 is a typical example of deposits of contaminant on the
supporting mesh before washing with petroleum ether. When it was washed
with petroleum ether in an ultrasonic bath, oil and fragile fragments of

TABLE 3--Interval of Pump Overhaul and the
Overhauled Pump Number.

Year	Overhaul Interval	Overhauled Pump Number
1977	3,000 hours	25
1978	3,000 hours	20
1979	3,000 hours	19
1980	3,000 hours	17
1981	3,000 hours	13
1982	4,000 hours	15
1983	5,000 hours	11
1984	6,000 hours	8
1985	15,000 hours	8

contaminant were washed off but some of contaminants like welding spatters or glue on the mesh were retained as shown in Figure 4. The mesh was again washed with toluene in an ultrasonic bath. Strongly glued foreign materials still remained on the mesh as shown in Figure 5.

The mesh was further washed with pyridine in an ultrasonic bath. Almost all of the foreign materials on the mesh were removed with Pyridine, as shown in Figure 6. Fractions extracted with toluene and pyridine were analyzed by infra-red (IR) Spectroscopy as shown in Figures 7 and 8.

The IR Spectra of the fraction extracted with toluene showed strong absorptions of -OH at 3200-3400 cm^{-1} and -COO at 1720 cm^{-1} which indicate the existence of oxidation products of hydrocarbon like fatty acids, and an apparent absorption of -COO-Me at 1600 and 1400 cm^{-1} which indicates reaction of fatty acid with metal ion [5]. The IR spectra of the fraction extracted with pyridine showed strong absorptions at 1600 cm^{-1}, 1400 cm^{-1}, and 1150 - 1100 cm^{-1} which indicate the existence of metal carboxylate [5]. The results of solvent extraction indicated that the contaminants of the mesh are organic materials and organic metal salts which are soluble in organic solvents.

DISCUSSION

Common Problems of the Three Case Studies

Existence of oil oxidation products and clogging of suction strainers were common to the hydraulic pump problems in the three cases studies. Suction strainers were used to protect hydraulic pumps but the clogged strainers caused problems to hydraulic pumps contrary to expectations.

Contaminants in Working Hydraulic Systems

This is one of the typical cases of the difference between test results in laboratories and in-service performance. In general filter performances are tested and evaluated by multi-pass test methods with new oil and ACFTD. However the majority of components of contaminants in working hydraulic systems are not solid particles like ACFTD but toluene soluble resinous matter, as shown in Table 2. Resinous matter has polarity and is adsorbed onto metal surfaces. It is sticky and attracts other solid and fiber contaminants which then stick to them. By this mechanism, suction strainers will be clogged.

The multi-pass filter test specification did not take the sticky contaminants into consideration. Even maintenance people do not like to deal with dirt or contaminants. Designers of hydraulic systems must also account for this problem.

New Finding of the Problems of Suction Strainers

A suction strainer is actually a full flow filter. Hydraulic oil passes through a suction strainer at very high velocity. This means that both oil and the strainer will be electrified by friction of oil with the strainer. Although suction strainers are made of metal, the mesh of pleat-type suction strainers are sometimes insulated at both end caps where resin is filled to seal. A strainer with openings at both the ends must be installed in a housing with O-rings insulating the strainer. The conductivity of used strainers before washing with solvent was measured as shown in Figure 9.

Some of the strainers were insulated by the coated resinous contaminants. When oil passes through insulated strainers, static electricity will be accumulated on the strainers and the accumulated electric charge will be discharged by sparks. Spark discharge of static electricity will crack oil molecules and produce free radicals. Free radicals will propagate autoxidation of oil and produce polymerized oil oxidation products (resinous matter). This cycle continues, as long as full-flow filters are used in hydraulic systems.

Investigation Conducted by the Japan Lubricating Oil Society

Japan Lubricating Oil Society investigated the lubrication problems by sending questionnaire to 2034 factories from November, 1994 to December, 1994. 511 replies were received in total [4].
The number of the answers to hydraulic pumps was 123. The majority of the answers to the questions about hydraulic pump problems was degradation of pump performance. The number of the answers to the questions about the causes of hydraulic pump problems was 111, of which the three major answers were (1) poor maintenance, (2) oil contamination and (3) improper design of the hydraulic systems. These conclusions were confirmed by the studies reported here.

Fig. 1   Metal suction strainer used for 1 year.

Fig. 2   The cut part of the strainer mesh.
The supporting large mesh was cleaned for analysis.

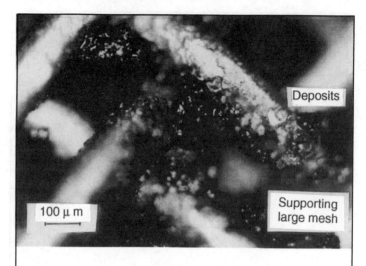

Fig. 3   Typical deposits on supporting mesh.

Fig. 4   Deposits on supporting mesh after
washing with petroleum ether.

Fig. 5   The mesh after washing with toluene.

Fig. 6   The mesh after washing with pyridine.

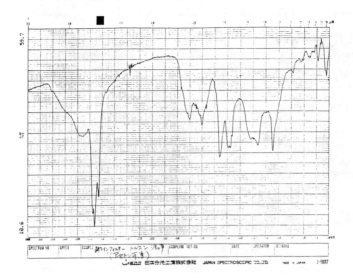

Fig. 7   IR Spectrum of toluene extracted fraction

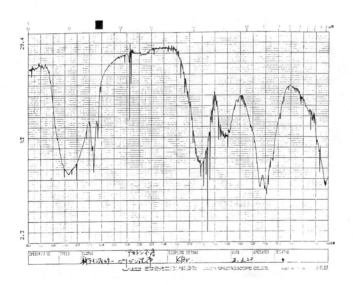

Fig. 8   IR Spectrum of pyridine extracted fraction

CONCLUSIONS

1. Pump problems are caused by clogged suction strainers with oil oxidation products.

2. Suction strainers have some possibility to generate static electricity by the friction of oil with the strainers. Spark discharge of static electricity propagate oil oxidation, through free radical formation.

3. It is recommended not to use suction strainers in hydraulic systems, if possible. Rather, it is recommended to that a comment be included in the pump test specification stating that the results are valid for test only and that it may not be reflective of in-service performance in factories.

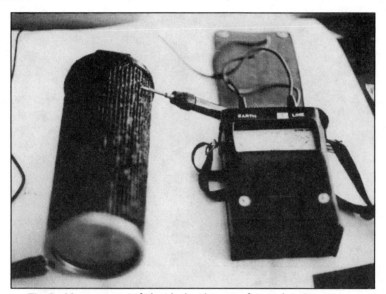

Fig. 9   Measurement of electrical resistance of a used strainer.

ACKNOWLEDGMENT

The author would like to express his thanks to Mr. Sasaoka of Kanto Seiki K.K., Mr. Naruse of Sumitomo Light Metal Industries Ltd. and Mr. Takayama of Toyoda Gosei K.K. for their valuable data.

REFERENCES

[1]   Sasaki, A & Sasaoka, M., Tobisu, T., Uchiyama, S., and Sakai, T., "The Use of Electrostatic Liquid Cleaning for Contamination Control of Hydraulic Oil", Lubr. Eng. **44**, 3, pp 251-256 (1988).
[2]   Naruse, K., "Maintenance Record Sumitomo Light Metal Industries Ltd. from 1977 to 1985" (Japanese).
[3]   Toyoda Gosei K.K., "Maintenance Records from Jan., 1989 to Aug., 1990" (Japanese).
[4]   Japan Lubricating Oil Society, "Investigation Report for Promotion of Efficient Lubrication Management" (Translation from Japanese), Tokyo Japan (1995).
[5]   Horiguchi, H., "IR Spectroscopy Chart Soran" (Japanese), Sankyo Shuppan, Tokyo, Japan, 1977.

Bench Test Development

Paul I. Lacey,[1] David W. Naegeli,[2] and Bernard R. Wright[2]

TRIBOLOGICAL PROPERTIES OF FIRE-RESISTANT, NONFLAMMABLE, AND PETROLEUM-BASED HYDRAULIC FLUIDS

REFERENCE: Lacey, P. I., Naegeli, D. W., and Wright, B. R., **"Tribological Properties of Fire-Resistant, Nonflammable, and Petroleum-Based Hydraulic Fluids,"** *Tribology of Hydraulic Pump Testing, ASTM STP 1310,* George E. Totten, Gary H. Kling, and Donald J. Smolenski, Eds., American Society for Testing and Materials, 1996.

ABSTRACT: Petroleum-based hydraulic fluids qualified under MIL-H-6083 are highly flammable. As a result, a fire-resistant polyalphaolefin (PAO)-based fluid, qualified under MIL-H-46170 and developed in the 1970s, and more recently, a nonflammable chlorotrifluoroethylene (CTFE) hydraulic fluid, were developed as potential replacements. However, the chemical and physical properties of both fluids are significantly different from those of conventional fluids. This paper outlines a laboratory study to define the interrelated parameters of degradation and wear of contacts lubricated with various hydraulic fluids, with particular reference to CTFE. The results rank the likely wear, corrosiveness, and oxidation properties of CTFE in comparison to currently used silicone-, PAO-, and petroleum-based fluids. In general, the antiwear characteristics of the CTFE hydraulic fluid were found to be similar to those of petroleum-based fluids but marginally inferior to those of PAO-based fluids. However, CTFE produced severe corrosion of brass at temperatures above approximately 135°C. As a result, it is believed that operation at very high temperatures is likely to cause unacceptable material removal from copper-based metals. Several surface treatment processes were identified to minimize potential side effects, evident under more severe operating conditions with CTFE.

[1] Senior Research Engineer, Automotive Products and Emissions Research Division, Petroleum Products Research Department, Southwest Research Institute, P.O. Drawer 28510, San Antonio, TX 78228-0510.

[2] Staff Scientist, U.S. Army Tank-Automotive Research, Development and Engineering Center Fuels and Lubricants Research Facility, Southwest Research Institute, P.O. Drawer 28510, San Antonio, TX 78228-0510.

KEYWORDS: Friction, wear, corrosion, hydraulic fluid, flammability, surface coating

---

INTRODUCTION

Hydrocarbon-based hydraulic fluids qualified under military specification MIL-H-6083, "Hydraulic Fluid, Petroleum Base, for Preservation and Operation," have been used as the recoil fluid in artillery and as the hydraulic fluid in most Army ground equipment since World War II. Military aircraft have used the equivalent MIL-H-5606, "Hydraulic Fluid, Petroleum Base; Aircraft, Missile, and Ordnance." Although both perform well in most applications, they are highly flammable. In the 1970s, PAO-based fluid qualified under MIL-H-46170, "Hydraulic Fluid, Rust Inhibited, Fire Resistant," (and MIL-H-83282, "Hydraulic Fluid, Fire Resistant Synthetic Hydrocarbon Base, Aircraft"), was developed as a replacement for fluids qualified under MIL-H-6083 (and MIL-H-5606). The PAO fluid proved to be less flammable with slightly better wear characteristics. It has since been widely used in military vehicles but remains a fire threat under certain conditions. Work was initiated in 1975 at the Air Force Materials Laboratory at Wright-Patterson Air Force Base, OH, to develop a thermally stable, nonflammable hydraulic (NFH) fluid [1]. A range of fluids was considered, including phosphate esters, silicones, cyclic esters, and fluorinated phosphonates. CTFE was ultimately selected, primarily due to its excellent fire resistance, although cost, availability, useable temperature range, viscosity-temperature properties, thermal stability, additive solubility, density, and compressibility were also considered.

CTFE is compatible neither with the existing fluids qualified under MIL-H-46170B and MIL-H-6083E nor with current hydraulic systems. A considerable amount of research has already been performed with CTFE, and a number of problems have been resolved. During development, the CTFE-based fluids failed to meet a number of target property requirements, including bulk modulus, wear rate, and viscosity index (VI) [2]. The relative change in viscosity with increasing temperature was initially controlled using a VI improver. However, VI improvers were not included in later fluids, as viscosity loss at high shear rate was experienced with the VI-improved CTFE. In addition, the properties of the baseline fluid alone were found to be sufficient after the requirements for VI were relaxed. Nonetheless, many contacts in current hydraulic systems are partially separated by a hydrodynamic film, and the viscosity and pressure-viscosity coefficient of the NFH fluid are appreciably less than that of the petroleum and PAO fluids [3, 4].

The net result of the reduced thick film lubrication is likely to be increased intermetallic contact between components that are in relative motion. In addition, discoloration and possible corrosion of other materials, such as bronze, have also been observed in full-scale pump loop tests [2]. The original CTFE basestock produced relatively severe wear when tested under boundary conditions and failed to meet the 1.0-mm wear scar diameter in the four-ball wear test required for petroleum-based oils qualified under MIL-H-6083 [2]. Many additives, such as tricresyl phosphate, that are commonly used in hydrocarbon-based oils are not effective in CTFE [2] due to a number of factors, including poor additive solubility. However, several additives, including molybdenum dialkyldithiophosphate, have produced a dramatic reduction in the wear rate observed with the base fluid. The formulated CTFE fluid used in the present work contains a fluorinated sulfonamide and a zinc dinonylnaphthalene sulfonate as antiwear and antirust agents, respectively.

APPROACH

The primary objective of this paper is to define both wear resistance and high-temperature stability of a CTFE-based hydraulic fluid qualified under MIL-H-53119, "Hydraulic Fluid, Nonflammable, Chlorotrifluoroethylene Base," using the oils described in Table 1 as a baseline for comparison. The baseline fluids include a fully formulated polyalphaolefin qualified under MIL-H-46170B; a silicone-based fluid qualified under MIL-B-46176, "Brake Fluid, Silicone, Automotive, All Weather, Operational and Preservative"; and a 100-mm^2/s silicone fluid composed of polydimethyl-siloxane that is not qualified under military standards. The CTFE fluid qualified under MIL-H-53119 is composed of approximately 99 percent CTFE basestock complemented by small amounts of corrosion inhibitor and antiwear

TABLE 1--Principal characteristics of test fluids.

Fluid Specification No.	Base Fluid	Fire Resistance	Kinematic Viscosity @ 100°C, mm^2/s	Specific Gravity	Pour Point, °C	Bulk Modulus, kPa, min.
MIL-H-6083E	Petroleum	Flammable	5.2	0.86	−59	1.45×10^6
MIL-H-53119	CTFE	NF^A	1.0	1.7	−60	1.24×10^6
MIL-H-46170B	PAO	FR^B	3.6	0.85	−54	1.38×10^6
MIL-B-46176	Silicone	FR	12.6	...	...	...
None	Silicone	FR	31.0	0.964	−65	...

^{A} Non-flammable
^{B} Fire resistant

additives. The formula for the CTFE fluid is $Cl(CF_2CFCL)_nCl$, with n normally equal to 3. Some higher molecular weight oligomers are also present. A more detailed description of the test methodology for corrosion and wear resistance follows.

Test Methodology for Evaluation of Oil Stability

High temperatures are generated within many modern highly loaded hydraulic systems. MIL-H-5606F, MIL-H-6083E, MIL-H-46170B, and MIL-B-46176 each require a minimum corrosiveness and oxidation stability according to Federal Standard 791C, Method No. 5307, "Corrosiveness and Oxidation Stability of Aircraft Turbine Engine Lubricants," and Method No. 5308, "Corrosiveness and Oxidation Stability of Light Oils (Metal Squares)." In the present study, the effects of oil degradation in a real environment are simulated using a modified version of ASTM Test Method for Corrosiveness and Oxidation Stability of Hydraulic Oils, Aircraft Turbine Engine Lubricants, and Other Highly Refined Oils (D 4636), which is an amalgamation of Federal Standard 791C, Method Nos. 5307 and 5308. Test coupons of C-93200 high-lead tin bronze, copper alloy NR 955, and 4140 steel were immersed in an oil bath to act as catalysts and to evaluate the fluid corrosivity. Metals which may be prone to corrosion by the MIL-H-53119 fluid were given particular attention.

The MIL-H-53119 specification requires that the fluid be suitable for use in the temperature range −54 to 135°C. Previous work with CTFE basestock has shown that the fluids degrade at temperatures above 232°C [1], although the presence of copper, iron, and tin accelerated the degradation process and produced measurable degradation at 175°C after 72 hours. This temperature also corresponds to the Advanced Field Artillery System (AFAS) operating envelope. Oxidation-corrosion tests were performed with each of the five formulated oils detailed in Table 1 at 135 and 175°C to reflect the predicted normal and absolute maximum operating temperatures, respectively, of a gun recoil mechanism [4].

Oil samples were periodically removed from the high-temperature test reservoir, and degradation was defined from measurements of viscosity, total acid number (TAN), and mass loss from the metallic coupons. This process indicates the interrelated effects of both metal type on lubricant degradation and acidic oil reaction products on the corrosion of specific metals. In addition, selected samples were examined using Fourier transform infrared (FTIR) spectroscopy and gas chromatography (GC) in an attempt to highlight any change in the molecular structure of the oil. In instances where appreciable corrosion was present on the metallic test coupons, surface analysis was performed using X-ray Photoelectron Spectroscopy (XPS) and Auger Electron Spectroscopy (AES).

Test Methodology for Wear Resistance

Many previous studies have examined the wear resistance of hydraulic fluid using the four-ball technique [3, 5]. Four-ball wear tests were performed on each of the fluids in this study in accordance with ASTM Test Method for Wear Preventive

Characteristics of Lubricating Fluid (Four-Ball Method) (D 4172). However, the unidirectional sliding of a highly loaded, counterformal AISI E-52100 steel contact in the four-ball test may not reflect the conditions present in many applications, such as a gun recoil brake mechanism.

The Cameron-Plint wear test apparatus (Figure 1) is intended primarily for the rapid assessment of the performance of lubricants and lubricant-metal combinations. In this test, an upper specimen slides on a lower plate with a pure sinusoidal motion, driven by a variable speed motor. The lower (fixed) specimen is mounted in a stainless steel oil bath attached to a heater block. The opposing (oscillating) specimen is loaded using a spring balance, with an applied load between 0 and 250 N. The resulting friction force may then be measured using a piezoelectric force transducer.

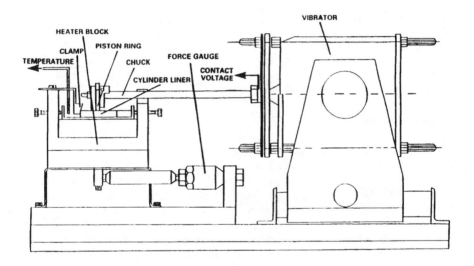

FIGURE 1--Schematic diagram of the Cameron-Plint wear test apparatus.

Hydraulic fluids are expected to perform satisfactorily in a wide variety of military systems. As a result, the present study attempted to compare fluid performance over the complete range of conditions available on the Cameron-Plint. The majority of laboratory-scale wear tests were performed using a hemispherically ended pin on a flat geometry. The upper C-93200 bronze specimen consists of a spherical end machined on commercially available rod stock, with a surface finish of approximately 0.3 μm. The metal was selected due to the susceptibility of C-93200 bronze to corrosion by CTFE fluid [4]. The opposing 4140 steel test flat is highly polished to a root mean square (rms) surface roughness of 0.04 μm using 1-μm diamond paste. This minimizes abrasion and adhesion and emphasizes the effects of corrosive wear.

The effects of elastic deformation on the unworn counterformal specimens were calculated using hertzian equations [6]. The apparent material loss due to elastic deformation is appreciably less than the true wear volume, making it negligible in practice. Prior to tabulation, the calculated wear volume was normalized using the fundamental equation discussed by J.F. Archard [7, 8]. The law for a contact may be expressed as follows:

$$K = \frac{VH_v}{3Ld}$$

where

K = Archard's wear coefficient,
V = wear volume, mm^3,
H_v = surface hardness, kg/mm^2,
L = applied load, kg, and
d = sliding distance, mm.

RESULTS

Oxidation-Corrosion Tests

The change in TAN measured during oxidation-corrosion tests at 135°C is shown in Figure 2. The TAN measured in tests performed at 175°C and corresponding viscosity measurements may be obtained in Reference 8. The TAN and viscosity of the highly stable silicone fluids remained unchanged throughout the duration of the 1100-hour oxidation-corrosion tests at both 135 and 175°C. In contrast, the petroleum- and PAO-based fluids experienced a break after an induction period of a few hundred hours at 135°C. An induction period of this kind, followed by rapid degradation, is probably due to exhaustion of the oxidation inhibitor additives. No induction period was observed at 175°C, and an immediate, severe increase in acid number was observed for the petroleum- and PAO-based fluids. The change in acid number for the petroleum-based oil was accompanied by a significant increase in viscosity at both temperatures, while a slight increase in viscosity was observed for the PAO oil at 175°C. No change in viscosity was observed for any of the remaining oils at either temperature.

No change in either TAN or viscosity of the CTFE fluid was observed at any temperature, although in previous work [1] at higher temperatures (302°C), the CTFE fluid experienced increased acid number and viscosity, especially in the presence of copper (Cu), iron (Fe), manganese (Mn), tin (Sn), titanium (Ti), and zinc (Zn). Samples of the new and degraded CTFE fluid were also examined using FTIR and GC. No variation was observed between the FTIR spectra of new oil and the spectra

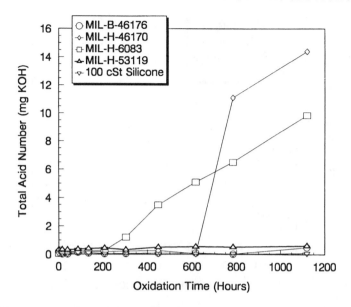

FIGURE 2--Total acid number measured during oxidation-corrosion tests
performed according to ASTM D 4636 at 135°C.

of oils oxidized at 175°C. However, considerable variation was apparent from the gas
chromatograms, indicating that some change in composition had occurred during
exposure to the severe oxidation environment.

Test coupons of C-93200 bronze, copper alloy NR 955, and 4140 steel were
immersed in the oils during the high-temperature oxidation-corrosion tests. No
corrosion of any metallic coupon was observed with either the hydrocarbon- or
silicone-based oils at 135°C, despite the high acid number achieved in some instances.
Similarly, very low levels of corrosion were observed with CTFE. As a result, the
corrosion measurements obtained at 135°C are not plotted in the present paper but
may be found in Reference 4.

The petroleum-, silicone-, and PAO-based oils did not produce severe corrosion
with copper or steel at 175°C, despite their relatively high acid numbers. (The 4140
steel received a sapphire blue coating in the silicone-based fluid; previous workers
have noted similar films [9].) In contrast, the C-93200 bronze metal was susceptible
to corrosion by each of the fluids at 175°C (Figure 3). Especially severe corrosion of
bronze was produced by the CTFE fluid, despite a stable TAN. Examination of the
corroded C-93200 specimen surface using a Scanning Electron Microscope (SEM)
shows a severely pitted topography [4]. CTFE also produced more severe corrosion of
NR 955 copper alloy at 175°C than the remaining fluids; however, the corrosion was
significantly less than that observed with C-93200. The 4140 steel was almost
unaffected by CTFE.

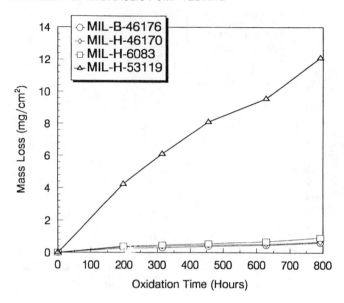

FIGURE 3--Mass loss measured on C-93200 bronze during oxidation-corrosion tests with fully formulated fluids at 175°C.

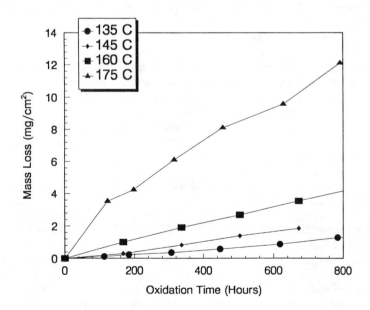

FIGURE 4--Corrosivity of MIL-H-53119 on C-93200 high-lead tin bronze at various temperatures.

Additional tests were performed at 145 and 160°C with the CTFE fluid qualified under MIL-H-53119 to define the effects of temperature on corrosion rate. The results are shown in Figure 4. As expected, the reaction rate of the bronze specimens is a nonlinear function of temperature and increases dramatically between 160 and 175°C. In each instance, the corrosion process is a linear function of time, with no initiation period required to deplete natural or artificial inhibitors. Oxidation-corrosion tests were also performed with copper, tin, lead, and zinc (>99.5 percent pure), which are the primary constituents in C-93200 bronze [4]. Severe corrosion of each specimen was observed, comparable to that of bronze.

The rate of degradation for conventional hydrocarbon fluids is highly sensitive to the availability of both oxygen and moisture. However, the severely aged CTFE fluid showed no increase in either acid number or viscosity, as would be expected for a conventional hydrocarbon. The previously detailed high-temperature corrosion test with C-93200 bronze was repeated in a nitrogen atmosphere. The remaining test parameters, including gas (nitrogen) flow rate, remained unchanged from the preceding tests. Results showed that corrosion of C-93200 bronze is significantly reduced by the elimination of oxygen at both 135 and 175°C. Some corrosion was observed at 175°C during the first 100 hours of testing, probably due to residual oxygen in the oil.

Wear Tests

Four-ball wear test according to ASTM D 4172--MIL-H-5606F and MIL-H-6083E each specify a maximum wear scar diameter of 1.0 mm for petroleum-based hydraulic fluids tested under the four-ball wear test, performed according to ASTM Test Method for Wear Preventive Characteristics of Lubricating Grease (Four-Ball Method) (D 2266). A revised procedure specifically for use with lubricating fluids (ASTM D 4172) is now available. Similarly, MIL-H-53119, which applies specifically to CTFE, requires a four-ball wear test result of less than 0.8 mm in diameter. (MIL-H-53119 also requires a 500-hour hydraulic pump loop test.) In the present work, each of the oils, except for those derived from a silicone base, produced acceptable wear in the ASTM standard test, as detailed in Table 2. (Note that the MIL-B-46176

TABLE 2--Results of four-ball wear tests performed according to ASTM D 4172.

Fluid Specification No.	Wear Scar Diameter, mm
MIL-H-6083E	0.64
MIL-H-46170B	0.38
MIL-B-46176	1.75
MIL-H-53119	0.61
100 mm^2/s silicone fluid	2.10

specification contains no minimum acceptable wear resistance.) Fully formulated CTFE produced a wear scar of 0.61 mm, which is in good agreement with previous studies in this area [2]. The PAO-based oil, qualified under MIL-H-46170B, is the most effective lubricant under these conditions. It should be noted that the four-ball procedure is at least partially affected by the extreme pressure (EP) properties of the fluid, which may not be representative of all applications. For example, EP characteristics are likely to be less relevant to the gun recoil brake mechanism, which typically consists of conformal contacts of softer metals.

Cameron-Plint tests--Laboratory-scale wear tests were performed at the conditions defined in Table 3 using the Cameron-Plint apparatus. A preliminary wear test series indicated that hydrodynamic lift has a significant effect on wear at higher sliding speeds [4]. This effect is amplified by the relatively large wear scar produced on the soft bronze specimens. As a result, all subsequent wear tests were performed at a mean sliding speed of 50 mm/s, which is the minimum speed possible on this apparatus. No tests were performed using the silicone fluid due to its relatively high viscosity of 30 mm^2/s, which makes it especially susceptible to hydrodynamic lift. In addition, the results of preceding sections indicate that its characteristics are broadly similar to those of the second silicone-based oil, qualified under MIL-B-46176.

TABLE 3--Conditions used in wear tests with counterformal contacts.

Parameter	Value
Speed, Hz	5
Speed, mm/s	50
Load, N	75
Pressure, Pa	75/(Contact area)
Contact Area, mm^2	$\pi \times$ (Wear scar diameter/2)^2
Temperature, °C	125
Duration, min	30
Amplitude, mm	4.76
Sliding distance, mm	85680

Figure 5 shows normalized wear rate as a function of temperature, with the remaining test conditions defined in Table 3. Wear rate with the CTFE fluid increased slightly with temperature. This result is unlikely to be solely due to decreased hydrodynamic lift, as no such effect is observed for the remaining oils, some of which are more viscous. At higher temperatures, the trend is temporarily reversed, and both friction and wear decrease, reaching a minimum at approximately 175°C. This decrease is repeatably accompanied by the formation of a strong contact resistance, confirming the presence of an effective chemical boundary film. Additional wear tests performed with degraded CTFE fluid (not plotted in Figure 5) indicate that the wear

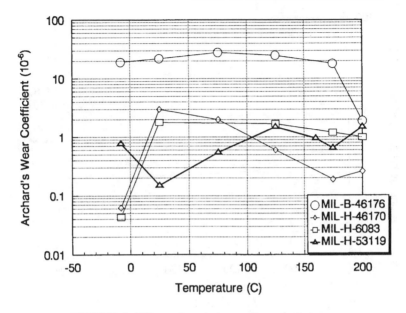

FIGURE 5--Effect of operating temperature on wear.

rate is approximately doubled by prestressing the oil, while a stronger contact resistance is also observed. Wear rate with the CTFE fluid increased at temperatures below ambient, possibly due to formation of small amounts of condensation in the test reservoir. Reference 1 indicates that the presence of moisture in CTFE greatly accelerates the corrosion rate of copper. In addition, the reaction mechanism is altered by moisture to form copper compounds that are more soluble in CTFE. Overall, however, the wear rate with CTFE was similar to that of the remaining fluids, except for MIL-B-46176, which produced very high wear over the complete temperature range. Similarly, most of the fluids produced a friction coefficient of approximately 0.2, except for MIL-B-46176, which had a friction coefficient of up to 0.5.

The effects of applied load on the measured wear rate with each of the test fluids are plotted in Figure 6. In general, a slight increase in the **normalized** wear rate is observed for each of the fluids at low loads. This result indicates that the wear rate at low loads was greater than that predicted to exist from consideration of contact forces alone (the true wear volume still decreased at low loads). Examination of the wear scar topography using an optical microscope indicated some adhesive transfer from the softer bronze specimen at higher loads. A relatively smooth topography was observed at low loads, typical of that caused by chemical corrosion. Chlorine compounds are occasionally added to lubricants to act as an EP agent which reacts with the rubbing surface at elevated temperatures, temperatures produced by either ambient conditions or by frictional heating. Such corroded surface layers may increase wear rate at low loads while acting as an EP agent at higher loads. A more detailed analysis of the surface chemistry present will be provided in the next section.

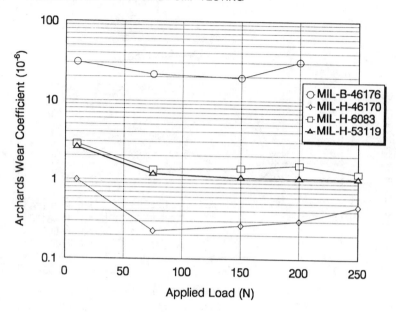

FIGURE 6--Effect of applied load on wear.

Chemical Analysis of Corrosion Mechanism

Surface analysis of C-93200 bronze test specimens was performed using AES, after sputtering to a depth of 33 Å. The relative concentration of the principal elements present on the surface of oxidation-corrosion test specimens exposed to CTFE at 135, 145, 160, and 175°C is shown in Figure 7. The data ignore any contaminants, such as carbon and oxygen, that remain following sputtering. Figure 8 shows the results of Auger analysis performed on wear scars produced during Cameron-Plint tests at a range of temperatures. Analysis was also performed on a "new" specimen that had been immersed in CTFE at room temperature to simulate the effects of unreacted CTFE remaining after the cleaning process.

No evidence of lead or zinc was found at the surface of any corroded or worn specimen[3], and more copper than tin was removed from the bronze surface as the CTFE corrosion temperature was increased. Traces of chlorine and fluorine were present on the surfaces from both the corrosion and wear tests. In each instance, the effect is temperature dependent, and neither chlorine nor fluorine was present on specimens immersed in the fluid at room temperature. The concentration of fluorine increased more rapidly than that of chlorine as temperature increased. This concentration could be derived from CTFE, the fluorinated sulfonamide antiwear

---

[3] The elemental composition of the C-93200 high-leaded tin bronze is 83 percent Cu, 7 percent Sn, 7 percent lead (Pb), and 3 percent Zn.

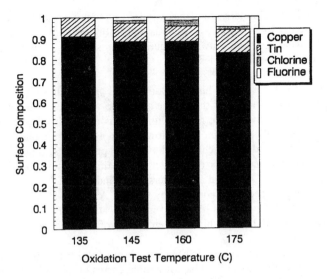

(Note: Results exclude contaminants such as carbon and oxygen.)

FIGURE 7--Surface composition of C-93200 bronze specimens from oxidation-corrosion tests at a sputtered depth of 33 Å.

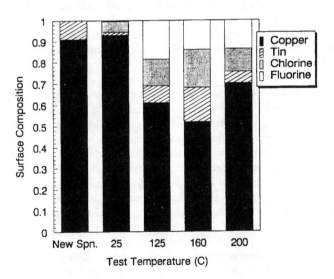

(Note: Results exclude contaminants such as carbon and oxygen.)

FIGURE 8--Surface composition of C-93200 material within a wear scar produced during tests lubricated with CTFE.

additive, or both. Similarly, the concentrations of both chlorine and fluorine were increased by the wear process, as summarized in Figure 8. The fluorine concentration outside the wear scar (not shown in Figure 8) was similar to that of the specimens from the oxidation-corrosion test.

Surface analysis from high resolution XPS spectra showed that the metallic constituents on the surface of the bronze specimen were present in an oxidized form over the complete temperature range, with cuprous oxide (Cu_2O), tin dioxide (SnO_2), and lead oxide (PbO) forms present. However, at 175°C, tin monoxide (SnO), lead difluoride (PbF_2), and lead dichloride ($PbCl_2$) are probably also present. Other workers have observed similar effects with these fluids. Using X-ray diffraction and X-ray fluorescence, Gupta and co-workers [1] found evidence of cuprous chloride (CuCl) on copper specimens corroded by CTFE. A related study [10] found no chlorine on bronze specimens immersed in CTFE at low temperatures using Energy Dispersive X-ray (EDX), but high concentrations of chlorine were present at 175°C.

X-ray fluorescence revealed traces of bromine in the hydraulic fluid and at increased concentrations in the black insoluble corrosion product present in the samples. Both the used and the virgin fluid contained approximately 40 ppm of bromine, compared to approximately 100 ppm of bromine in the reaction products. The black insoluble product appears to be an oxy-fluorocarbon containing small amounts of heavier elements such as copper, zinc, bromine, and chlorine entrapped in the structure. The product has a lower density than that of the CTFE hydraulic fluid, probably because oxygen has displaced halogen atoms in the molecule. Bromine was probably present as an impurity in one of the raw materials, possibly in the form of a bromotrifluoroethylene polymer similar to the chlorine derivative used in the synthesis of CTFE.

A definitive description of the high-temperature corrosion mechanism present between CTFE and bronze has not been achieved. It is likely that the corrosion process is initiated by the breaking of carbon-bromine (C-Br) and carbon-chlorine (C-Cl) bonds in the hydraulic fluid. The trace quantity of bromine in the fluid may play a disproportionate role in the corrosion mechanism due to the relatively low dissociation energy of the C-Br bond. Furthermore, increased concentrations of bromine are present in the reaction products. The corrosion process seems to be driven by a metal-catalyzed auto-oxidation reaction in which a free radical mechanism is initiated by breaking chemical bonds. The corrosion is initiated when a bromine derivative of the hydraulic fluid reacts with the metal. The reaction forms a CTFE free radical that in turn reacts with oxygen to form a peroxyl radical, CTFE-OO•. The peroxyl radical oxidizes the metal and forms an alkoxy type radical which then decomposes to form a carbonyl group on the CTFE molecule and a chlorine atom radical. The chlorine atom radical causes more oxidation of the metal and continues the free radical chain. Additional oxidation-corrosion tests using a CTFE fluid free of bromine are required to confirm the suggested corrosion mechanism.

Evaluation of Surface Treatment Procedures for C-93200 Bronze

Background to surface treatment evaluation--A brief test series was performed to define the effectiveness of various surface coating processes in minimizing corrosion and wear of bronze by CTFE. No metal or fluids other than C-93200 bronze and CTFE were evaluated. The surface treatment processes summarized in Table 4 were evaluated under contact conditions identical to those used in the preceding sections. The PVD Titanium Nitride did not adhere to some of the wear test surfaces, and irregular discoloration of the treated surfaces was apparent even at room temperature. No difficulty was encountered during application of this process to the test coupons intended for use in the oxidation-corrosion test, probably due to their smooth surface texture.

TABLE 4--Surface treatment procedures evaluated in this study.

Treatment Type	Thickness, m $\times$ 10^{-6} Range	Tol.	Attributes	Rockwell Hardness
Nickel/Polymer	5-75	$\pm$ 5.0	Friction/wear/ corrosion	65
IBD[A] Chrome	2-3	...	Wear/corrosion	82
Chrome	$\approx$25	...	Wear/corrosion	30
Sulfamate Nickel	10-15	$\pm$ 2.5	Corrosion	22
PVD[B] Titanium Nitride	1-3	$\pm$ 0.5	Friction/wear	85
Modified electroless nickel[C]	30	$\pm$ 2.5	Wear/corrosion	70

[A] Ion Beam-Deposited
[B] Physical Vapor Deposition
[C] Composite electroless nickel containing fluorocarbon particles
    [not polytetrafluoroethylene (PTFE)]

Oxidation-corrosion tests with surface-treated specimens--The normalized material removal rate during oxidation-corrosion tests at 135°C is plotted in Figure 9. A second test series at 175°C was also performed. (A detailed summary of these results may be obtained in Reference 4.) As expected, severe corrosion was present on the untreated specimens. Material removal was significantly reduced by each of the surface coatings, although the IBD Chrome and PVD Titanium Nitride were only

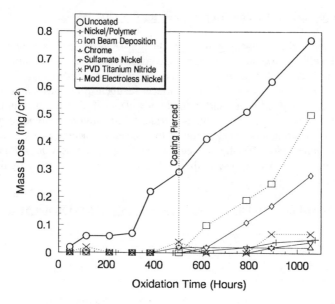

(Note: The broken lines indicate that the surface coating has begun to flake.)

FIGURE 9--Mass loss from surface-treated C-93200 specimens immersed
in CTFE hydraulic fluid at 135°C.

partially successful, particularly at 175°C. The remaining surface coatings (chrome, sulfamate nickel, and modified electroless nickel) survived completely untarnished and appeared similar to the pretest finish.

The entire surface of each specimen was completely covered by the protective coating prior to the initiation of the oxidation-corrosion test. During practical applications, some penetration of the thin surface coating is likely to occur due to accidental scratching or wear. The effect of such damage was simulated by indenting the surface of each specimen through the protective coating at 500 hours, as denoted by the vertical broken line in Figure 9. At high temperature (175°C), the material removal rate is not affected by the indentation, as in many instances severe corrosion was already present. At low temperatures, surface corrosion of the less well-adhered coatings was greatly increased by the localized penetration of the surface coating. The increased corrosion is apparent on the IBD Chrome and PVD Titanium Nitride. Slightly increased corrosion of the nickel/polymer-coated specimen was also observed at low temperature. However, the nickel/polymer coating was successful at high temperature with only slight discoloration at conclusion of the test.

Wear tests with surface-treated specimens--The results of Cameron-Plint wear tests performed using surface-treated specimens are provided in Figure 10. The contact materials were reversed (compared to previous tests) to allow wear testing of

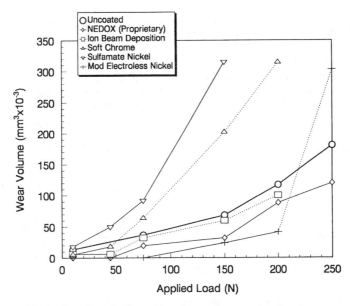

(Note: The broken lines indicate that the surface coating has been removed
due to wear. No wear tests were performed with PVD Titanium Nitride due
to poor adhesion with bronze substrate.)

FIGURE 10--Wear test results obtained with surface-treated
C-93200 bronze specimens.

various surface treatment processes intended for use with C-93200 bronze. Once
again, a counterformal contact geometry was selected to allow accurate wear
measurement. However, the C-93200 material (which is the treated specimen) was
machined to form the test flat rather than the reciprocating ball contact. Previous
experience has shown that the surface coating would be almost immediately removed
if applied to the hertzian contact on the test ball. The wear volume was measured
from the test flat using a Talysurf surface profilometer. The remaining test conditions
are identical to those previously described in Table 3.

Relatively large variations in wear rate were observed among the coated
specimens. Indeed, measured wear rate was apparently increased by application of
both the soft chrome and sulfamate nickel surface treatments. The chrome coating did
not adhere well to the bronze surface and was eliminated at loads above 50 N (as
denoted by the broken line in Figure 10), leaving a deeper wear scar. In contrast, the
sulfamate nickel coating showed good adhesion but had low hardness (22 HRC)
combined with a high application thickness (>10 μm). As a result, deformation of the
adhered sulfamate nickel layer gives apparently high wear, even at low applied loads.
Similarly, the ion beam-deposited layer is removed at loads above 50 N but is
sufficiently thin to make little difference in the perceived wear rate. However, the

comparatively hard ion beam-deposited layer is effective at low loads prior to subsurface deformation and removal of the surface coating.

The hard nickel/polymer and modified electroless nickel surface coatings, both of which consist of a nickel/polymer composite, appeared to be the most effective overall surface treatment processes. Both coatings were previously shown to eliminate surface corrosion and may be seen to minimize wear in Figure 10. No flaking or removal of the nickel/polymer coating was present even under the severe test conditions, while penetration of the modified electroless nickel coating only occurred in the most highly loaded tests.

DISCUSSION

The primary objective of the present study was to compare the corrosion and wear characteristics of a nonflammable CTFE hydraulic fluid with those of more conventional fluids. Most hydraulic systems contain a number of metals and disparate contact conditions. As a result, the complete program [4] considers a number of metals and a similarly broad range of contact variables. Previous workers have concluded that consideration of a wide range of contact parameters is required, even for the successful characterization of fluids intended for use in a well-defined system [11]. The present study concentrates on those parameters found to be most critical (i.e., the C-93200 bronze material, low sliding speeds, and relatively high temperature).

Previous work has indicated relatively severe corrosion of bronze by unformulated CTFE at 175°C, which corresponds to the AFAS operating envelope [1, 10]. The results of the present study indicate that formulated CTFE fluid will corrode bronze and each of its elemental constituents (i.e., elementally pure copper, lead, tin, and zinc) at temperatures as low as 135°C. Previous studies have also noted degradation of copper by CTFE [1]. Operation at lower temperatures will greatly reduce the level of corrosion observed, as will use of appropriate surface treatment procedures. The level of degradation appears more severe than that observed in previous studies with alternate metals [1]. Chlorine and trace quantities of bromine in the hydraulic fluid seem to be the main cause of corrosion. A number of surface treatment processes that reduce or even eliminate corrosion of the bronze substrate were identified. In addition, two of the processes provide greatly improved wear resistance, due largely to increased surface hardness.

It is recognized that the test severity necessary to achieve oil degradation in a practical time period necessitates use of a severe environment. For example, the laboratory tests were performed in a high-temperature environment saturated with dry air, which was shown to accelerate the degradation process, although most hydraulic systems are closed. Fluid temperatures of 121°C are common in currently used equipment, with temperatures of up to 177°C observed in some military prototype testing [10]. Indeed, the test conditions were considerably less demanding than some previous studies performed at higher temperatures in the presence of moisture [1], with a corresponding reduction in test duration. To put this in perspective, both the petroleum- and PAO-based fluids eventually suffered severe degradation in the present laboratory study but normally provide acceptable field service.

In general, the oil degradation process of the hydrocarbon-based fluids is very different from that of CTFE, which showed no change in TAN, viscosity, or composition as defined by FTIR (although some degradation was apparent using GC). Moreover, a distinct induction period was present prior to the breakpoint of the hydrocarbon-based fluids, while chemical corrosion of the bronze material occurred immediately after commencing the test with CTFE. As a result, no direct comparison may be made between the high-temperature stability of the CTFE and hydrocarbon-based oils, although the results of the present study indicate that corrosion of bronze and a number of other elements by CTFE fluid would commence immediately upon initiation of high- temperature operation above approximately 135°C. The effect of chemical reaction between CTFE and bronze in practical applications is unclear and will depend to a great extent on the operating environment. Previously, however, full-scale pump loop tests have indicated surface corrosion with a similar but possibly not identical fluid [2].

Lubricant decomposition components can be either beneficial or detrimental to wear protection. Simultaneous corrosion and mild mechanical contact may result in a combined effect that is greater than the additive effect of each process taken alone, although formation of an effective chemical film is necessary to reduce adhesive wear and scuffing. The wear tests performed in this study indicate that the previously described surface corrosion does indeed affect the observed wear mechanism. In general, however, the laboratory wear tests performed using both the Cameron-Plint and four-ball apparatus indicate that the lubricity of the CTFE fluid was similar to that of the hydrocarbon-based oils and appreciably better than that of the silicone fluid. In no instance was the lubricity of the CTFE oils less than that of the MIL-H-6083E baseline fluid, which has given excellent field service for many years [2]. In addition, the poor viscosity index of the CTFE fluids reduced hydrodynamic lift during laboratory testing compared with the PAO- and silicone-based oils, both of which have excellent viscosity retention at higher temperatures [11].

During consideration of the results, it should be noted that the lubricating qualities of polydimethylsiloxanes, or silicone fluids, are typically poor [12]. This result is partially due to the low pressure-viscosity coefficient of silicone, causing inferior elastohydrodynamic film formation (distinct from hydrodynamic). In addition, dimethylsilicones do not normally form boundary films in a manner similar to hydrocarbon surfactants, such as fatty acids, although weak films have been reported at higher temperatures [9]. Finally, the boundary lubricating characteristics of the silicone fluids may not normally be greatly improved using additives, a result which seriously restricts the use of silicone fluids in systems in which hydrodynamic or full-fluid lubrication is likely to fail [12].

During the present study, the normalized wear coefficient for the hydrocarbon and CTFE fluid ranged between 0.05 and 2.0 × 10^{-6}, which typically corresponds to a lubricant of average quality in a contact of dissimilar metals[4] [6]. In general, the most that may be expected from any practical laboratory wear test procedure is an accurate qualitative ranking in comparison to other fluids of known quality [11]. Overall, the results indicate that the PAO-based oil provides marginally the best boundary wear protection for the fluids and conditions considered. Nonetheless, the CTFE fluid should provide good wear protection under boundary lubricated conditions at all but the highest operating temperatures. Higher temperatures will promote surface corrosion in bronze alloys, which may be augmented by sliding wear. However, the temperature required to produce corrosion with CTFE would **eventually** also cause severe degradation of the petroleum and PAO fluids.

CONCLUSIONS

The following conclusions may be drawn from the laboratory tests performed during the course of the study:

a.   The level of wear observed at normal operating temperatures with CTFE oil qualified under MIL-H-53119 was comparable to petroleum-based fluids qualified under MIL-H-6083E and so should be acceptable in practical applications.

b.   The PAO-based oil qualified under MIL-H-46170B normally provided marginally better wear protection than both the petroleum- and CTFE-based fluids.

c.   Severe friction and wear were present for the silicone-based fluids in the absence of hydrodynamic lift.

d.   CTFE produced severe corrosion of bronze and elementally pure copper, lead, tin, and zinc at temperatures above approximately 135°C. The rate of material removal increased disproportionately at temperatures above this value. Corrosion begins immediately following immersion in the fluid.

e.   CTFE fluid provided good lubricating qualities and corrosion resistance with ferrous metals.

---

[4] Considerable disagreement exists in the literature with regard to the value of Archard's wear coefficient that corresponds to acceptable lubrication [8, 13-15].

f.   No change in either acid number or viscosity was observed for the CTFE fluid during oxidation-corrosion tests in the absence of moisture at temperatures up to 175°C.

g.   Corrosion of bronze by CTFE may be due to preferential reaction with near-surface material by chlorine, fluorine, and trace quantities of bromine.

h.   Increased acid number and viscosity were observed for the formulated hydrocarbon fluids during high-temperature oxidation-corrosion tests after an appreciable initiation period. However, metallic corrosion was comparatively mild compared to the CTFE-based oil.

i.   Of the fluids studied, MIL-B-46176 (derived from a silicone basestock) provided optimum thermal stability.

j.   Both surface corrosion and wear of bronze lubricated with CTFE may be greatly reduced through use of nickel/polymer or modified electroless nickel surface treatment processes.

k.   Corrosion of bronze by CTFE is greatly reduced in an oxygen-free environment.

ACKNOWLEDGEMENTS

This work was performed by the U.S. Army Tank-Automotive Research, Development and Engineering Center (TARDEC) Fuels and Lubricants Research Facility (TFLRF) located at Southwest Research Institute (SwRI), San Antonio, TX, during the period July 1992 to January 1995 under Contract No. DAAK70-92-C-0059. Ms. C. Van Brocklin and Mr. A. Rasberry served as project technical monitors at the U.S. Army Tank-automotive and Armaments Command (TACOM), Mobility Technology Center-Belvoir (MTCB). Mr. R. Espinosa and Ms. L. Marazita served as the project technical monitors at the U.S. Army Armament Research, Development and Engineering Center (ARDEC) within the U.S. Army Armaments, Munitions and Chemical Command (AMCCOM), Ft. Belvoir, VA. Major funding was provided by ARDEC, and limited funding was provided by the U.S. Army TACOM, MTCB. The authors would also like to acknowledge the efforts of TFLRF personnel, including Mr. J.J. Dozier and Ms. J.R. Rocha for performing the laboratory work, and Ms. M.M. Clark for editing the final draft of the paper.

REFERENCES

[1]   Gupta, V.K., Warren, O.L., and Eisentraut, K.J., "Interaction of Metals With Chlorotrifluoroethylene Fluid at Elevated Temperatures," Lubrication Engineering, Vol. 47, No. 10, 1991, pp. 816-821.

[2]   Snyder, C.E., Gschwender, L.J., and Campbell, W.B., "Development and Mechanical Evaluation of Nonflammable Aerospace (−54°C to 135°C) Hydraulic Fluids," Lubrication Engineering, Vol. 38, No. 1, 1982, pp. 45-51.

[3]   Gschwender, L.G., Snyder, C.E., and Sharma, S.H., "Development of a −54°C to 175°C High Temperature Nonflammable Hydraulic Fluid for Air Force Systems," to be published.

[4]   Lacey, P.I., Naegeli, D.W., Wright, B.R., "Evaluation of Corrosiveness, Oxidation, and Wear Properties of Hydraulic and Recoil Fluids," Interim Report TFLRF No. 287 (AD A301987), prepared by U.S. Army TARDEC Fuels and Lubricants Research Facility (SwRI), Southwest Research Institute, San Antonio, TX, November 1995.

[5]   Klaus, E.E. and Perez, J.M., "Comparative Evaluation of Several Hydraulic Fluids in Operational Equipment, a Full-Scale Pump Test Stand and the Four-Ball Wear Tester," SAE Paper No. 831680, Society of Automotive Engineers, Warrendale, PA, 1983.

[6]   Wear Control Handbook, M.B. Peterson and W.O. Winer, Eds., American Society of Mechanical Engineers, New York, 1980.

[7]   Rabinowicz, E., Friction and Wear of Materials, John Wiley, 1965.

[8]   Archard, J.F., "Contact and Rubbing of Flat Surfaces," Journal of Applied Physics, Vol. 24, 1953, pp. 981-988.

[9]   Jemmett, A.E., "Review of Recent Silicone Work," Wear, Vol. 15, 1970, pp. 143-148.

[10]   Gupta, V.K. and Eisentraut, K.J., "Interaction of Alloys with Chlorotrifluoroethylene Fluid at 177°C," Lubrication Engineering, Vol. 47, No. 12, 1991, pp. 1028-1034.

[11]   Mizuhara, K. and Tsyura, Y., "Investigation of a Method for Evaluating Fire Resistant Hydraulic Fluids by Means of an Oil Testing Machine," Tribology International, Vol. 25, No. 1, 1992, pp. 37-43.

[12]   Tabor, D. and Willis, R.F., "Thin Film Lubrication with Substituted Silicones: The Role of Physical and Chemical Factors," Wear, Vol. 11, 1968, pp. 145-162.

[13]   Hirst, W., "Wear of Unlubricated Metals," Proceedings of the Conference on Lubrication and Wear, Institution of Mechanical Engineers, London, 1957, pp. 674-681.

[14]   Rabinowicz, E., "New Coefficients Predict Wear of Metal Parts," Product Engineering, Vol. 29, 1958, pp. 71-73.

[15]   Holm, R., Electric Contacts, Almquist and Wiksells, Stockholm, 1946.

Martin Priest [1] , Christopher N. March [1] and Paul V. Cox [2]

A NEW TEST METHOD FOR DETERMINING THE ANTI-WEAR PROPERTIES OF HYDRAULIC FLUIDS

REFERENCE: Priest, M., March, C. N., and Cox, P. V., **"A New Test Method for Determining the Anti-Wear Properties of Hydraulic Fluids,"** *Tribology of Hydraulic Pump Testing, ASTM STP 1310,* George E. Totten, Gary H. Kling, and Donald J. Smolenski, Eds., American Society for Testing and Materials, 1996.

ABSTRACT: Existing pump test procedures for determining the anti-wear properties of hydraulic fluids are expensive to undertake, have a long test duration and require a large quantity of test fluid. This paper describes a project to develop an alternative low cost bench screening procedure of short duration, requiring only a small quantity of test fluid and using simple reproducible test specimens. A high level of correlation is demonstrated between the new test method and existing vane pump test procedures for a range of hydraulic fluids including mineral oils and water-glycol fluids. The new test procedure is highly suited to the hydraulic fluid development process requiring only ten litres of fluid per test and a maximum total test duration of six hours.

KEYWORDS: hydraulic fluids, anti-wear properties, vane pumps, standard test procedures, mineral oils, water-glycols

In recent years much development has taken place in design, material selection and fluid additive technology to improve the efficiency and reliability of hydraulic equipment. A substantial proportion of this effort has concentrated on the performance of the highly loaded sliding contacts which are essential elements of many types of hydraulic units such as vane pumps, gear pumps and piston pumps. Such contacts are lubricated solely by the system fluid and therefore depend heavily on the performance of the hydraulic fluid and in

[1] Research Engineer and Manager respectively, The Industrial Unit of Tribology, Department of Mechanical Engineering, University of Leeds, Woodhouse Lane, Leeds, LS2 9JT, UK

[2] Chairman, BSI Sub-Committee MCE/18/6 - Hydraulic Fluids, Castrol Technology Centre, Whitchurch Hill, Pangbourne, Reading, Berkshire, RG8 7QR, UK

particular on the properties of the additives. One of the most important properties of a hydraulic fluid in this regard is its anti-wear performance.

Currently, standard test procedures in the UK and elsewhere to evaluate the anti-wear properties of hydraulic fluids are based on full scale testing using production hydraulic equipment such as the vane pump. Examples of such tests are BS 2000 Part 281: 1983 in the UK and ASTM Test Method D 2882 in the USA which both use the Vickers V104C vane pump. Such pump test procedures are expensive to undertake, have a long test duration and need a large quantity of test fluid. Although viewed as essential by the hydraulics industry to demonstrate the performance of hydraulic fluids in service conditions, the use of such large scale testing during the early formulation stages of fluids is expensive and slows and restricts the pace of fluid and additive development. This situation has been exacerbated with tests based on the Vickers V104C pump by technical difficulties in both completing the test sequence, and thereby obtaining meaningful results, and in establishing acceptable repeatability and reproducibility data. This has led to numerous laboratory specific variations on the test method and general moves to abandon the test.

The UK hydraulics industry has identified the need for a low cost bench screening procedure of short duration, requiring only a small quantity of test fluid and using simple reproducible test specimens to evaluate anti-wear properties of fluids at an early stage of development without recourse to expensive pump testing. As a first step in this direction an industry initiated study was undertaken to determine the range of contact conditions which represent the critical operating limits in many types of hydraulic unit and to establish the requirements that should be met by a suitable screening procedure for fluids of all levels of performance from premium grade mineral oils to high water based fluids. It was concluded that vane pumps impose the severest anti-wear requirements on hydraulic fluids. Although other types of pumps and components may require different characteristics, the vane pump is seen as the primary performance measure in this aspect of fluid properties. As a consequence, the specification of a screening test was defined on the basis of vane pump operating conditions and in particular those encountered in the standard test BS 2000 Part 281: 1983. Another major conclusion of this study was that none of the common tribology bench tests had the capability to form the basis of a realistic screening test and therefore a new test rig was needed.

This paper describes a BSI project to develop a new test method fulfilling these requirements and demonstrates correlation of this test with existing pump test procedures. Standard anti-wear performance tests based on the Vickers V104C vane pump (BS 2000 Part 281: 1983) and the Vickers 35VQ25 vane pump (Sperry Vickers procedure) have been undertaken to provide validation data for the new test rig on a range of hydraulic fluids including mineral oils and water-glycol fluids. The future development of the new test method is also outlined.

TEST RIG DESIGN

The design of the test rig took as its starting point the Vickers V104C pump and its operating conditions in the UK standard anti-wear performance test BS 2000 Part 281: 1983. The geometry of the V104C vane pump internal components, which are referred to as the cartridge, is illustrated in Figure 1. It basically consists of a rotor with twelve forward facing slots which house single piece vanes that are free to move in out of the slots and slide against the contoured inner surface of the cam ring. As the rotor turns, the volume between adjacent vanes changes according to their circumferential position. Hydraulic fluid is drawn into these volumes through axial ports in side plates, which form the axial walls of the volumes, when they are expanding and is outlet through further axial ports when the volumes are small and the fluid is compressed. There are flow passages in the side plates which supply fluid at the outlet pressure to the back of the vanes forcing them radially outwards against the cam ring and it is these highly loaded sliding interfaces between the vanes and the cam ring which are used to assess the anti-wear properties of the hydraulic fluid in standard pump tests by measuring the weight loss of the vanes and cam ring by weighing them before and after a specified test cycle. In BS 2000 Part 281: 1983 the pump is run with a maximum outlet pressure of 14.0 MPa (2000 psi) with mineral oils and synthetic fire resistant fluids and 10.5 MPa (1500 psi) with other types of fire resistant fluids for a total test duration of 250 hours.

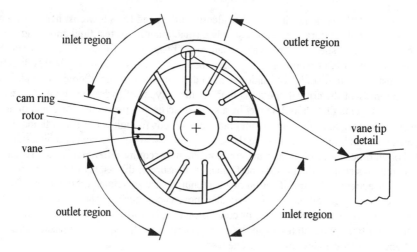

FIGURE 1—Vickers V104C Vane Pump Cartridge

The tip of the vane is shown as new in Figure 1 with a chamfer on the trailing edge. When the pump is run, however, this chamfer soon develops into a convex profile due to wear of the vane tip as it slides against the cam ring. The effect of sliding against this contoured ring is that the curvature changes constantly as does the attitude of the vane and the point of contact on the vane surface. Measurements of the convex profile of the vanes from a V104C test on a good fluid with low wear have shown that this convex profile can be approximated by a circular arc with a typical radius of 0.5 mm.

The new test rig, which is called the IUT AW-2 tribotesting machine, models this geometry using a pin on disc configuration with circumferential line contact as shown in Figure 2. The disc rotates at a constant speed and simulates the cam ring sliding past the vanes. The pin oscillates through an angle of 30° and simulates the vane oscillating against the cam ring. Hydraulic pressure which forces the vanes radially outwards against the cam ring in the vane pump is represented by static loading between the two specimens. The use of cylindrical components with a rotating motion about their own axes means that the contact geometry remains unchanged as the specimens wear, as long as gross wear is avoided, and ensures that contact stresses at a given load do not change with time.

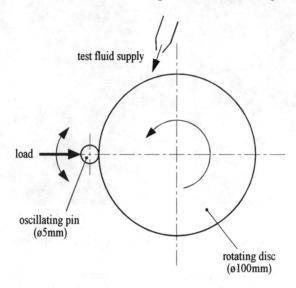

FIGURE 2—IUT AW-2 Test Rig Concept

The disc is 100 mm in diameter and manufactured from 535A99 alloy steel (BS 970 : Part 1 : 1983) and the pin is 5 mm in diameter and manufactured from BM2 tool steel (BS 4659: 1971). Both components are hardened and tempered to 60–63 HRC and ground to an axial surface roughness of 0.2–0.4 Ra (BS 1134 : Part 1, 0.8 mm cut-off) on the outside diameters. These material specifications for the disc and pin correspond to those of the cam ring and vanes respectively from the V104C vane pump. The length of the line contact at the interface of the specimens is 15mm.

Figure 3 shows the interior of the rig with the specimens in situ. The disc is located on a fixed cantilevered drive shaft and the pin is mounted at the centre of an oscillating shaft driven by a four–bar linkage which can be seen on the left. This oscillating shaft supports the rear of the pin to prevent bending under load and is mounted between self-aligning rolling element bearings in two load plates which rotate about a fixed pivot to ensure the discs remain in contact as they wear and also have a degree of independence to allow the pin to self–align with the disc and therefore eliminate the possibility of edge loading. The required normal load at the disc interface is applied by weights via a load arm acting on the load plates as shown in Figure 4. The disc rotates at a constant speed for a given test condition and is powered by a variable speed a-c electric drive. The pin is powered independently by a variable speed d-c drive through the four-bar linkage.

FIGURE 3—Interior of the IUT AW-2 Test Rig with the Specimens In Situ

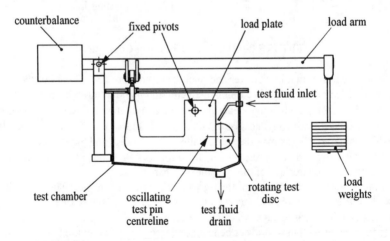

FIGURE 4—Schematic View of the IUT AW-2 Test Rig

The test fluid is fed on to the surface of the disc above the test specimen interface and the disc then transports the fluid down into the interface. The used fluid drains under gravity into a thermostatically controlled bath and is then recirculated back to the test disc using an integral pump supplied with the bath temperature control unit.

The pin can be raised out of the test chamber, with the lid removed, for the purposes of examination and dismounting by rotating the load plates using the external lifting handle which can be seen in the photograph of Figure 5. A pneumatic cylinder also acts on this lifting handle which gently lowers the specimens together at the start of the test once the drives are up to test speed and the test fluid is circulating correctly and also separates the specimens at the end of the test prior to stopping the motors and fluid circulator. The intention is to eliminate unrepresentative wear or scuffing of the test specimens during starting and stopping.

The anti-wear performance criteria for the test rig is weight loss of the test pin over a specified test cycle supported by assessment of the surface condition of the test pin and disc at the end of the test.

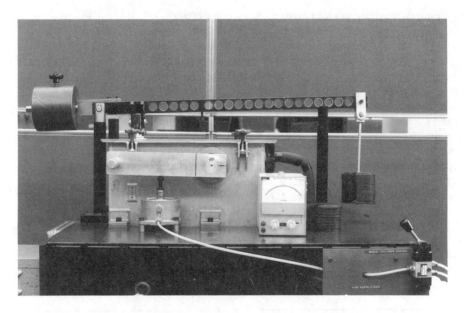

FIGURE 5—Side Elevation of the IUT AW-2 Test Rig

TEST FLUID SPECIFICATION

To assist in the development of the new test procedure it was decided to prepare four ISO VG 32 (ISO Standard 3448) mineral oil hydraulic fluids for vane pump and rig testing with varying degrees of zinc dialkyldithiophosphate (ZDDP) type anti-wear additives to give the following target performance with respect to the vane pump test procedure BS 2000 Part 281: 1983 (V104C) and the Sperry Vickers procedure (35VQ25)

Fluid 1 :   passes V104C and 35VQ25
Fluid 2 :   passes V104C, borderline 35VQ25

Fluid 3:   borderline V104C, fails 35VQ25
Fluid 4 :   fails both V104C and 35VQ25

To investigate the validity and repeatability of the new test procedure with a fluid not used in the development process a further mineral oil, which will be referred to as fluid 5, was also tested. This was an ISO VG 46 mineral oil with ZDDP type anti-wear additive.

Two proprietary ISO VG 46 water-glycol HFC (ISO Standard 6743/4) fluids were also included in the test programme with the aim of demonstrating the suitability of the rig to test this type of fluid and to establish appropriate operating conditions. Vane pump test procedures such as BS 2000 Part 281: 1983 use a reduced load when testing this type of fluid.

TEST PROCEDURE DEVELOPMENT AND RESULTS

Initial trials on the test rig led to the establishment of an outline test procedure to use during the first stages of testing on fluids 1 to 4. In summary this procedure involved starting the test rig with the specimens separated, running the specimen drives and fluid circuit until system stability was achieved, lowering the test specimens together using the pneumatic cylinder with just the load from the load plates and associated mechanism (load stage 1), running for fifteen minutes and then increasing the load at fifteen minute intervals by adding 0.5 kg weights to the load arm until the desired test load of 3.0 kg (load stage 7) was reached. It should be noted that the load arm is balanced by a counterweight such that it applies no load due to its own mass. The test was then run for a set duration, initially four hours was selected, giving a total test duration of 5.5 hours, or until severe wear was observed. At the end of the test period the specimens were separated using the pneumatic cylinder and the rig stopped. The pin and disc were thoroughly cleaned using a 1.1.1 trichloroethane based solvent and hexane at the start and finish of the test. The operating conditions of the test are given in Table 1 along with the corresponding data for the V104C and 35VQ25 vane pump tests.

TABLE 1—Comparison of Operating Conditions in the First Phase of Rig Testing

Parameter	V104C Vane Pump Test	35VQ25 Vane Pump Test	IUT AW-2 Test Rig
cam ring/disc sliding speed, m/s	4.2 – 5.2	10.4 – 11.4	9.4 (1800 rpm)
vane/pin oscillation frequency, Hz	48	80	6 (360 rpm)
contact load per unit length, kN/m	0 – 28	0 – 16	39 (load stage 7)
maximum Hertzian contact stress, GPa	0.02 – 1.60	0.17 – 1.70	0.78
inlet temperature, °C	65 (ISO VG 32)	90 – 96	55 – 60 (60°C at bath)

The contact loads per unit length and Hertzian contact stresses for the vane pumps were determined with the aid of measurements from pump components and data presented in [1]. Comparison of sliding speeds shows the rig to be operating close to the speed of the 35VQ25 pump rather than the rather slower V104C. The oscillation frequency of the test pin is, by design, considerably less than that of the vanes in both the pumps. Test pin oscillation is designed to distribute wear evenly such that the radius of curvature is kept as constant as possible and to move the high temperature region associated with the contact across the surface to prevent local overheating. Increasing the frequency to that of the vanes would not greatly change this behaviour but would make the design of the rig more complicated and expensive. Contact load per unit width is higher in the rig than the maximum seen in both the pumps and the resulting Hertzian contact stress is about the mean of that encountered in the pumps. With regards to fluid temperature, it is difficult to closely compare inlet temperatures for the rig and the pumps due to the unknown thermal characteristics of the two types of device. It is reasonable to observe however that the rig temperature conditions are more akin to the V104C pump rather than the 35VQ25.

TABLE 2—Results from the First Phase of Rig Testing

Fluid	Test Sequence Completed ?	Pin Weight Loss, mg	Failure Load Stage	Failure Time, minutes
1	no	17.19	3	43
	no	18.60	3	43
2	no	239.90	1	7
	no	155.36	1	5
3	yes	0.05	...	...
	yes	0.23	...	...
4	no	6.93	3	34
	no	7.60	3	34

Results from the first phase of rig testing are given in Table 2. A marked difference between the performance of the fluids is evident with no fluid apart from fluid 3 even surviving the initial step loading phase up to the test condition of load stage seven. The first test on fluid 1 lasted up to load stage three before the test was terminated because of excessive fuming, load arm vibration and metal-to-metal contact noise. The second test exhibited the same behaviour and was allowed to run for the same time to enable comparison of pin weight loss data. In both cases the pin was scuffed and the disc scored. The weight losses obtained are two orders of magnitude greater than those for fluid 3 measured after a full 5.5 hours running. Fluid 4 gave comparable performance to fluid 1 although the tests did not last quite as long. As with fluid 1 the pin was scuffed and the disc scored. Fluid 2 on the other hand behaved completely differently not even surviving the first load stage. From the start of both tests the pin drive mechanism vibrated excessively leading in a few short minutes to severe fuming and termination of the test. No metal-to-metal contact noise was noted, however, and the test pins at the end of both tests had deep well defined, fine finish wear scars, with no evidence of overheating, as if material had simply been machined away. The discs were barely marked with just minor scratches but the pin weight loss figures were very high.

During the first phase of testing on the rig there was no filtration of the test fluid during the test sequence. The only means to trap particulates in the fluid was a pair of magnets placed in the thermostatically controlled bath which were intended as a crude experiment to see if there was significant debris circulating in the fluid during the test. Despite the undoubted inefficiency of this approach, in terms of the nature and size of material that would be trapped, a surprisingly large quantity of debris was collected by the magnets. As a consequence of this and the fact that the test fluids did not appear to be performing in accordance with their intended performance specification, it was decided to introduce a filter circuit. A separate filtration circuit was constructed that draws fluid from the thermostatically controlled bath, filters it using a low volume filter unit with a rating of $3\mu m$ at $\beta x = 200$ (ISO 4572, BS 6275 [modified for silt control]) and then returns it to the bath. One test on each of fluids 1 to 4 was then undertaken to investigate the influence of fluid filtration on anti-wear performance and the results are given in Table 3. Comparison with the first phase of testing (Table 2) shows that the performance of fluid 1 has improved greatly from failure at load stage 3 to a completed run at load stage 7 although partial scuffing of the pin and partial scoring of the disc occurred. The performance of fluid 2 improved to a lesser degree from failure at load stage 1 to failure at load stage 3 with pin scuffing rather than the previous failure mode which was characterised by a deep well defined, fine finish wear scar on the pin. The performance of fluids 3 and 4 was essentially unchanged. It was concluded from these tests that effective filtration of the test fluid is clearly an important factor in experimentally determining anti-wear performance of particular hydraulic fluids and therefore the filtration circuit was retained.

TABLE 3—IUT AW-2 Test Rig Results with Filter Circuit

Fluid	Test Sequence Completed ?	Pin Weight Loss, mg	Failure Load Stage	Failure Time, minutes
1	yes	3.87	...	...
2	no	20.24	3	41
3	yes	0.22	...	...
4	no	6.99	3	34

It seemed appropriate at this stage in the development of the test procedure to investigate the sensitivity of the results to changes in operating parameters. A parametric study was therefore undertaken concentrating on fluids 1 and 3 as fluids 2 and 4 were proving very troublesome in both rig tests and vane pump tests, simply producing a series of failures. As a consequence of this study it was decided decrease the disc drive speed to 1350 rpm and to increase the maximum load stage to stage 9 (4.0 kg of load weights). All other parameters were retained at their original values including running times of fifteen minutes at each load stage and four hours at the maximum load stage giving a total test duration of 6 hours. The chosen operating conditions for mineral oils are given in Table 4 along with the corresponding data for the V104C and 35VQ25 vane pump tests and the results for fluids 1 and 3 at these operating conditions are given in Table 5. Fluids 2 and 4 were not actually tested at these exact conditions but from tests performed at other similar conditions, it is assumed that they would fail.

TABLE 4—Final Operating Conditions for Mineral Oil Hydraulic Fluids

Parameter	V104C Vane Pump Test	35VQ25 Vane Pump Test	IUT AW-2 Test Rig
cam ring/disc sliding speed, m/s	4.2 – 5.2	10.4 – 11.4	7.1 (1350 rpm)
vane/pin oscillation frequency, Hz	48	80	6 (360 rpm)
contact load per unit length, kN/m	0 – 28	0 – 16	49 (load stage 9)
maximum Hertzian contact stress, GPa	0.02 – 1.60	0.17 – 1.70	0.87
inlet temperature, °C	65 (ISO VG 32)	90 – 96	55 – 60 (60°C at bath)

TABLE 5—IUT AW-2 Test Rig Results at the Final Operating Conditions

Fluid	Test Sequence Completed ?	Pin Weight Loss, mg	Failure Load Stage	Failure Time, minutes
1	yes	0.21	...	...
2	———————failure assumed———————			
3	yes	0.29	...	...
4	———————failure assumed———————			
5	yes	0.37	...	...
	yes	0.38	...	...
	yes	0.13	...	...
	yes	0.10	...	...
	yes	0.32	...	...
	yes	0.23	...	...
	yes	0.29	...	...
	yes	0.26	...	...
	yes	0.26	...	...
	yes	0.44	...	...
	yes	0.17	...	...
6	yes	1.01	...	...
	yes	0.10	...	...
	yes	0.88	...	...
7	no	18.35	3	32
	yes	−0.06	...	...
	yes	0.30	...	...

Also included in Table 5 are the results of eleven test runs on fluid 5 which show good repeatability with all tests running to completion with similar specimen surface conditions and pin weight losses ranging from 0.10 mg to 0.44 mg with a mean of 0.27 mg. This mean is similar to the results obtained for fluids 1 and 3. Successful tests on fluids 1, 3 and 5 were all characterised by pins with polished wear scars and discs with signs of minor abrasion.

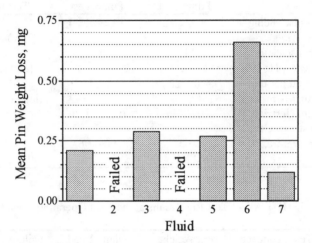

FIGURE 6—Summary of IUT AW-2 Test Rig Results

Prior to testing fluids 6 and 7, a parametric study was carried out to establish suitable operating conditions for this type of fluid. This resulted in a decrease in fluid bath temperature from 60°C to 50°C and a reduction in maximum load stage from stage 9 to stage 4 (1.5 kg of load weights) with a consequent decrease in the overall test duration to 4.75 hours. Three tests were then undertaken on fluid 6 and fluid 7 using the developed procedure and Table 5 includes the results. Fluid 6 produced three completed runs but with a wide range of pin weight losses. Fluid 7 gave two completed runs, one of which gave a negative weight loss (i.e. a weight gain), but the first test on this fluid failed with fuming and rapid wear leading to a scuffed pin. No obvious explanation has been found for this anomalous test. In all five successfully completed runs the pins had a similar surface appearance characterised by a deep turquoise/black coating over the entire wear scar worn through in places to reveal pin metal beneath. There was also a light brown fine metallic grain effect at several locations on the wear scar. The disc surfaces were similar to those obtained with successful mineral oil tests but also had this light brown fine metallic grain effect in places. One problem envisaged prior to the tests on these fluids was water evaporation and certainly there was visual evidence to suggest water loss in that a layer of moisture collected on the underside of the perspex lid of the test rig. Routine analysis of the fluids after the each test, however, indicated no problems when compared to a new fluid sample apart from small increases in viscosity within the ISO grade of the fluids. In the final test on fluid 7 the volume of fluid drained from the rig at the end of the test was measured and found to be approximately 4.8 litres. Given an initial fill of 5.0 litres this shows a volume loss of 4% although a small proportion of this may be accounted for by fluid trapped in crevices, etc. in the rig.

Figure 6 shows the mean pin weight losses obtained for fluids 1 to 7 for direct comparison with the vane pump test results in the next section.

VANE PUMP TEST RESULTS

Anti-wear performance tests based on the Vickers V104C and 35VQ25 vane pumps were undertaken on fluids 1 to 4 to provide validation data for the new bench test rig using existing test stands in five industrial laboratories. V104C vane pump tests were performed in accordance with BS 2000 Part 281: 1983 with some in-house variations relating to cartridge preparation and assembly in an attempt to ensure that the test method examines the anti-wear performance of the fluid in the vane/cam ring contacts and not undesirable wear or failure of other parts of the pump. Tests based on the larger 35VQ25 vane pump are now in use in industry where more severe conditions are required but there are no national standards. A test method developed by Sperry Vickers which has been adopted as an industry standard formed the basis of testing for this project. As the 35VQ25 procedure is regarded as being more severe than the V104C, it was decided to undertake the V104C programme first and only apply the 35VQ25 procedure to those fluids that achieved passes in the V104C tests.

The mean total ring and vane weight losses for fluids 1 to 4 are shown in Figure 7. It is worth noting that of the 25 test runs attempted, 15 (60%) were aborted for reasons other than failure at the vane/cam ring interface giving no meaningful results. This observation supports anecdotal evidence regarding the poor reliability of test methods using the V104C pump. Based on an industry adopted pass criterion for a high performance fluid of less than 50 mg total ring and vane weight loss, only fluids 1 and 3 achieved passes and were therefore eligible for 35VQ25 testing. The mean total ring and vane weight losses for fluids 1 and 3 in the 35VQ25 tests are given in Figure 8.

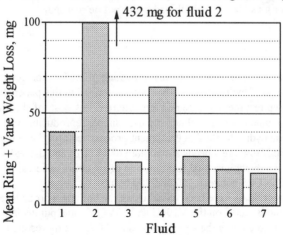

FIGURE 7—Summary of V104C Vane Pump Test Results

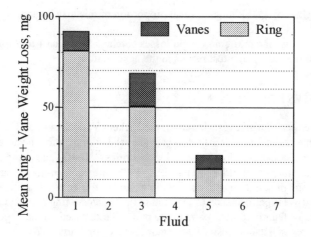

FIGURE 8—Summary of 35VQ25 Vane Pump Test Results

Figures 7 and 8 also include vane pump test results provided by the manufacturer's of fluids 5 to 7 from their own test programmes. For fluid 5, the exact test procedures followed are not known but it is reasonable to assume, with a mean total ring and vane weight loss of 27 mg in the V104C test and 16 mg mean ring weight loss and 8 mg mean total vane weight loss in the 35VQ25 test, that the fluid passes both procedures. The V104C vane pump test results for fluid 6 and fluid 7 are from tests undertaken in accordance with BS 2000 Part 281 : 1983 for a water-glycol (HFC), 250 hours at 10.5 MPa (1500 psi), and the fluids exhibit an almost identical anti-wear performance. Results were also provided from a short duration run at the full mineral oil load, 24 hours at 14.0 MPa (2000 psi), that clearly show fluid 7 to be superior to fluid 6 with weight losses of 39 mg and 78 mg respectively. This short duration vane pump test could not be simulated in the IUT AW-2 test rig as the mineral oil load could not be reached.

DISCUSSION

The development of a new bench test screening procedure for determining the anti-wear performance of hydraulic fluids has been described in this paper. In comparison to existing vane pump test methods the new procedure is low cost, of short duration, requires only a small quantity of test fluid, and uses simple reproducible test specimens. It is highly suited to the hydraulic fluid development process requiring only ten litres of fluid per test, whereas the V104C test needs a minimum of 75 litres and the 35VQ25 typically needs 410 litres, and a maximum total test duration of six hours, whereas the V104C lasts 250 hours and the 35VQ25 between 150 and 250 hours.

Seven different fluids have been tested using the both new procedure and existing vane pump test methods based on the V104C and 35VQ25 vane pumps. An attempt has been made in Table 6 to classify the vane pump and IUT AW-2 rig results for all seven fluids tested in terms of pass/fail criteria. Appropriate criteria for the vane pump tests have been extracted from the standard methods where available and from industry practice where no criteria are given as in the case of BS 2000 Part 281: 1983. The IUT rig pass/fail criteria are tentatively proposed and are based on our experience of fluid performance on the rig to date. Below is a summary of the criteria

- V104C Vane Pump Test (BS 2000: Part 281 and equivalents)

  pass   :   ≤ 50 mg total ring and vane weight loss
  borderline :   > 50 mg and ≤ 100 mg total ring and vane weight loss
  fail   :   > 100 mg total ring and vane weight loss

- 35VQ25 Vane Pump Test (Sperry Vickers procedure and equivalents)

  As the Sperry Vickers procedure with subjective judgement when insufficient test runs are available.

- IUT AW-2 Test Rig

  pass   :   ≤ 1.0 mg pin weight loss
  fail   :   > 1.0 mg pin weight loss
  When three test results are available on a particular fluid (e.g. fluid 6) then one failure may be disregarded.

Full correlation is obtained for fluid 3, 5, 6 and 7 with passes and fluid 2 with failures. It is worth noting that fluid 2 failed in both V104C and IUT AW-2 rig tests whereas the target performance for this fluid was for a pass in the V104C test and, therefore, by implication in the IUT AW-2 rig test. For fluid 1 there is full correlation between the IUT AW-2 rig and the V104C vane pump tests with passes but the 35VQ25 tests must be regarded as borderline on the basis of the tests undertaken. Fluid 4 was a clear failure on the IUT AW-2 rig but gave borderline performance in the V104C vane pump test. This high level of correlation between the IUT AW-2 rig and vane pump test methods is most encouraging.

TABLE 6—Correlation between the Vane Pump Tests and IUT AW-2 Test Rig Results

Fluid	V104C Vane Pump Test	35VQ25 Vane Pump Test	IUT AW-2 Test Rig
1	Pass	Borderline	Pass
2	Fail	...	Fail
3	Pass	Pass	Pass
4	Borderline	...	Fail
5	Pass	Pass	Pass
6	Pass	...	Pass
7	Pass	...	Pass

A standard test method has been prepared based on the work reported in this paper and has been published as a British Standard Draft for Development [2].

FUTURE DEVELOPMENTS

Work is ongoing on this test procedure to further investigate the repeatability of the results and the validity of the method to all types of hydraulic fluids that would currently be tested using vane pump methods. Current emphasis is on mineral oils with "ashless" type anti-wear additives as opposed to the ZDDP type tested in this paper.

The introduction of a pass/fail criteria with greater emphasis on visual assessment of the specimen surfaces rather than pin weight loss is also under consideration at the request of industrial laboratories who do not have the facilities to weigh the pins to the required resolution of 20 µg [2].

More long term plans include the manufacture of further test rigs to establish reproducibility data and automation of the test procedure

CONCLUSIONS

• A low cost bench test screening procedure of short duration, requiring only a small quantity of test fluid and using simple reproducible test specimens has been developed.

• This procedure has been validated against existing standard vane pump anti-wear performance tests.

• A standard test procedure has been prepared.

• Future developments have been identified.

ACKNOWLEDGEMENTS

The authors would like to thank members of BSI R&D Steering Panel MCE/18/6/1 for providing valuable guidance to this work and BSI, London and members of the British Fluid Power Association (BFPA) for funding the work and for permission to publish this paper.

REFERENCES

[1]   Broszeit E., Steindorf H. and Kunz A., "Prüfung von Hydraulikflüssigkeiten mit Flügelzellenpumpen", Tribologie und Schmierungstechnik, Vol. 37, No.4, July/August 1990, pp.202–209.

[2]   "Determination of Anti-Wear Properties of Hydraulic Fluids using the Oscillating Pin on Rotating Disc Method", BS DD 225 : 1995, BSI, London, October 1995

Jürgen Reichel[1]

IMPORTANCE OF MECHANICAL TESTING OF HYDRAULIC FLUIDS

REFERENCE: Reichel, J., **"Importance of Mechanical Testing of Hydraulic Fluids,"** *Tribology of Hydraulic Pump Testing, ASTM STP 1310,* George E. Totten, Gary H. Kling, and Donald J. Smolenski, Eds., American Society for Testing and Materials, 1996.

Abstract: Anti-wear properties of hydraulic fluids are important because hydraulic pump and motor wear is costly. Hydraulic fluid performance specifications represent minimum requirements. International hydraulic fluid performance standards are being developed by ISO/TC28/SC4 committee as draft (ISO DIS 11158 "Specifications for Mineral Oil Hydraulic Fluids"). Performance specifications for non-mineral oil hydraulic fluids are also being developed.

Typically, both the user and fluid manufacturer have insufficient information relating to the anti-wear properties of a new fluid to be used in hydraulic equipment, such as axial piston pumps, vane pumps or radial piston motors. Therefore, pump lubrication and operation requirements, preferably pre-existing in pump manufacturer's specifications, must be determined. The required fluid lubrication properties may be determined by either laboratory pump tests or by a field trial, often at the expense of the customer. More preferably, the lubrication properties of the hydraulic fluid should be determined under mechanical conditions equivalent to field practice. In this paper, the use of both the vane pump test and the FZG Gear Test to predetermine the "recommended" hydraulic fluid lubrication performance will be discussed. In this way, fluid performance may be determined at significantly lower cost than more expensive large scale hydraulic pump and motor tests which slower and more energy consuming.

KEYWORDS: Hydraulic, pump, hydraulic fluids, testing, lubrication, anti-wear, FZG Gear Test, pump tests

---

[1] Dipl.-Ing., Research Engineer, DMT-Gesellschaft für Forschung und Prüfung mbH, Franz-Fischer-Weg 61, D-45307 Essen, Germany

INTRODUCTION

Mineral oil based hydraulic fluids are the second-largest sales volume of industrial lubricants. Motor oils are the largest sales volume of industrial lubricants. The performance of fluid power systems depends on both the physical and lubrication properties of these fluids. Therefore hydraulic fluids should be regarded as an important design element in the fluid power system.

Both pump tests and bench tests may be performed on hydraulic fluids. The objectives of this work include:

1. Mechanical Testing of Hydraulic Fluids

In addition to pump testing, mechanical (bench) testing of hydraulic fluids in oil testing machines, such as four-ball, pin-on-V-block, Timken anti-wear test machines or testing in standard gear test rigs are run for determination of anti-wear properties. When standard vane pump testing is performed the criteria of failure is the total weight loss on both the cam rings and vanes. When testing is conducted using a standard gear testing rig such as FZG gear test, the testing criteria is the number of load increases to failure, or the failure stage. In both cases, pump tests or bench tests, the performance is compared to minimum standard specifications.

2. Testing of Axial-Piston Pumps

Wear properties of roller bearings and drives of axial-piston pumps may be determined for hydraulic fluid, especially non-mineral oil fluids including vegetable oil, aqueous fire-resistant fluids and synthetic oil based fluids using either changing or constant load. The load capacity of the critical components is a function of both the hydraulic fluid being tested, the operation pressure of the pump and the sequence ant amount of pressure variation, if any. The criteria for failure is determined by the pump manufacturer.

3. Evaluation of Pump Sensitivity to Solid Contamination

Determination of the wear properties of hydraulic pumps exposed to a hydraulic fluid and contaminated with solid particles, (contamination sensitivity test) is also important. The pump damage and the effects on the pump behavior are analyzable [1]. Leakage increase with respect to individual components (tribosystem, cylinder block/control plate, rotary pistons, valves, etc.) is typically plotted as a function of volume flow versus pressure, and versus the type and amount of induced contamination. The objective of such tests is to increase wear resistance, to develop low-wear designs and to select optimal materials for tribosystems of hydraulic pumps and motors

4. Pump Testing for Fluid-Development

Simple pump tests are desired to determine of particular properties of hydraulic fluids under increased load. Recently, it has been necessary to develop pump testing protocol to evaluate the changing properties of rapidly biodegradable hydraulic fluids. The objective is to develop fluids with longer lifetimes and to improve the performance of these, and other new fluids, by varying either the additive system, base fluid or both.

For mineral oil based hydraulic fluids, viscosity as a function of temperature, wear properties under extreme pressure conditions, the oxidation stability, corrosion effects on steel and non-ferrous metals, and elastomer compatibility are important. Hydraulic oils containing additives for improved wear and corrosion protection are tested in oil test machines by means of simple standard friction material pairs to determine anti-wear performance. One of the advantages of these tests is the relatively low cost of the test specimens and the relatively short testing times. Test results obtained using FZG gear test machine for the development of rapidly biodegradable hydraulic fluids are provided in **Figure 1.**

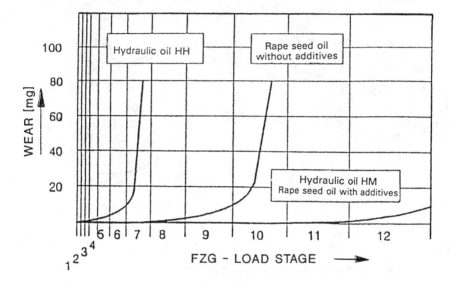

Figure 1   Evaluation of the anti wear properties of a rapidly biodegradable hydraulic fluid using the FZG gear test

Minimum anti-wear requirements for mineral oil based hydraulic fluids are specified per DIN 51524 and require standard testing in oil testing machines. These machines and methods have become standard in Europe. Examples of this standards include: "Mechanical Testing of Hydraulic Fluids in the Vane Pump" DIN 51389 and "Testing in the FZG gear test rig to method A 8,3-90" DIN 51354 part 2. The FZG test machine is schematically illustrated in **Figure 2.**

International standards, such as ISO/DIS 11158 "Specifications for mineral oil hydraulic fluids" and the standards of the American Automobile Manufacturers Association (AAMA) integrated these requirements (in draft standards). For special base fluids, predominantly mineral oil fluids are used, and their properties are the basis for the design of highly-performing pumps and motors of high longevity. However, the utility of this test procedures is limited to mineral oils since the test results obtained with other classes of hydraulic fluids have little, or no, correlation with the wear obtained in industrial hydraulic pump applications.

The objective of this paper is to discuss the significance of mechanical testing of hydraulic fluids and the development of axial-piston pump testing procedures.

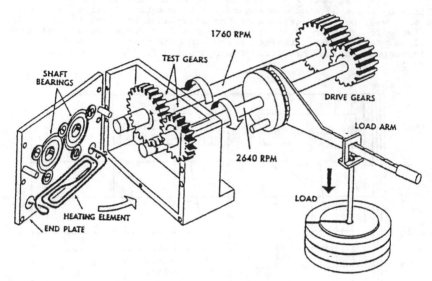

Figure 2   Schematic illustration of the FZG gear test machine

DISCUSSION

A. MECHANICAL TESTING OF HYDRAULIC FLUIDS

For many applications, mineral oil hydraulic fluids are absolutely excluded. For example, fire-protection , healthy and safe workplace concerns, product compatibility, and environmental protection requirements demand the use of mineral oil-free alternative hydraulic fluids. Currently, there is increasing use of fire-resistant fluids or environmentally acceptable hydraulic fluids. However, only some of these fluids are covered by pump performance standards. Some properties of fire-resistant fluids or rapidly biologically degradable fluids are significantly different than mineral oils. If thermal or anti-wear properties are not met, pump failure due to excessive wear can happen. Therefore, it is becoming increasingly important to identify a potential oil testing machine for preliminary screening of non-mineral oil hydraulic fluids. This test machine would make use of relatively simple standardized material pairs which would permit the identification of lubrication failure modes that correlate with pump wear failures.

1. Task Identification

Pump manufacturers and users typically have insufficient information on the wear properties of non-mineral oil derived hydraulic fluids used in gear pumps, vane pumps, axial- or radial-piston pumps which are adequately lubricated with most mineral oil based fluids. **Figure 3** provides a schematic illustration of these different pump types. From the viewpoint of the pump manufacturers, lubrication of certain points of

friction in pumps and machinery operation should be assessed. There are
two possibilities: Field trials are usually conducted at the user's expense. There is a greater probability of success if the pumps and motors
have been tested under mechanical conditions of field use.

2. Vane Pump Tests

Vane pump testing using the Vicker's V 105 C vane pump is standardized by ASTM 2882 and DIN 51389. This testing procedures have served
the industry well and their use has led to substantial anti-wear fluid
formulation improvements, particularly for HFC fire-resistant fluids.
Furthermore, cost-effective tests for assessing a fluids "life" under
wear conditions that may be encountered in field use.

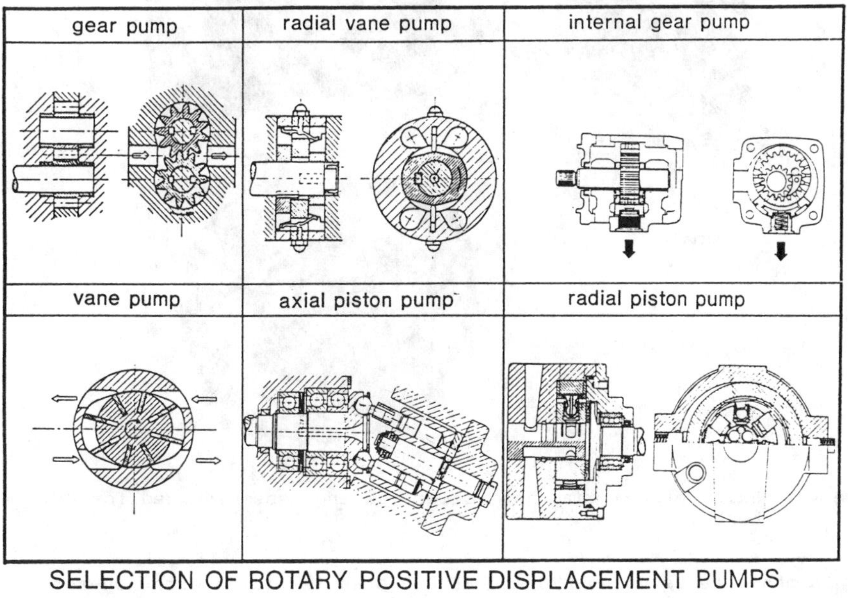

| gear pump | radial vane pump | internal gear pump |
| vane pump | axial piston pump | radial piston pump |

SELECTION OF ROTARY POSITIVE DISPLACEMENT PUMPS

Figure 3   Schematic Illustration of various common hydraulic pump
designs

3. Axial-Piston Pump Tests

Field practice has shown that while axial-piston pumps may be successfully operated using HFC fluids, roller bearing failures reduce
their operational lifetimes. Axial-piston pump tests on HFC fluids [2]
on test rigs at DMT in Essen, Germany, have shown that even after three
sets of roller bearing assemblies had failed due to material fatigue
(see Figures 4 and 5), the moving parts (pistons, cylinders, and controls) of bent-axis axial-piston pump had not yet reached the end of
their useful lives, even after 6000 hours of operation under continuous
load. The HFC fluid had also not reached the end of its useful life when
the test was completed. The continuous acoustic measurement of body

sound on the pump test rig allowed for the detection of impending roller bearing failures. This would not have been possible in actual field trials.

High-speed axial piston pumps of swash plate design and low speed radial-piston motors exhibited, even after successful testing on the FZG gear test rig, wear on steel-on-steel friction points when run on synthetic polyol ester hydraulic fluids. Piston pump testing showed that this unexpected wear increase was due to fluid contamination with small amounts of water, < 1%. Subsequent analysis showed that the non-ferrous metals, such as brass, metal components had undergone, and the fluid contained small amounts of zinc. This problem was resolved by changing the additive.

cylinder / control plate

transmission

roller bearing

Figure 4   Axial-piston pump and roller bearing assembly used for DMT tests.

In other tests using HFC and HFDU hydraulic fluids with non-supercharged type axial-piston pumps on variable-load test rigs, the control plates exhibited cavitation-induced damage. This problem was remedied by design modifications of the pump which was originally designed for use with mineral oil hydraulic fluids.

In the pump tests run by DMT, a pump and a motor were tested simultaneously by connecting them hydraulically and mechanically to each other in the "regenerative circuit" shown in **Fig. 6**. This test may be run in open or closed circuit. By solenoid valve control the motor's stroke is varied over a part or its full range, and the pump responds - in pressure control mode - by changing volume per stroke. With this test, the fluid volume can be matched to the requirements. A fluid volume of 80 liters (l) and a variable volume flow by ranging between 50 and 150 l/min may be run in closed circuit at an operation pressure of 420 bar (6000 psi). In all these cases, only the total of motor and pump losses needs to be catered for in terms of drive energy. These losses are approximately equal to 15 % of the total hydraulic performance output. In addition, the cooling of the system, the hydraulic control energy, and, if necessary, system fluid top-off needs to be monitored.

bevel ball bearings in axial piston units

fatigue after 1639 h   fatigue after 1562 h

fatigue after 1652 h   fatigue after 1396 h

Figure 5   Bevel ball bearing assembly run using an HFC hydraulic fluid
in an axial-piston pump.

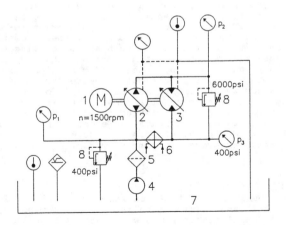

1) drive motor   5) filter—unit 25—μm
2) pump swash plate design   6) heat exchanger
3) motor bent shaft design   7) reservoir 80 L
4) boost pump   8) pressure relieve valve

Figure 6   Hydraulic schematic for a regenerative test circuit for
testing hydraulic fluids.

B. COMMENTS AND RECOMMENDATIONS

Due to the variety of material friction pairs and components that may undergo wear, such as roller bearings, plain bearings, pistons, sliding elements, cylinders, control plates and adjustment systems, the development of a hydraulic fluid test rig or apparatus using a single axial-piston pump, in a standard analogous to the ASTM 2882 or DIN 51389 van pump test, - is impossible. Test results cannot simply be reported in terms of material losses of certain components, since wear may occur by different fluid-related mechanisms such as lubrication and cavitation induced failure.

The piston pump tests reported here are "standard" tests for hydraulic fluids and do not relate to their universal use in all hydraulic pumps or even just piston pumps. These tests are substantially more expensive than the standard methods for the determination of anti-wear properties of hydraulic fluids in the vane pump per ASTM 2882 and DIN 51389 or the variable-load test on the FZG gear test rig per DIN 51354 part 2, which have shown excellent correlation to the wear expected in industrial hydraulic systems using mineral oils.

Axial-piston pump tests, however, will be absolutely necessary in the future for designs which are commonly used in hydraulic drive technologies requiring the use of non-mineral oil derived hydraulic fluids, such as fire-resistant and rapidly biodegradable hydraulic fluids. Therefore, the potential utility of these fluids in various piston pumps using various bearing designs must be assessed. This assessment is product-specific and is referred to as "suitability" testing. Potential utility of non mineral-oil derived hydraulic fluids cannot be assessed by laboratory bench testing equipment only. Also, due to potential use hazards and equipment costs, the suitability of non-mineral fluids cannot often be tested in field tests according to the customer's specific needs.

It is recommended that testing and assessment criteria of an axial-piston pump test on the variable-load test rig be summarized in a technical report (TR) which should specify, in practice-oriented form, fluid volumes, load cycles, test temperature, test pressure, and duration of test. Fluid properties, such as aging, viscosity loss, anti-wear properties, filtration, and behavior in contact with non-ferrous metals, steels, seal materials, water, and mineral oil should be assessed. This technical report should provide recommended threshold values (the values which control the useful life of the fluid such as the viscosity changes and changes in the neutralization number). After the development of a successful (standard) suitability test, fluid formulators and users will have available equipment manufacturer's recommendations, and hopefully approval, for using a specific hydraulic fluid in a particular hydraulic pump or motor.

CONCLUSIONS

Various possibilities for determining wear properties with hydraulic fluids in standard testing machinery, and a method for testing standard-production axial-piston pumps with various hydraulic fluids, have been introduced. For verification of the minimum specifications for hydraulic fluids, the vane pump and the FZG gear test rig have provided excellent wear correlations with industrial hydraulic equipment. Many non-mineral oil hydraulic fluids such as fire-resistant and rapidly biologically degradable fluids, or synthetic hydraulic fluids cannot be used in axial piston systems based upon only the results obtained with these tests. Additional testing is always required. Based on experience, proposals for energy-saving testing of these fluids on the variable-load test rig were discussed. These tests are based on axial piston-pump systems, or other displacement systems, and are focused on the determination of the industrial suitability of non-mineral oil derived hydraulic fluids. Standardization of these methods is only possible after the issuance of a technical report, and should be used to develop minimum testing requirements.

REFERENCES

[1] Lehner, S. and G. Jacobs: "Contamination Sensitivity of Hydraulic Pumps and Valves". Tribology of Hydraulic Pump Testing ASTM STP   , George E. Totten, Gary H. Kling, and Donald M. Smolenski, Eds., American Society for Testing and Materials, Philadelphia, 1996

[2] Reichel, J. "Fluid Power Engineering with Fire Resistant Hydraulic Fluids", Lubrication Engineering , Vol. 50, No. 12, 1994, PP 947-952

John G. Eleftherakis,[1] Rod P. Webb [2]

CORRELATING FLUID LUBRICATION CHARACTERISTICS TO PUMP
WEAR USING A BENCH TOP SURFACE CONTACT TEST METHOD

REFERENCE: Eleftherakis, J. G. and Webb, R. P., **"Correlating Fluid Lubrication Characteristics to Pump Wear Using a Bench Top Surface Contact Test Method,"** *Tribology of Hydraulic Pump Testing, ASTM STP 1310,* George E. Totten, Gary H. Kling, and Donald J. Smolenski, Eds., American Society for Testing and Materials, 1996.

ABSTRACT: One measure of a good test procedure ultimately answers the question, "How well does the procedure duplicate field observed results?" This is especially true with test methods for hydraulic components such as pumps due to the unique wear processes at play in today's high performance hydraulic systems.

This paper discusses the development of a benchtop, surface contact pin and journal wear test method and the research conducted to determine its usefulness in predicting the wear characteristics of fluid system components. Discussed are the limitations discovered in the various other machines and methods used for assessing the antiwear characteristics of fluids (i.e., lack of fluid circulation system, lack of contamination control and lack of variable speed control of the wearing surfaces). These limitations are emphasized when applied to hydraulic system related testing.

Further discussed in the paper are two surface wear studies--one on the lubrication characteristics of hydraulic fluids and the other a wear assessment of fuel pumps. Both studies are shown to correlate well with actual field experience.

These studies are the basis to justify further efforts to correlate results from the bench test proposed with hydraulic pump testing. Such correlation would minimize the need to conduct the more expensive and time consuming pump testing.

[1] President, Fluid Technologies, Inc., P.O. Box 669, 1016 E. Airport Road, Stillwater, OK 74076.
[2] Marketing Manager, Fluid Technologies, Inc., P.O. Box 669, 1016 E. Airport Road, Stillwater, OK 74076.

KEYWORDS: Lubricity, Hydraulic Pump, Wear

INTRODUCTION

Many test methods exist to assess the performance of hydraulic pumps. The specific method used in a particular test program will depend upon the objectives of the program. Obviously, the method and conditions used to assess the durability of a pump will be different from the method and conditions needed to test contaminant sensitivity of the same pump. The various methods, however, fall in the general categories listed in Table 1.

TABLE 1--Test Categories of Hydraulic Pumps

PERFORMANCE TESTS	STRUCTURAL TESTS
Pump Filling Characteristics	Pump Structural Integrity
Steady-State Efficiency Characteristics	Constant Pressure Endurance
Dynamic Response Characteristics	Cyclic Endurance
Thermal Stability Test	
Pump Fluidborne Noise Generation	
CONTAMINATION TESTS	**ENVIRONMENTAL TEST**
Pump Break-In and Wear Generation	High/Low Temperature
Pump Contaminant Sensitivity	Performance

Even the simplest test method requires significant resources to perform and, in the final analysis, may not correlate to actual pump performance. Unless such correlation exists, in most cases, the test data has limited value. This paper discusses a test method which assesses pump performance based upon the wear characteristics of the pump material.

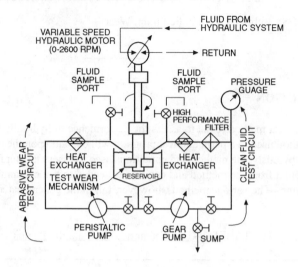

Fig. 1   Schematic of FTI Wear Test System.

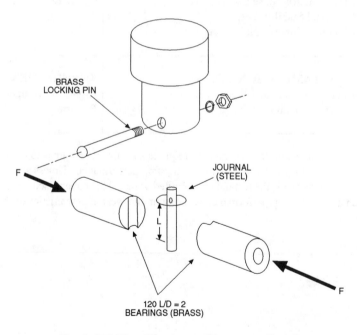

Fig. 2   FTI Wear Test Journal/Bearing Assembly.

LUBRICITY: THE KEY TO MAXIMUM PUMP PERFORMANCE

Studies have shown the relationship of lubricity to pump performance [1]. In fact, if lubricity is studied in terms of the many conditions that cause it to degrade (i.e., solid contamination, temperature extremes, additive loss, etc.) it is the most important condition that impacts performance of a properly designed and applied pump.

Several lubricity tests have been used to assess fluid lubricity characteristics. These were previously identified and considered for assessing hydraulic fluid lubricity [2] but were found not to directly duplicate hydraulic system conditions [3]. Proper assessment of hydraulic system components requires:

1) the ability to control the solid particle contamination level through controlled contaminant injection and, if necessary, filtration;
2) the ability to control temperature; and
3) the ability to control the sliding speed of wear by varying drive speed.

The amount of lubricity, and thus wear, is directly related to the level to which these factors are controlled. The FTI Wear Test (formally the Gamma Test) described below controls these variables to a high level.

HISTORY OF THE FTI WEAR TEST

As noted above, several standard test methods exist to assess fluid lubricity primarily for lubricating fluids. One such method is the Falex Lubricant Tester (ASTM D-2670). This tester is the foundation of the FTI Wear Test. The Falex tester uses a "V" block bearing and circular journal (under load) submerged in fluid. The journal is rotated at a constant speed which creates wear on the journal and bearing set. However, the "V" block configuration often results in non-uniform wear.

The FTI Wear Tester shown in Figures 1 and 2 modifies the Falex tester, fully described in ASTM 2670, in the following ways:

1) The ability to produce a more linear wear rate to the journal and bearing using 120° bearing halves rather than a "V" block.
2) The ability to simulate a recirculating fluid supply.
3) The ability to control temperature of the test fluid.
4) The ability to control contaminant levels.
5) The ability to create both boundary layer and hydrodynamic lubrication by varying the rotation speed of the journal.
6) The ability to control load.

TEST PROCEDURE AND REPORTING

Following is the test procedure for the FTI Wear Test and example results:

1) Clean the fluid reservoir and circulating system. Remove all residue fluid.
2) Install the test journal and bearings.
3) Fill the reservoir with 500-550 ml of test fluid. The fluid must cover the load jaws so that the journal and bearings are completely immersed.
4) Circulate and heat the fluid and adjust the temperature to the specific test temperature plus or minus 2°F.
5) Conduct a 2 minute break-in period by applying a load equal to one-half the specified test load and at the specified speed.
6) Increase the applied load to the specified level and record the initial wear reading.
7) Readjust gear tooth load wheel every two minutes to maintain a constant load.
8) Record number of gear tooth advances versus time.

The wear readings (as read from gear wheel movement which correlates to bearing wear) in terms of the number of gear teeth advanced versus time are tabulated and plotted. The gear tooth reading is taken in the same manner as in ASTM D-2670. Figure 3 is a representative sample of FTI Wear Test results.

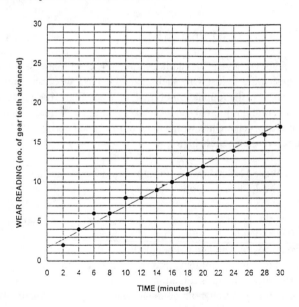

FIG. 3--Sample Results of FTI Wear Test.

RECENT TESTING USING THE FTI WEAR TEST

ATF Testing

The test history of the FTI Wear Test dates back to the Fluid Power Research Center at Oklahoma State University. There is a long history of its use for assessment of the lubricity characteristics of Automotive Transmission Fluids [4]. Testing has most recently been conducted to assess the lubricity of fuel mixtures [5] and the wear characteristics of compressor piston materials[6].

The test approach described above was followed to assess the lubricity characteristics of Dexron II® Automotive Transmission Fluid. Separate tests were run for fluid containing 2 different additive packages (AP1 and AP2) at temperatures of 100°F and 160°F at the following conditions:

Test Load:	100 lb.
Rotational Speed:	290 rpm
Journal Material:	Steel
Bearing Material:	Brass

Additionally, 1 percent distilled water was added to both formulations of Dexron II®, and tests were again run at 100°F and 160°F. Testing with water was performed for correlation of field tests and to investigate a hypothesis that some fluids provide better wear protection of steel-brass material combinations with added water [4].

A number of interesting points were immediately noticeable:

1) Dexron II® with AP2 showed less wear in all experiments.
2) Water addition reduces, or does not change (with tolerance) wear rates.
3) High temperature operation for AP1 dramatically increases the wear rate; while, for AP2, the wear rate with water was reduced or changed by an insignificant amount.

The findings that Dexron II® with AP2 reduced wear rate was not surprising because this result has been observed by others. Additionally, this result confirms that the test results are in agreement with other findings.

The most interesting results are from Dexron II® with 1 percent water. Both AP1 and AP2 showed a decrease in wear rate when tested at 100°F. In fact, a 34 percent reduction is experienced with AP1 and a 55 percent reduction with AP2. It is believed that a chemical reaction-based process is occurring between the steel-brass combination and is promoted by the water.

The fact that Dexron II® with AP2 shows less wear rate at high temperatures and reduced, or almost unchanged, wear rate with water is a clear indication of the superior operational characteristics of this formulation, which can be observed from the test results. Although unpublished, pump wear tests using Dexron II® with and without water followed the same wearing conditions.

Fuel Pump Testing

In the fuel lubricity test program, the objective was to correlate lubricity test results with actual fuel pump testing [5]. The fuel samples chosen for the program were from the Auto/Oil Air Quality Improvement Research Program [7]. This set of 15 fuels had carefully controlled chemical formulations and extensive field use within the study conducted in Reference [7]. Table 2 presents the properties of those fuels as reported in Reference [7].

TABLE 2   Fuel Properties—Actual

AO CODE	ID	AROMATICS %	MTBE %	OLEFINS %	T_{∞} °F	RVP PSI	BENZENE %
A	AVG	32.0	0	9.2	330	8,7	1.5
B	CERT	29.9	0	4.5	309	8.7	0.5
C	AMot	43.8	15.4	3.3	286	8.7	1.3
D	amOT	20.7	0	22.3	357	8.5	1.5
E	AMOT	43.7	14.8	17.2	357	8.7	1.4
F	amot	20	0	3.2	279	8.8	1.4
G	AmOt	44.3	0	17.4	286	8.8	1.5
H	aMOt	20.2	14.6	20.2	286	8.5	1.5
I	AmoT	42.9	0	4.1	353	8.9	1.3
J	aMoT	21.4	14.9	4.0	358	8.6	1.3
K	Amot	45.7	0	4.9	294	8.8	1.5
L	AmOT	47.8	0	17.7	357	8.5	1.4
M	aMOT	18	14.5	21.8	356	8.7	1.5
N	aMot	21.4	13.9	5.7	292	8.8	1.4
O	AMOt	46.7	14.6	19.3	283	8.6	1.4
P	amOt	20.3	0	18.3	284	8.5	1.5
Q	amoT	21.5	0	4.8	357	8.6	1.5
R	AMoT	46	15.2	4.0	354	8.4	1.4

A/a   HIGH/LOW AROMATICS
M/m   HIGH/LOW MTBE
O/o   HIGH/LOW OLEFINS
T/t   HIGH/LOW T_{∞}

Figures 4 and 5 give FTI Wear Test results from standard 30 minute tests for Auto/Oil Fuels (AO) C and D. Each fuel was tested three times for repeatability. The Lubricity Index (LI) for each of the fuels are given in Figure 6 and represents the average slope (or wear rate) using a least squares fit for each fuel. Note the good spread of data and discrimination of differing results. As represented in Figures 4 and 5, the LI is defined as the number of gear teeth advance on the loading wheel of the FTI Wear Test Machine per unit of time; or, more literally, a direct measure of the wear rate of the test bearing on the test journal.

Analysis of the data given in Figure 6 results in an analytical expression for the Lubricity Index (LI)--see equation (1). Stepwise regression of the data provides the following (refer to Table 1 for reference fuel chemistry),

$$LI = 0.296206\ (A\%) + 1.087215(M\%) - 0.080913\ (T) + 31.457115 \qquad \text{Eq (1)}$$

$$R_2 = 0.81694$$

where

A%=percentage composition of Aromatics
M%=percentage composition of MTBE
$T = T_{90}$ temperature

Fuel pump tests were then conducted on gerotor type pumps to measure the wear particle generation with the different Auto/Oil fuel. No test contaminant was added to the fuel because it was desired to measure only the wear debris generated by normal operation of the fuel pumps-not accelerated durability, or the like. The generated wear debris was allowed to recirculate within the fuel circulation system of the qualified test stand. Thus, accurate measurements of generated wear debris were made will all parameters being unchanged except for the fuel used.

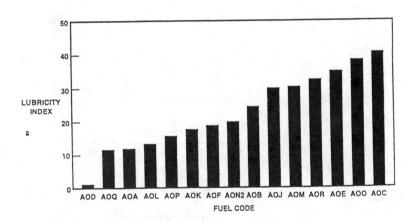

FIG. 6--Lubricity Index Measured for Test Fuels
Gamma Lubricity Test Results

The rubbing interfaces in the pump are as follows: pump gear elements are powder metallurgy, sintered iron with the inner gear rotation on a hardened steel shaft and the outer, ring gear outer diameter riding on an aluminum die cast pump housing. The ends of the gear pump elements are rotating against Teflon™ coated phenolic/fiberglass sealing discs.

Figures 7 and 8 show the particle counts (per m*l*) greater than 2 and 5 microns, respectively, for a Gerotor fuel pump. AO fuel C and D (AOC and AOD) were used for pump testing because of their dramatically differing lubricity characteristics as measured by the FTI Wear Test Machine. The wear debris generated using AOC is markedly more than that generated using AOD, as predicted by the lubricity test results. Multiple tests with many other Reference Fuels followed these same results, providing very good correlation between the FTI Wear Test results and the fuel pump tests.

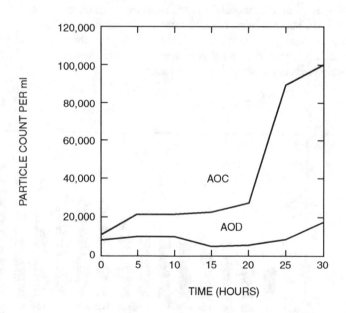

Fig. 7   Gerotor Lubricity Test
2 Micron Comparison

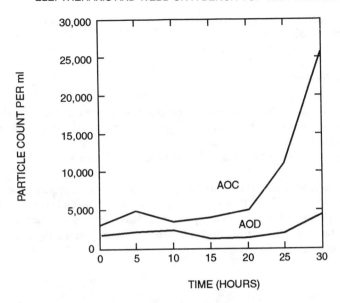

Fig. 8   Gerotor Lubricity Test
5 Micron Comparison

These two studies provide strong support that the FTI Wear Test can correlate very well to observed results with pumps. As described below, further study is underway to provide a more direct correlation between the FTI Wear Test and specific hydraulic pump wear.

CURRENT TESTING UNDERWAY

The FTI Wear Test is currently being used to assess hydraulic pump failure mode differences. Pump testing is being conducted at four sites--three laboratory sites and one field site. All sites are testing the same type of pump from the same manufacturer using used fluid controlled to a stated cleanliness level. Premature pump failure was experienced at two of the sites.

Upon further investigation, it was found that the failure modes of the pumps at the two sites were substantially different. It was further discovered that each of the four sites were using test fluid from different geographical locations. The difference in the test fluid lubricity characteristics may account for the difference in pump failure modes and the frequency in the failures.

To help confirm this theory, the FTI Wear Test is being used to assess the lubricity characteristics of the fluid. The journal and bearing set is made of the same material used in the pump internal surfaces. Testing is currently underway and results will be forthcoming. However, this is another example of how the FTI Wear Test can be used to assess the lubricity characteristics of a fluid system component, specifically a hydraulic pump.

CONCLUSION

Fluid lubricity is vital to hydraulic pump performance. The test method proposed in this paper provides strong correlation between lubricity and pump performance. Substantial data exists which supports such correlation. This data is presented from lubricity test on Automatic Transmission Fluid and a fuel pump correlation study.

Based upon the strong correlation from these tests, a hydraulic pump correlation study is underway. This study will attempt to show correlation between FTI Wear Test results with both field and laboratory pump performance results. If strong correlation is shown, it will provide further support that the FTI Wear Test is a less expensive, accurate method to predict hydraulic pump wear characteristics.

Richard K. Tessmann[1], David J. Heer[2]

THE GAMMA WEAR TEST FOR HYDRAULIC FLUID QUALIFICATIONS

REFERENCE: Tessmann, R. K. and Heer, D. J., **"The Gamma Wear Test for Hydraulic Fluid Qualifications,"** *Tribology of Hydraulic Pump Testing, ASTM STP 1310,* George E. Totten, Gary H. Kling, and Donald J. Smolenski, Eds., American Society for Testing and Materials, 1996.

ABSTRACT: The qualification of fluids for use in a hydraulic system has been a major concern to component manufacturers for several years. The primary issue in a hydraulic system is the compatibility of the fluid with the main pump in the system. There are several bench-type wear tests for the evaluation of fluids, however, most of these are not intended specifically for hydraulic fluids but were developed for lubrication fluids. Therefore, many engineers have relied upon the pump tests which were available in order to select the fluid to be used in their hydraulic system. The testing of a hydraulic pump to qualify a hydraulic fluid is very time consuming and expensive.

This paper reviews several of the more wide-spread bench-type tests which are used to evaluate hydraulic fluids. However, the main purpose of the paper is to describe a new test system for evaluating the anti-wear characteristics of hydraulic fluids called the Gamma Wear Test System.

KEYWORDS: Fluid qualification, fluid testing, gamma wear testing, lubrication, hydraulic fluid

---

[1] Vice President, FES, Inc., 5111 North Perkins Road, Stillwater, OK 74075
[2] Laboratory Manager, FES, Inc., 5111 North Perkins Road, Stillwater, OK 74075

INTRODUCTION

Hydraulic fluids, whether mineral based or one of the fire resistant types, ~~are~~ is the life blood of the hydraulic system. The properties and characteristics of the system fluid can make all the difference in the world to the satisfactory operation or failure of the components. Lobmeyer [1] reported at the 1977 SAE Earthmoving Industry Conference that hydraulic fluid, can be just as important for improving the system service life as good filtration and the use of contaminant insensitive components. Therefore, the qualification of fluids for use in a hydraulic system has been a major concern to component manufacturers for many years.

Different hydraulic fluids are needed under various operating conditions. However, an effective fluid for any hydraulic system should enhance efficiency, reduce wear, and resist degradation [2]. To an experienced system engineer, it is obvious that a low compressibility (high bulk modulus) is necessary to achieve these goals. In addition, the ability of the system fluid to reduce friction is a very important consideration. The degradation of the fluid with both time and operation is a factor which must be considered in the selection of hydraulic fluids. However, one of the most critical characteristics of a fluid is the ability to reduce wear and promote long component life.

Without the use of additives to enhance their properties, the fluids available for use in hydraulic systems would have a very limited working range. Although there are a large number of parameters and characteristics which describe a hydraulic fluid, most of these properties are well defined. The antiwear and lubrication properties of a hydraulic fluid needs a qualification procedure which is effective and economical.

There are several test methods which have been developed to assess the antiwear properties of liquids. Unfortunately, most of these procedures are intended to evaluate fluids other than those used in hydraulic systems. The bench type wear test methods are designed for lubricating oil. The test procedures normally utilized for hydraulic fluids use a designated pump and consist of an entire hydraulic system. In order to avoid the cost and complexity associated with a complete hydraulic system and to eliminate the shortcomings of standard bench type wear tests, researchers at the Fluid Power Research Center, formerly at Oklahoma State University, developed and introduced a bench test entitled the Gamma Wear Test System. The term Gamma was adopted to separate this test from such tests as the Shell Four Ball Test, the Timken Wear Test, etc. This paper will describe the Gamma Wear Test System and present data obtained from tests on hydraulic fluids.

TECHNICAL BACKGROUND

The factors which are critical in considering the wear rate of fluid power components are [3]:

- Area of surface in contact
- Surface roughness and geometry
- Material properties
- Applied load

- Sliding speed of surfaces
- Fluid properties

All of these factors are very important in determining the wear rate of hydraulic components and, in fact, directly or indirectly affect the durability, reliability, and life of the hydraulic system. A considerable amount of research has been conducted on these factors and some theoretical formulations have been developed [4].

One of the most comprehensive effort directed toward the evaluation of the antiwear properties of hydraulic fluids resulted in the development of standards (e.g. ASTM D-2882) which utilized a complete hydraulic system involving a pump such as the vane pump. Not only are these tests complicated and involve an abundance of equipment, they are also built around a particular component (pump) which may not be available over the long term. Investigators have relied upon several bench-type tests to appraise the influence of antiwear additives [5], along with a broad range of pump tests focused on a variety of pump designs.

The selection of the test to be used in evaluating a hydraulic fluid depends to a great extent on who is doing the selection. A fluid supplier would much prefer the bench tests since they are quicker to run and less expensive. However, the pump manufacturer and the system user will insist on a pump test using the particular pump. The decision on the part of these players is understandable since there has been very little reported on the correlation between bench tests and pump tests. In fact, most reports indicate a lack of such correlation.

The most commonly used bench-type wear tests that have been used for hydraulic fluids are shown in Table 1. The Shell tests, as shown in Fig. 1, involve three stationary balls and one rotating ball which are located in a cup of the test fluid.

The configuration for the Falex wear tester, as shown in Fig. 2 , consists of two stationary Vee blocks which are loaded against a rotating journal. The bearing blocks and the journal are immersed in a retaining cup of the fluid under test.

The Timken Wear Test concept consists of a stationary rectangular bar loaded against rotating cup, as shown in Fig. 3. As is the case with the Falex tester, the Timken tester wear specimens are submerged in a cup of test fluid.

It should be noted that in none of these bench-type wear test methods is the temperature of the fluid controlled. In addition, the contamination level of the fluid being tested is not under control. Instead, both of these parameters are allowed to vary in relationship to the wear process occurring during the test.

In order to make a rational selection of hydraulic fluid from a wear protection standpoint, a test procedure was needed with the following characteristics:

- It should be a bench type test that is as simple as possible
- It should provide for fluid circulation in order to control both temperature and contamination level
- It should provide a specimen configuration which will simulate a real world tribological processes
- It should provide a wide range of loading and relative speed between the wearing surfaces

TABLE 1 -- Common Bench Type Wear Tests

Type of Machine	Type of Friction	Media	Remarks
Shell (four-ball) E.P. ASTM D2783	Sliding	Stationary balls - Rotating ball	Assess the load-wear index and weld point under severe conditions.
Shell (four-ball) Wear ASTM D2266	Sliding	Stationary balls - Rotating ball	Assess the anti-wear properties of test sample and wear scar of ball under light load condition.
Falex wear properties test ASTM D2670	Sliding	Stationary V-blocks - Rotating journal	Assess the wear properties of fluid by the number of ratchet-type mechanism to keep constant load over a fixed test time.
Falex E.P. properties test ASTM D3323	Sliding	Stationary V-blocks - Rotating journal	Assess the load-carrying properties of fluid by increasing the load.
Timken E.P. test ASTM D2782	Sliding	Stationary block - Rotating cup	Assess the load-carrying capacity of lubricating properties by determining the minimum load at which scoring or seizure occurs and the maximum load that it cannot occur.

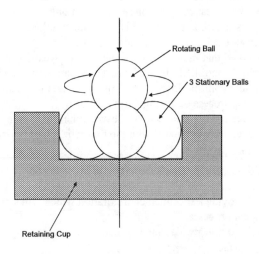

FIG. 1 -- Shell Four Ball Tester

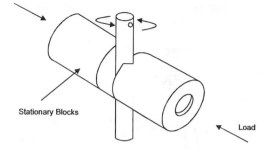

FIG. 2 -- Falex Specimen Configuration

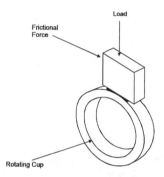

FIG. 3 -- Timken Wear Specimen

To incorporate these characteristics, eliminate the short comings of the more standard bench type wear tests and avoid the complexity of a complete component test, the authors, developed and introduced a bench test entitled the Gamma Wear Test System. This system incorporated all of the characteristics which are needed for basic fluid selection and some characteristics which expand its capability. For example, the Gamma Wear Test System can be used with different material combinations and various loading conditions.

THE GAMMA WEAR TEST SYSTEM

After carefully exploring the bench type wear testing apparatus available, many features of the Falex Wear Test machine loading system were selected. The bearing loading mechanism used in the Gamma Wear Test System is illustrated in Fig. 4. This type of loading mechanism permits the wear specimen to be loaded from a fraction of a pound to several hundred pounds by simply using various calibrated springs. The wear specimen incorporated in the Gamma Wear Test System are not of the Vee-block design as normally found in the Falex tester. The Gamma System uses a journal identical in dimensions to that of the Falex, however, the bearings are constructed to approach the configuration of a journal bearing, as shown in Fig. 5. The schematic of the circulation system for this new wear test method as shown in Fig. 6.

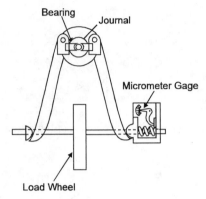

FIG. 4 -- Loading Mechanism for the Gamma Wear Test System

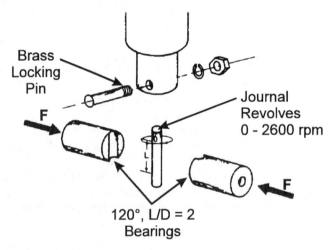

FIG. 5 -- Journal and Bearing Pairs for Gamma Wear Test System

As a result of the unique features shown in Figs. 4, 5, and 6, the Gamma Wear Test System provides complete freedom in the selection of bearings and journal mating pairs. The bearings are designed with the critical 120° bearing surface needed to provide linear wear test results and a stable wearing condition. In this system, the mating pairs can be subjected to various speeds, loads, temperatures, cleanliness levels, and circulating flow.

In Fig. 6 it can be seen that a variable speed hydraulic motor was used to rotate the test wear pin at speeds up to 4500 RPM. In the Gamma Wear Test System, the wear bearings and the accompanying spindle are placed in small (500 milliliter) reservoir and submerged in the test fluid. There are two circulating circuits shown in Fig. 6. In the circuit to the right, a gear pump causes the test fluid to circulate through a filter and a heat exchanger to remove any particle which have become entrained in the fluid and to regulate the test temperature. The circuit to the left utilizes a peristaltic pump to eliminate any interference between the pump and the contamination level of the test fluid. This peristaltic pump causes the test fluid to flow through a heat exchanger to control the temperature. With the circuit designated contaminant environments can be subjected to the test specimen by entraining given contaminants in the test fluid.

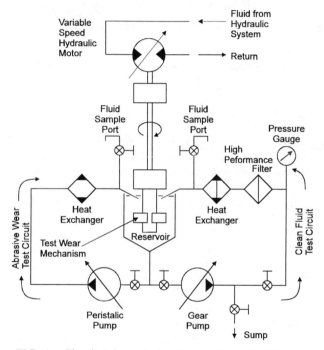

FIG. 6 -- Circuit Schematic for Gamma Wear Test System

The primary advantage of the Gamma Wear System lies in the ability to simulate any lubrication regime. As shown by the Stribeck Curve [6, 7] shown in Fig. 7, the lubrication regime will depend upon the viscosity of the fluid, the relative velocity between the wear surfaces, and the load imposed on the wearing interface. Therefore, by changing the speed of rotation or the load applied to the bearing, any lubrication regime can be imposed during the test. Such a broad range of control parameters permits the controlled testing with the Gamma Wear Test System to duplicate almost any conceivable operating condition that may be experienced in a specific application.

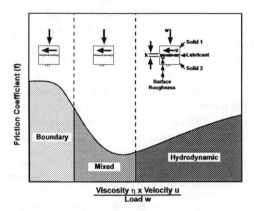

FIG. 7 -- The Stribeck Curve showing the Three Basic Lubrication Regimes

TEST RESULTS USING THE GAMMA WEAR TEST SYSTEM

Tests conducted at both the Fluid Power Research Center and FES, Inc. have assessed the antiwear characteristics of many different fluids at various temperatures, speeds, and load conditions, using a broad spectrum of bearing materials. As shown in Fig. 8, the Gamma Wear Test can distinguish between the wear created by various particle size distributions entrained in the test fluid. These particle size distributions were obtained from AC Fine Test Dust by elutriation. In this effort, a test on clean MIL-L-2104 was conducted to provide a baseline. Sequential tests were then conducted using the Gamma Wear Test System on the same fluid with 300 mg/l of 0-10, 0-20, 0-30, 0-40 micrometer size ranges of AC Fine Test Dust created by elutriation from the full distribution of the standard test dust. These tests revealed that each different particle size distribution provided unique wear characteristics which increased with increasing particle size range.

In addition, the Gamma Test can effectively distinguish between various fluids. Fig. 9 illustrates the results of Gamma tests conducted on a widely used hydraulic fluid (MIL-H-5606) and a lubrication fluid often used in hydraulic systems (MIL-L-2104). As can be seen in this figure, the test is capable of revealing the difference in wear protection provided by various fluids used in many hydraulic systems. The Gamma Wear test has been employed to evaluate the characteristics of high water content fluids as shown in Fig. 10. The test results shown in this figure reveal that the more additive used with the water phase of the high water base fluid, the better the anti-wear performance. The effect of fluid temperature is illustrated in Fig. 11 using a steel journal and brass bearings with a base stock fluid containing one percent ZDDP antiwear additive tested at various temperatures. Finally, Fig. 12 reveals the influence of high unit loading using a steel journal and steel bearings with both an antiwear fluid (MIL-L-2104) and a straight base stock oil.

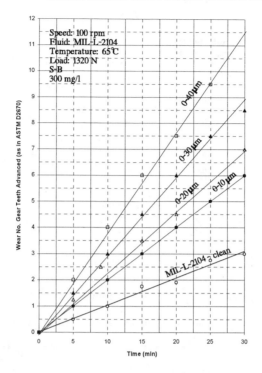

FIG. 8 -- Effect of Particle Size Distribution on the Anti-Wear Characteristics of
MIL-L-2104 Class 10

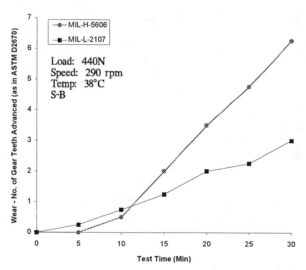

FIG. 9 -- Antiwear Characteristics of MIL-H-5606 and MIL-L-2104 Class 10

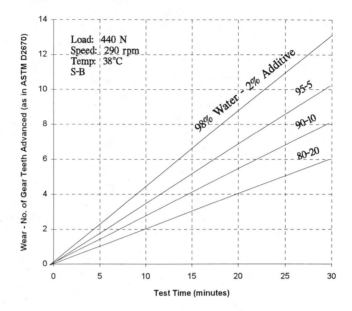

FIG. 10 -- Antiwear Characteristics of High Water Content Fluids

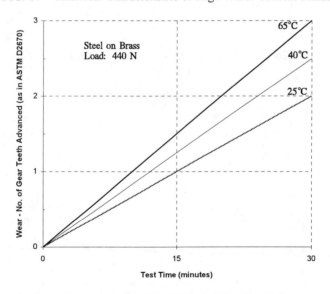

FIG. 11 -- Antiwear Characteristics of Bask Stock Oil with 1% Antiwear Additive

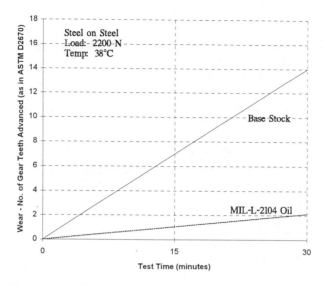

FIG. 12 -- Antiwear Characteristics of Base Stock versus Antiwear Oil

CONCLUSION

The users of hydraulic systems are relying much more today than ever before upon the system fluid to protect components, improve operation, and increase service life. In addition, the severity of the expected duty cycle and the demand of fire resistant characteristics means that the hydraulic system design needs more and better information concerning the performance capability and limitations of hydraulic fluids. The oil suppliers have accomplished an excellent job in providing most of the necessary performance data on their products. However, in the case of the antiwear properties, such data has been severely limited by the test procedures available.

It is believed that the Gamma Wear Test System eliminates the problems associated with the bench type standard tests in use today. In addition, it is simple enough to recommend it as a screening test to minimize pump testing. The broad range of operating conditions which can be used with this tester produces data which is extremely valuable to the hydraulic component and system designer as well as those who use the high performance hydraulic systems on the market today.

References

[1]   "Lobmeyer, Raymond J., et al, "Hydraulic Fluid Contaminant Sensitivity Test"; No. 770543, Earth Moving Industry Conference, Peoria, IL, Society of Automotive Engineers, Central Illinois Section, (18-20 April 1977).

[2]   Chu, J. and R.K. Tessmann, "Additive Packages for Hydraulic Fluids"; The Basic Fluid Power Research (BFPR) Journal, 1979, 12:111-117.

[3]   Rainwater, R.L., "An Experimental Study of Wear and Effects of Anti-Wear Fluids Development of a Wear Test Model"; Sixth Annual Fluid Power Research Conference, Fluid Power Research Center, Oklahoma State University, Stillwater, OK, 1972..

[4]   Marnoun, M.M., "Fundamentals of Modern Theory of Friction and Wear"; 72-W-4 Sixth Annual Fluid Power Research Conference, Fluid Power Research Center, Oklahoma State University, Stillwater, OK, 1972.

[5]   Chu, S. Arrington, and R.K. Tessmann, "A Bench Wear Test for Hydraulic Fluids"; The BFPR Journal, 1980, 13:189-194.

[6]   Hong, I.T., "Effect of System Filtration Characteristics on Tribological Wear of Rotating Elements"; PhD Dissertation, Oklahoma State University, Stillwater, OK, 1980.

[7]   Stribeck, R., "Die Wasentlicken Eigenschaften der Glut - Und Rollenlager"; VDI-Zeitschrift, 46 (1902), 1341, 1342, 1463.

[8]   Fitch, E.C., "Fluid Contamination Control", FES, Inc., 5111 North Perkins Road, Stillwater, OK, 1988.

Joseph M. Perez[1]

A REVIEW OF FOUR-BALL METHODS FOR THE EVALUATION OF LUBRICANTS

REFERENCE: Perez, J. M., ''A Review of Four-Ball Methods for the Evaluation of Lubricants,'' *Tribology of Hydraulic Pump Testing, ASTM STP 1310,* George E. Totten, Gary H. Kling, and Donald J. Smolenski, Eds., American Society for Testing and Materials, 1996.

ABSTRACT: A review of four-ball wear tester and test methods developed at Penn State by Prof. Emeritus E.E. Klaus (Deceased) and students of Klaus. The test methods include standard methods for fluid and additive evaluation, microliter tests to evaluate tribochemical reactions in the contact zone, Ball-on-Three Flat tests to evaluate materials and fluids, scuffing test methods and sequential four-ball tests, one of which was developed to screen hydraulic fluids for full-scale pump stand tests and field vehicles. Applications range from room temperature to over 425 °C. The variety of methods demonstrate the versatility of the FBWT, including applications related to the tribology of hydraulic fluid systems.

KEYWORDS: Hydraulic Fluids; Bench Tests; Four-ball; Sequential Wear Tests; Tribochemistry; Ball-on-three flats; Scuffing Test

This review will focus on hydraulic systems and pump stand tests, and describe some four-ball wear test methods developed at Penn State, one of which was developed to simulate pump stand wear. There are over 234 friction and wear devices described in the literature [1]. The four-ball wear tester (FBWT) is one of the more widely used devices. Invented in the 1930's [2], the FBWT has evolved over the last fifty years as one of the primary tools for the evaluation of friction and wear of lubricants at temperatures from room temperature to vapor-phase lubrication above 400 °C. Applications range from standard quality control tests to materials and additive development.

The FBWT is a simple test that engineers over the years have placed anywhere from no

[1]Adjunct Professor, Tribology Group, Chemical Engineering Department, The Pennsylvania State University, University Park, PA 16802

reliability to complete reliability in the test results. Researchers such as Klaus, Feng, Fein and Rounds were champions of the method. They understood how to use the tool and its limitations. They have published numerous papers [3-10] over the last 40 years describing applications of the method. E.E. Klaus' 1950 PH.D. thesis at Penn State [11] is a basic bible on the use of the instrument. It contains evaluations of most lubricants and additives available in the 50's and was one of the first attempts to correlate the method with full-scale pump tests. The following sections will briefly discuss some of the variations of the methods that were developed by Klaus or students of Klaus. Most of the developments were industry driven.

Four-ball Wear -- The typical wear vs. load pattern for lubricants in the four-ball testers is shown on Figure 1. If additives are controlling wear, it would correspond to the first segment of the plot, up to point A. The use of the four-ball to evaluate fluids and additives usually involves testing to establish the limits of this regime. In this regime, wear scars corrected for the Hertz elastic indentation, normally are below 20 mm in diameter, even at 392N (40 kgf) loads. Test repeatability in the first segment is good, $\pm$ 5%. Above this regime, the wear scar increases rapidly until it becomes large enough to support the load once again, point B. In a typical curve, point A can be defined as the point of scuffing. At this point, the additives fail to control the wear rate. Usually, a change in the coefficient of friction also occurs at this point. In this transition regime, the test repeatability is poorest, $\pm$15 %. In the final segment, B-C, wear increases to failure. This regime is better evaluated in the EP tester.

   Essentially all long life operating systems, such as pump stand tests and vehicle systems, function in the load range below point A. That means the location of point A and the wear rates below this value are of primary interest. The load at which point B occurs depends on both the material and the lubricant being tested.

EXPERIMENTAL METHODS.

Instrumentation.

   The machines used at Penn State over the past 40 years include the Roxana Shell Four-ball and EP Testers, the GE/Brown modification by the Roxana Machine Works and more recently the Falex version. The features of one of the current wear testers include pneumatic loading for adjustable loads, an air bearing to improve friction accuracy, a controlled temperature heating block, various configurations of test cups, atmospheric gas control, a variable drive motor to control spindle speed and elimination of the old mercury moat temperature contacts.

Methods used.

   The methods described in this paper include the standard methods of using the four-ball test, sequential wear tests, the ball-on-three flats, a scuffing test and micro-four ball tests. The geometry of all of the tests including the four-ball and ball-on-three flats is the same, Figure 2. The load is applied axially through a chuck ball to three specimens locked in place. The chuck or rotating ball is nestled on the three locked specimens and the test speed is normally 600 RPM.

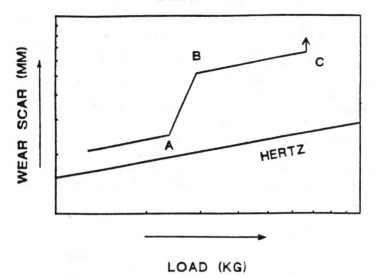

LOAD (KG)

Fig.1--Typical Four-Ball Wear vs Load

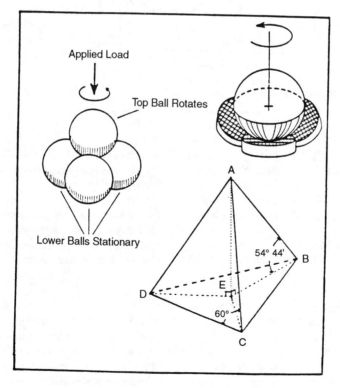

Fig.2--Four-Ball Test Geometry

The standard ASTM methods are used primarily for screening and quality control. The sequential and scuffing tests were developed in studies to correlate the method with an industry full-scale hydraulic fluid pump stand test and off-highway equipment field tests [12-14]. The use of three flats to replace the fixed balls in the test was developed to enable testing of ceramic materials not readily available as 1.27 cm diameter ball-bearings [15]. The micro-sample tests were developed to enable research evaluations of fluids and additives, especially for development of high temperature lubricants [16]. Each of the methods will be described and examples of their application presented.

The typical FBWT conducted in the studies described ranged in temperature from ambient to 204 °C. However, modifications of the four-ball enabled evaluations of liquid lubricants at temperatures up to 425 °C. An extreme high temperature modification, Figure 3, enabled evaluation of vapor phase lubricants and advanced materials [17,18] over a temperature range of 400 to 1000 °C.

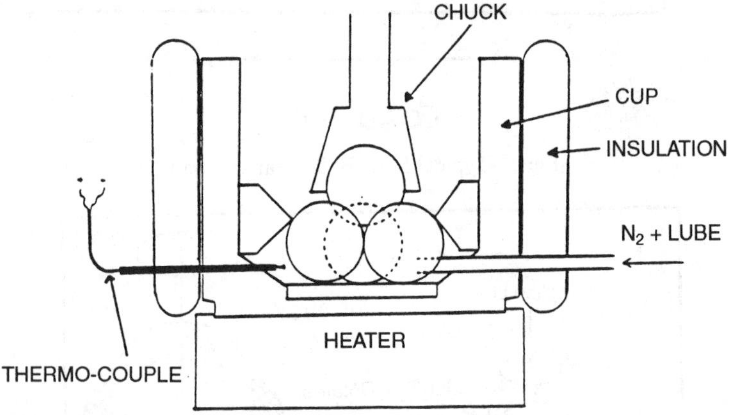

Fig. 3--Schematic of Vapor-Phase Unit

ASTM Standard Methods [19]--A brief comparison of the standard four-ball methods is shown in Table 1. In all there are three ASTM methods, one to evaluate lubricants on the FBWT, ASTM Standard Test Method for Wear Preventive Characteristics of Lubricating Fluid (D 4172-94), one to evaluate liquid lubricants in the Four-Ball Extreme Pressure (FBEP) tester, ASTM Standard Test Method for Measurement of Extreme-Pressure Properties of Lubricating Fluids (D 2783-88) and a FBEP method to evaluate greases, ASTM Standard Test Method for Measurement of Extreme-Pressure Properties of Lubricating Grease (D 2596-93). The details of the methods can be found in the Annual Book of ASTM Standards [19].

Sequential Wear Tests--There are several variations of the sequential wear test depending on whether the lubricant or film forming tendencies of the additives are of interest. The first use of sequential test was at the PRL in the late 50's to evaluate the mechanism of lubrication [20]. The method was later adopted and modified to study hydraulic systems [12-14]. The third modification, developed at NIST [15], was used to study tribochemical reactions in the contact zone and used to evaluate high temperature liquid lubricants and materials [21].

Table 1-- Comparison with ASTM Methods [19]

CONDITION	PSU	D-4172-94	D-2783-88	D2596-93
TEMP.,°C	75± 2	75±2	18.33-35.0	27±8
SPEED, rpm	600±30	1200±60	1760±40	1770±60
TIME, min	30/60/90	60±1	0.166	0.166±0.2
LOAD, N/kgf	Varies	147/392±2	Varies	Varies

The test developed to study hydraulic fluids correlated the performance of various types of petroleum base hydraulic fluids and their performance to a full-scale hydraulic fluid pump stand test that was correlated to field performance of various equipment hydraulic systems. This sequential test is conducted by running three separate 30-minute segments without changing the fixed balls. The initial study was extended to other types of hydraulic fluids and finally to the evaluation of new and used fluids from pump stand tests and vehicle systems.

The sequential test used to evaluate hydraulic fluids is a sequence of three thirty minute runs using the same wear surfaces throughout the sequence, Table 2. The first and second test segments are run with the test fluid to demonstrate "run-in" and "steady state" wear respectively. The third thirty minute segment is run with a non-additive white oil. This segment demonstrates the effectiveness of the surface finish and/or the nature of the chemical coating on the wear surface, if present. The wear values reported, incremental values of wear (delta wear scars), are used to evaluate the performance of the fluid.

Table 2--Sequential 4-Ball Test Procedure

1. Clean Unit, Add Oil and Assemble
2. Heat to Desired Temperature
3. Apply Load and Run for 30 min.
4. Cool & Clean w/o Removing Balls
5. Measure Wear Scars
6. Add Fresh Oil & Reassemble
7. Repeat Steps 2 to 5
8. Add Paraffinic White Oil
9. Repeat Steps 2-5
-- End of Test--

Four types of hydraulic fluids with the characteristics similar to one of the types shown on Figure 4 were evaluated. These represent different types of lubricants. A non-additive mineral oil would tend to show similar wear rates in the SS and SF segments of the test. An extreme pressure, EP, type additive would show low wear typical of chemical reactivity with the bearing surface in the SS segment and increased wear in the final SF segment, indicating that no lasting surface coating or mating was achieved in the SS segment. An effective anti-wear type additive would form a surface film and would tend to exhibit less wear increase than either the non-additive or EP type additives in the third segment of the test. The SF wear would be an indication of how effective the film formed was in the SS segment of the test. An R&O type oil would perform similar to the mineral oil in the sequential test.

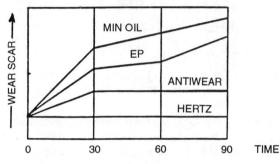

Fig.4-- Typical Hydraulic Fluid Wear [12]

The first, or "run-in segment", wear is determined by subtracting the Hertzian (elastic) contact area from the scar measured at the end of the 30-minute segment. The wear for the "steady-state" and "surface conditioning" segments are determined by subtracting the initial and final wear scar measurements of the segment. This system of considering the wear difference improves the evaluation of four-ball wear data. The test is run at different temperatures and loads depending on the conditions of the system the test is attempting to simulate.

Additional information on the mechanism involved may be obtainable by varying the test temperature and the load and observing the type of changes that occur in the individual segments of the sequential test.

Scuffing Test--The scuffing test was developed to better evaluate the load carrying ability of used fluids. In the scuffing test, the loads are increased in increments until scuffing occurs. By measuring the friction, a quick evaluation of the loads required for scuffing can be approximated. The scuffing test does use a new set of ball-bearings to measure wear for each subsequent load. An increase in the wear scar diameter of 0.20 mm or more in any 30-minute segment of the test is an indication that scuffing was occurring (Point A exceeded). The test was used to show the loss of load carrying additives in a pump stand test and in vehicle systems. The depletion of additive with time of fluid use is easily determined[14].

BTF Test--The BTF sequential step loading test has been used to establish scuffing and failure loads. In this test the loads are increased incrementally every few minutes without changing the fluid or the flats. Once points A and B (Figure 1) are identified, 30 minute tests are conducted to confirm the transition zones. The test was used for ceramic materials that were not readily available as 1.27 cm diameter balls [15].

Micro-four Ball Tests--The micro-four ball test methods were developed to study the chemical interactions in the contact zone. The use of micro-liter amounts of lubricant in thin films results in less heat transfer and higher temperatures in the contact zone, accelerating oxidation reactions and accumulation of products as shown in Figure 5. Perez et al used the method to study the formation of reaction products of tricresyl phosphate [22]. Chao [23] modified the procedure to study the effects of fluid volatility on friction and wear. The amounts of lubricant used vary from 1 uL to 6 uL.

The standard 6 uL test is used with the four-ball set-up. In this test, the balls are fully flooded with a 10 ml charge of a purified white oil and a 60 minute run-in test is conducted at 75 °C using a 392N (40 kgf) load. At the end of this run-in period, the ball pot and chuck ball are cleaned with solvents without removing the balls from the pot or chuck. The wear scars are measured to insure that the run-in wear falls within established limits (0.65±0.03mm). If acceptable, the chuck ball is also thoroughly cleaned and dried and the unit is reassembled. A 6 uL sample, or less, of the test lubricant is rapidly added to the scar area on the chuck ball, Figure 6. The chuck ball is carefully rotated on the three fixed balls without any applied load to insure the lubricant film is well distributed. The load is applied and the test resumed. The test is terminated when the fluid film can no longer control wear. This is indicated by a sharp increase in friction. The reaction products can be analyzed by any of several instrumental techniques.

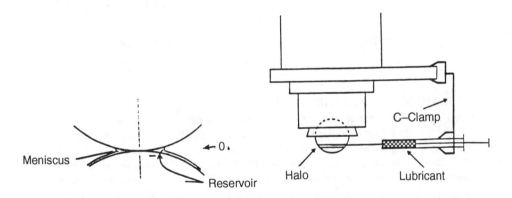

Fig. 5--Micro-Test Chemical Environment   Fig. 6--Fluid Application-Micro-Test

Tricresyl phosphate, a common anti-wear additive, was studied using a combination of methods, including the 6 uL test [22]. The friction traces were used to establish chemical trends and select samples for analysis. The surface reaction products were examined using FTIR. The additive depletion was used to establish required effective concentrations.

The method was used to screen high temperature lubricants developed for advanced diesel engines [21]. Conventional mineral oil base lubricants will control friction and wear for less than 30 minutes in the 6 uL FBWT. The test duration was extended to several hours by using selected additives and base fluids in lubricants designed for severe applications.

Typical results showing the effect of sample size on test duration is shown on Figure 7. Fluid loss can be determined by analysis of the final products which are dissolved from the balls and analyzed by gel permeation chromatography, GPC, Figure 8.

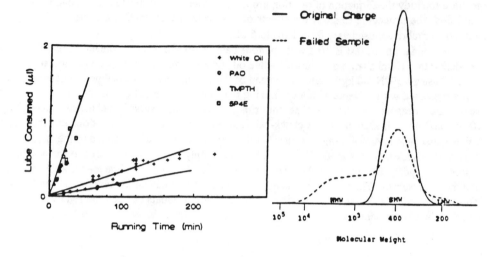

Fig.7--Lubricant Depletion   Fig.8--GPC Analysis of Used Lubricant
    Micro-Wear Test   6 ʋl Test

HYDRAULIC PUMP STAND TESTS.

A series of studies were conducted [12-14] involving a full-scale hydraulic pump stand test, fluids from off-highway vehicle hydraulic systems and the four-ball wear test. In the initial studies, the purpose was to reduce the number of costly tests conducted on the pump stand. The pump stand was correlated to field tests over a number of years through comparisons of vehicle performance with laboratory data. The wear mechanism in the pump system is similar to the four-ball test in that the initial part of the test involves a "break-in" period in which the wear is initially high and decreases as the bearing surfaces, vanes and cam ring, become mated. If the hydraulic fluid does not provide adequate lubrication, pump failure in the form of welding and seizure between the vane tips and cam ring is experienced. The pump stand differs from the four-ball sequence in that the pump stand test, as currently run, does not have a comparable third step to characterize the effectiveness of the film formed.

The pump stand test procedure involved conducting a series of tests at various temperatures and pressures to determine if the hydraulic fluid performance fell below or above established field failure criteria. The sequential wear test method was used initially to rank four fluids of known performance in the system. Various loads and temperatures were used to establish performance and scuffing data for the fluids, Table 3.

Table 3--<u>Typical Sequential Four-Ball Test Data</u>.

| FLUID ID No. | ----------------------SS + SF------------------------ ---------- Wear Scars, Δ mm ---------------- | | | PUMP STAND PERFORMANCE |
	75°C (167°F) 98N(10kgf)	75°C (167°F) 98N(10kgf)	75°C (167°F) 392N(40kgf)	
4550	0.03	0.08	0.1 0	EXCELLENT
4551	0.06	0.19	0.05	BORDERLINE
4552	0.16	0.24	0.12	FAILED
4553	0.18	0.31	0.30	FAILED

In subsequent studies, several fire resistant type fluids were evaluated and new and used fluids were also studied to establish the feasibility of extending the drain intervals on the systems. The results in both studies were consistent with the first study and the sequential test can be used to obtain pass - fail results of a number of fluids. A summary of some of the scuffing test results for new and used fluids are found in Table 4.

Table 4--Sequential Four-Ball Tests for New and Used Fluids

| TYPE FLUID (Comparable to fluids in Table 3) | PUMP STAND TEST RESULT | --------- SEQUENTIAL TESTS ---------- | | |
| | | 40kg WEAR, (Δ SCAR,mm) | SCUFFING TEST FAILURE LOAD, kgf | |
			NEW OIL	USED OIL
4550	EXCELLENT	PASS (0.16)	>120	100
4551	BORDERLINE	BL (0.19)	60	------
4551	PASSED	PASS (0.10)	80	60
4551	PASSED	BL (0.06)*	50	60
R&O	FAILED	SCUFFED	40	-------

* Winterized Fluid Containing Kerosene

SUMMARY.

In summary, the FBWT is a simple and useful tool to screen materials, lubricants, and to evaluate new chemistry. The tester can be used to understand the role of lubricants in various tribological systems. Even though the four-ball methods fundamentally involve only sliding wear, it is possible to correlate the test methods with full-scale mechanical systems. The key to establishing a correlation depends on the understanding of the function of the lubricant in the system and selecting the four-ball conditions that best simulate the results.

ACKNOWLEDGEMENTS.
This paper is dedicated to Professor E.E. Klaus. His untimely death has left a significant void in the tribology of lubricants. Over the past forty years, a large number of organizations from both industry and government contributed to the support of Professor E.E. Klaus and his students at Penn State. Their support has been deeply appreciated.

REFERENCES.
[1] Friction and Wear Devices, Publ. American Soc. Of Lubrication Engineers (now STLE), Park Ridge, Il (1976)], second edition.

[2] Boerlage, G.D., US Pat.2,019,948 (1935).

[3] Klaus,E.E., Tewksbury, E.J. and Fenske, M.R.,"Critical Comparisons of Several Fluids as High Temperature Lubricants", J.Chem. Eng. Data, 6,1, 99-106, January (1961).

[4] Klaus, E.E. and Fenske, M.R., "Chemical Structure and Lubrication", Div. Pet. Chem., ACS, Atlantic City, NJ, Sept (1956).

[5] Fein,R.S.,"Transitional Temperature Distribution Within a Sliding Hertzian Contact",ASLE Trans.,3,34-39.(1960).

[6] Fein,R.S. and Kreuz,K.L.,"Chemistry of the Boundary Lubrication of Steel by Hydrocarbons",ASLE Trans.,8,29-38,(1965).

[7] Klaus, E.E. and Bieber, H.E.,"Effects of P32 Impurities on the Behavior of Tricresyl Phosphate-32 as an Antiwear Additive",ASLE Trans.,8,12-20,(1965).

[8] Rounds, F.G.,"Some Effects of Amines on ZDTP Antiwear Performance as Measured in 4-Ball Wear Tests", ASLE Trans.,24,4, 431-440 (1981).

[9] Rounds, F.G.,"Additive Interactions and Their Effect on the Performance of Zinc Dialkyl Dithiophophate",ASLE Trans.,21,2,91-101,(1978).

[10] Rounds, F.G.,"Soots From Used Diesel Engine Oils - Their Effects on Wear as Measured in 4-Ball Wear Tests",SAE Paper No.810499,Detroit, MI,(1981).

[11] Klaus, E.E.,Wear and Lubrication Characteristics of Some Mineral Oil and Synthetic Lubricants, PH.D. Thesis, The Pennsylvania State University, University Park, Pa 16802, (1952).

[12] Perez, J.M. and Klaus, E.E., "Comparative Evaluation of Several Hydraulic Fluids in Operational Equipment, A Full Scale Pump Test Stand and the Four-Ball Wear Tester," SAE Paper No. 831680, October 1983, San Francisco, CA.

[13] Perez,J.M., Klaus, E.E., and Hansen,R.C.,"Cooperative Evaluation of Several Hydraulic Fluids in Operational Equipment, A Full-Scale Pump Stand Test and the Four-Ball Wear Tester. Part II-Phosphate Esters, Glycols and Mineral Oils", Lubr. Eng., 46,4,249-255,April, 1990.

[14] Perez,J.M., Klaus, E.E. and Hansen,R.C., "Cooperative Evaluation of Several

Hydraulic Fluids in Operational Equipment, A Full-scale Pump Stand Test and the Four-Ball Wear Tester. Part III- New and Used Mineral Oils", Lubr. Eng., 52, 5, 416-422 (1966).

[15] Gates, R.S.,Yellets, J.P., Deckman,D.E. and Hsu, S.M., Considerations in Ceramic Friction and Wear Measurements , ASTM STP No.1010, Selection and Use of Wear Tests for Ceramics (1988).

[16] Gates, R.S. and Hsu, S.M., " Development of an Oxidation Wear-Coupled Test for the Evaluation of Lubricants", Lubr. Eng., 40,(1),27-33,(1984).

[17] Klaus, E.E., Vapor Delivery-A Technique Designed for High Temperature Lubrication , USDOE, Energy Conservation Utilization Technologies (ECUT), Report No. DOE/EC-88/3, May-June 1988.

[18] Klaus, E.E.,Jeng,G.S.,and Duda, J.L., A Study of Tricresyl Phosphate as a Vapor Delivered Lubricant ,Lubr. Eng.,45,11,717-723(1989).

[19] American Society of Testing Materials, Book of ASTM Standards, 1916 Race Street, Philadelphia,PA 19103

[20] Klaus, E.E., Tewksbury, E.J., Angeloni, F., Perez,J.M., and Spooner, D.,Wright Paterson Air Development Division(WADD) Technical Report, 55-30, Pt VIII, February (1960).

[21] Hsu, S.M. and Perez, J.M., "High Temperature Liquid Lubricant for Diesel Application", SAE Paper No. 910454, Detroit, MI (1991).

[22] Perez,J.M., Ku, C.S., Pei, P., Hegemann, B.E. and Hsu, S.M. "Characterization of Tricresylphosphate Lubricating Films by Micro-Fourier Transform Infrared Spectroscopy," Tribology Conference, Baltimore,Md., October 1988, Trib. Trans., 33 (1990), 1, 131-139.

[23] Chao, K.K., A Study of Tribology By Using A Microsample Test, PH.D. Thesis, December (1990).

Author Index

Subject Index

A

Additives, 230, 361
  effects, 165
Aging, 230
  fluid, 220
Aluminum, 277
Analyzing devices, 140
ASTM standards, 3, 156, 165
  D 2882, 85, 96, 106, 118, 129,
     200

B

Ball-on-three flats, 361
Bearing/sealing parts, 49
Bench screening procedure, 314
Bench testing, 65, 349, 361
Bench top surface contact test,
  338
Biodegradable fluids, 208, 220
Biodegradable hydraulic
    pressure media, 230
Brass, 291

C

Cam rings, 106
Cartridges, 118
  preparation, 106
Cavitation, 65, 277
Chlorofluorocarbon, 38
Chlorotrifluoroethylene, 291
Cleanliness, 277
Clogging, 277
Coating, surface, 291
Contact loading, 200
Contacts, steel on steel, 156
Contact test method, surface, 338
Contamination, 247, 338
  sensitivity, 261
Copper, 291
  containing metals, 186
Corrosion, 291
  tests, 186

D

Denison pump, 186
Deposits tests, 186
DIN standards, 85
DU-bearing wear, 208

E

Elastohydrodynamic, 3
  lubrication, 21
Electrostatics, 277
Energy efficiency, 165
Energy transfer, 65
Esters, 230
  synthetic, 208
Extrusion molding, 277

F

Failure mechanisms, 3
Film attributes, 3, 38
  thickness, 21
Fire resistance, 21
Flow degradation, 247
Flow rate, 106
Flow tests, 186
Flywheel testrig, 220
Four-ball methods, 361
Friction, 3, 38
  coefficient, 165, 208
  force, 49
  reduction, 85
  test, 165
FZG gear test, 329

G

Galling, 106
Gamma Wear Test System, 349
Gear pumps, 176
Gear test, 329
Glycol and water fluids, 31